Edwin Kiel (Ed.)

Drive Solutions

Edwin Kiel (Ed.)

Drive Solutions

Mechatronics for Production and Logistics

With 310 Figures and 51 Tables

 Springer

Editor

Dr. Edwin Kiel
Lenze AG
Hans-Lenze-Str. 1
31855 Aerzen, Deutschland
Kiele@Lenze.de

Translation of the German book „Antriebslösungen" by E. Kiel (Hrsg.)

ISBN 978-3-540-76704-6   e-ISBN 978-3-540-76705-3

DOI 10.1007/978-3-540-76705-3

Library of Congress Control Number: 2007939488

© 2008 Springer-Verlag Berlin Heidelberg

Typesetting: supplied by the editor
Production: LE-TEX Jelonek, Schmidt & Vöckler GbR, Leipzig, Germany
Cover: eStudioCalamar S.L., F. Steinen-Broo, Pau/Girona, Spain

Printed on acid-free paper

9 8 7 6 5 4 3 2 1

springer.com

For Elisabeth Belling

our companion for the past 60 years

Preface

This book wants to fill a gap. There are many reference books covering electric motors and drive control systems [e.g. GiHaVo03, Le97, Sch98, SchHo05b], but to date there is nothing describing the use of electronically controlled drives in production and logistics systems. Equivalent technical literature exists only for machine tools [notably We01a, We01b].

The basis of this book is the extensive application knowledge of Lenze. Its genesis dates back to 2003, when we began to classify the vast number of applications that are implemented with our products. Until then we regarded each application as an individual case that required a customised solution. But this did not allow for an industrial approach to the engineering process. By thorough analysis we were able to identify and describe twelve classes of drive functions that are used in many industries. The initial outcome of this classification process, in 2005, was a promotional publication, "Drive Solutions", which provided an eye-catching double-page description of each of these twelve application groups [Le06]. This publication proved extremely popular because it gives an orientation about drive technology applications in mechanical engineering.

The decision was then taken to explore these twelve drive solutions in greater technical depth. And since there was no specialist literature available in this field, we decided to publish the research in the form of a reference book. The result of this exercise is here before you.

Of course, writing a reference book comprising over 500 pages is not a task to be undertaken lightly. A motivational trigger was needed to turn this project into a reality, and Lenze's 60th anniversary in 2007 provided that trigger.

Alongside its products, Lenze has always offered its customers technical advice on drive applications, and so it was a natural decision to choose "Drive Solutions" as the title of the book. In addition, in 2006 we began to use the term "Mechatronics", describing the cooperation between mechanical engineering, electronics and IT, for our solutions. Mechatronic drive solutions are used in production and increasingly also in logistics centres for achieving a high level of automation.

At this point I want to thank Dr. Erhard Tellbüscher, the CEO of Lenze AG, who provided the inspiration and the freedom that was needed to carry out this undertaking.

The contents of this book represent the combined work of a total of 26 authors. I am grateful to them for sacrificing their free time in order to structure and then write their individual sections. They are, in addition to me, Volker Bockskopf, Andreas Diekmann, Sabine Driehaus, Martin Ehlich, Frank Erbs, Dr. Carsten Fräger, Thorsten Gaubisch, Olaf Götz, Ralf-Torsten Guhl, Martin Harms, Torsten Heß, Dr. Sven Hilfert, Markus Kiele-Dunsche, Detlef Kohlmeier, Karsten Lüchtefeld, Sebastian Lülsdorf, Ralf Scharrenbach, Rolf Sievers, Mathias Stöwer, Uwe Tielker, Markus Toeberg, Johann Peter Vogt, Karl-Heinz Weber, Hans-Joachim Wendt and Stefan Witte.

An essential part of any good reference book is its illustrations. Michael Hartung, Andreas Mletzko, Stephan Riese and Jutta Stuckenbrock did sterling work in this area. Jörg Baldig and Sandra Schindler sourced additional illustrations from other companies.

Additional support for this book was provided by Volker Arlt, Jörg Baldig, Nicole Funck, Harald Hilgers, Josef Lackhove, Cristian M. Popa and Andreas Tolksdorf.

Detailed technical revision was undertaken by Dr. Jörn Steinbrink of the Leibniz-Universität Hannover and Professor Karl-Dieter Tieste of the Fachhochschule Braunschweig-Wolfenbüttel. In particular, our former colleague Professor Tieste immersed himself thoroughly in the project and offered many valuable suggestions for improvements to the contents.

We should also like to thank Springer-Verlag, especially Eva Hestermann-Beyerle and Monika Lempe, and Steffi Hohensee from the company Le-Tex for their help and cooperation.

The English translation of the originally German edition was organized by the Technical Documentation team of Lenze. The Lenze UK sales organisation organised a thorough proof reading. Thanks to all who helped to ensure a good English translation and style.

Finally I want to single out three more colleagues without whom this book would not have been completed. The first is Stephan Riese, who despite having only recently joined the company took over a great deal of the organisational work without complaint.

Karl-Heinz Weber has the highest applications expertise of any of my colleagues and as such was the brains behind the drive solutions and the many usage examples. He worked on four chapters himself and provided extensive support to the authors of the other applications-related chapters.

Lastly, my special thanks go to my colleague Hans-Joachim Wendt, who offered guidance and encouragement throughout the project and helped to shape its final form. The depth and maturity of the book would be the poorer without his help.

Hameln, October 2007   Edwin Kiel

List of authors

Volker Bockskopf. Studied mechanical engineering at the University of Paderborn, postgraduate studies and completing the "Quality and environmental management" course at the Fresenius Academy in Dortmund.
1995-2001: Implementation of quality and environmental management systems. Implementation of process management systems.
Joined Lenze in 2001: International process management, specialising in the protection of the environment and workplace safety.

Andreas Diekmann. Studied electrical engineering (specialising in power engineering) at the University of Applied Sciences of Bielefeld.
2003-2004: Electrical engineering design, PLC programming.
Joined Lenze in 2004: Technical product support for servo drives, product training, software development for standardised applications.

Sabine Driehaus. Studied electrical engineering at the University of Applied Sciences of Osnabrück and the University of the West of England (Bristol).
Joined Lenze in 2001: Technical product support, key account management, service and applications support in the USA, application development for frequency inverters.

Martin Ehlich. Studied electrical engineering at the Ruhr-Universität Bochum.
1997-2001: Worked as a drive engineering project manager for a manufacturer of machine tools.
Worked for Lenze from 1990-1996 and rejoined us in 2001: Software development for digital inverters. Project manager for the development of digital servo inverters. System integration for new products, advance development of inverters, system architecture.

Frank Erbs. Studied electrical engineering at the Leipzig University of Applied Sciences.
1998-2001: Control technology, control cabinet configuration for an engineering consultant specialising in innovative water engineering techniques.
Joined Lenze in 2001: Technical product support for servo drives, product training, key account management.

Dr. Carsten Fräger. Studied electrical engineering at the University of Paderborn and the University of Hannover.
1994 Received Dr.-Ing. degree from the University of Hannover.
Joined Lenze in 1989: Servo technology application. Head of motor development. Head of product management for servo drives.

Thorsten Gaubisch. Studied information technology (specialising in process and automation technology) at the University of Applied Sciences and Arts of Hannover.
Joined Lenze in 2003: Technical product support for frequency inverter technology, product training.

Olaf Götz. Studied electrical engineering (specialising in automation) at the University of Paderborn.
Joined Lenze in 1998: Technical product support, drive dimensioning and product configuration. Development of PC-based tools for dimensioning drive systems.

Ralf-Torsten Guhl. Studied engineering (specialising in drive engineering) at the Otto von Guericke University of Magdeburg .
1993-1999: Mechanical engineer for an engineering consultant specialised in gearboxes.
Joined Lenze in 1999: Head of gearbox development.

Martin Harms. Studied electrical engineering (specialising in power engineering) at the University of Hannover.
1995-1999: Sales and application of electrical drive engineering. Design, configuration, commissioning of drive and automation systems.
Joined Lenze in 1999: Application engineer for electrical drive engineering. Technical product support for servo drives (specialising in cam drives).

Torsten Heß. Qualified as an industrial foreman in electrical engineering (measuring and control equipment) at the Technical Military Academy in Aachen. Studied electrical engineering (specialising in process information technology and automation) at the University of Applied Sciences and Arts of Hannover.
1986 - 1994: Weaponry service and maintenance team leader with the German army.
Joined Lenze in 1999: Technical product support for servo drives.

Dr. Sven Hilfert. Studied electrical engineering at the Otto von Guericke University of Magdeburg, specialising in electrical drives and electrical automation.
1999 Received Dr.-Ing. degree from the University of Essen.

Joined Lenze in 1999: Development of servo motors. Head of three-phase AC motor development.

Dr. Edwin Kiel. Studied electrical engineering at the Technical University of Braunschweig, Germany.
1994 Received Dr.-Ing. degree from the Technical University of Braunschweig, Germany.
1984-1995 Working at the Institute of Applied Microelectronics (IAM) in Braunschweig, Germany, at last head of engineering.
1995-1998: Head of engineering for a drive manufacturer.
Joined Lenze in 1998: Head of development. Head of innovation management.

Markus Kiele-Dunsche. Studied electrical engineering at the University of Paderborn.
1999-2001: Employed by a pump manufacturer developing drive solutions for pumps.
Joined Lenze in 2001: Technical product support for frequency inverters, development of PC-based tools for dimensioning drive systems.

Detlef Kohlmeier. Studied electrical engineering at the University of Applied Sciences and Arts of Hannover.
Joined Lenze in 1993: Configuration and commissioning of test systems. Development of reliability strategies and process management. Head of management systems.

Karsten Lüchtefeld. Studied electrical engineering (specialising in automation) at the University of Applied Sciences of Osnabrück.
Joined Lenze in 2001: Technical product support for servo drives, software development for standardised applications.

Sebastian Lülsdorf. Studied electrical engineering and automation at the Technical University of Munich.
1988-2000: Software product development for a manufacturer of machine tool control systems. Technical project management, software product development, Internet security for a computer manufacturer.
Joined Lenze in 2001: Development of engineering software for drive-based and controller-based systems, specification, usability.

Ralf Scharrenbach. Studied electrical engineering (specialising in automation) at the University of Cooperative Education in Mannheim.
1994-2001: Service for electrical drive engineering, fieldbus technology and PC software applications. Innovation management for engineering, fieldbuses, drive engineering for a drive manufacturer.
Joined Lenze in 2001: Customer solutions project management, key account management. Head of robotics & handling.

Rolf Sievers. State-qualified engineer, Stadthagen.
Joined Lenze in 1974: Service and development of DC converters. Development of frequency inverter technology. Technical product support, application support for frequency inverter technology.

Mathias Stöwer. Studied electrical engineering (specialising in information technology) at the University of Applied Sciences and Arts of Hannover.
Joined Lenze in 1992: Development of DC converters, servo inverters and intelligent servo technology. Development of engineering software for drive-based and controller-based systems, specification, usability.

Uwe Tielker. Studied electrical engineering at the Technical University of Lippe at Lemgo.
1986-1988: Commissioning of a telecommunications system for an electrical engineering company. Development of control systems for moving staircases for a lift manufacturer.
Joined Lenze in 1989: Customised development/application of analogue DC converters and hardware accessories. Development of digital DC converters. Application, configuration, commissioning of DC and servo drive systems. Technical support for key accounts in the development of new machines. Head of technical product support.

Markus Toeberg. Studied electrical engineering (specialising in drive engineering) at the Technical University of Lippe at Lemgo.
Joined Lenze in 2001: Technical product support for servo drives, product training, key account management.

Johann Peter Vogt. Studied electrical engineering (specialising in electrical machinery and drives) at the University of Hannover.
1997-2001: Development of torque motors for a ventilation and drive manufacturer.
Joined Lenze in 2001: Technical product support for servo drives specialising in cam drives, development of PC-based tools for dimensioning drive systems.

Karl-Heinz Weber. Studied electrical engineering (specialising in power electronics) at the Technical University of Lippe at Lemgo.
Joined Lenze in 1978: Application, configuration, commissioning, servicing of drive systems and equipment. Applications project management, customised development projects, product specifications. Head of system engineering. Set up the customer training facility and applications lab. Head of technical application basics for drive systems.

Hans-Joachim Wendt. Studied electrical engineering (specialising in power engineering) at the University of Applied Sciences of Bielefeld.

Joined Lenze in 1988: Development of frequency inverters. Product management and marketing. Product range management, innovation management.

Stefan Witte. Studied electrical engineering (specialising in power engineering) at the University of Applied Sciences and Arts of Hannover.
Joined Lenze in 2003: Technical product support for safety engineering, product training, key account management.

Table of Contents

1 Introduction

Highly automated factories, requiring only limited manual intervention, are a fundamental feature of our lives today. As a consequence, the supply of most everyday goods involves only minimal manpower. This book explores how controlled drives in these factories and in logistics systems ensure that products are manufactured and delivered to the end user in large volumes and with maximum efficiency.

Highly automated production and logistics. Most people in industrialised nations take for granted the possession of cars, books, computers, consumer electronics and communications devices. At the same time, the workplace for most people has shifted from farms and factory floors to offices and shops. This is a huge step forward from pre-industrial times, when the majority of the workforce had to be employed solely for the purposes of food production.

This industrial transformation has come about as a result of the fact that production and transport processes have become increasingly mechanised and automated, meaning that they no longer require direct human intervention and are taken over by machines. In modern factories the human workforce concentrates largely on planning and monitoring processes, while fewer and fewer operations are performed by hand.

Factories manufacture their products in huge numbers because people's needs are generally very similar. Many people eat the same yogurt, read the same newspaper or book, and drive the same car (with only minor individual modifications). The more consumers a factory can reach with its products, the higher its production rate can be, and the more economic it is too as a rule. Economic limits are set only by the size of the potential market (number of consumers), individual demand (how often I buy a yogurt, a paper, a book or a car) and people's willingness to buy the same products.

Today a factory can sell its products in the industrialised markets (Western Europe, USA, Japan) – and increasingly also in the developing regions – to upwards of 100 million people, without hindrance from political borders or inefficient transport routes. Of those 100 million people, if just one percent wishes to buy a car with a lifetime of 15 years built on an assembly line operating two shifts per day, then that assembly line has to build that car at a rate of one every 2.7 minutes (the VW Golf is built in

Wolfsburg, Germany, at a rate of one every 0.3 minutes). The same calculation applied to a yogurt that is eaten every day for breakfast by the same number of people would require a production rate, with round-the-clock operation, of 14 units per second or 50,000 units per hour (high-speed lines currently produce 30,000 pots per hour). These examples show why the production of many consumer and utility goods in highly automated factories is measured today in units per minute or units per second.

How is this achieved? Even in the earliest factories it was known that it is far more efficient to perform the same operation repeatedly than to combine multiple operations before moving on to the next product. Although this principle of division of labour is efficient, it soon becomes boring and monotonous to the worker. It is quite a different case with machines. It is much easier to build a specialist machine for a single operation than a universal machine for a variety of different tasks. That is why most production machines are designed according to the division of labour principle, breaking down the overall work process into a series of individual steps that are performed by individual workstations. The operating rate of these workstations corresponds to the production rate of the plant. This basic principle of automated mass production has shaped the appearance of modern factories: a continuous flow of products that moves as if by magic from one station to the next, before being despatched at the end of the production process.

Similarly, the onward logistics, in other words the transfer of goods from production to the end user, can no longer be managed efficiently without automation. For some time now, most internal transport and storage operations have been performed by conveyor belts and storage and retrieval systems rather than by humans. The process of putting together deliveries (order picking) is now increasingly automated too. The bottom line is that virtually all consumer and utility goods today are produced in high volumes in highly automated factories and then delivered to the customer via equally highly automated logistics and distribution centres. For many day-to-day consumer goods (such as foodstuffs), the first time they are moved by a person is when they are taken off the supermarket shelf by the customer. Prior to that they will have been moved by hundreds of controlled drives: during production, packaging, stacking onto transport pallets and storage and retrieval.

Controlled drives in highly automated machines and production plants. One of the most important functional elements in highly automated machines and plants are controlled drives, which function as actuators and whose motion sequence is determined by a control system. Such automated drives replicate the operations performed by a person in a manual process. The interplay of brain and muscle is replaced by the interplay of software and an energy transformation that it controls, generally from electrical into mechanical energy. The importance of drives for factories and machines is illustrated by the fact that the first factories were only made possible by the central provision of drive energy via water power or a steam engine ("power machine"); this drive energy was then distributed mechanically by the "transmission" to the "work machines". This transmission is now provided by distributed electric motors.

A variety of different drive principles can be found in the field of engineering. This book concentrates solely on *electromechanical energy transformation by controlled electric motors*, however. The main reason for this is that electric motors can be controlled very effectively, making them ideal for tasks that require precision motion control. Since three-phase AC motors are used almost exclusively for such tasks nowadays, they are all that are considered here.

The task of controlling the flow of energy to the electric motor to enable it to execute the desired function is undertaken by *inverters*. They consist of two key elements: the *power electronics*, which are responsible for the electrical energy transformation, and the *control electronics*, with the software that controls this energy transformation. The software has to accomplish two tasks here:

1. Defining the motion response of the drive (*motion control*).
2. Controlling the motor and its response at any given time in accordance with the desired response (*drive control*).

The drive control software is generally located in the inverter, while the motion control software can be located in either the inverter or a higher-level control system. While many positioning movements are implemented directly in the inverter, in the case of coordinated multi-axis movements this software function is executed by a special motion control system.

In many applications the optimal operating point of an electric motor does not correspond to the requirements of the work machine. In such cases *gears* adjust the mechanical operating point. In addition, drive elements (shafts, clutches, spindles, toothed belts, rack and pinions) are used to couple the rotary motion of the motor or gear to the effective point in the work process in the most efficient way. Some of the drive elements also convert the rotary motion into a linear motion.

Mechatronic drive solutions. Controlled drives connect the automation control systems, which nowadays are software-based systems, to the machine and hence to the mechanical system. The combination of IT, electronic and mechanical systems is known as "*mechatronics*", a made-up word that was coined in 1969. So we refer to a controlled drive that executes a function in a machine or plant as a "mechatronic drive solution". This book describes how these mechatronic drive solutions are constructed and how they execute their function in the machines.

As diverse as the products that are produced by highly automated methods are the machines that are used to make them. The focus on product ranges and production processes means that machine manufacturers are often highly specialised, and the high production rate of their machines limits their market. For example, if one production line can meet 1% of the need of 100 million people, the entire world market consists of 1 billion people (in the developed countries) and a machine operates for 10 years, then only 100 new machines are needed each year for production worldwide. If the market is split between two equally-sized suppliers (to avoid any monopoly), then each can only sell 50 machines or plants per year to meet global demand. Sales can only be increased if global demand rises, i.e. if more people live in developed countries.

Despite this diversity in the machines and their limited numbers, there is a level at which commonalities can be found between drives in different machines without losing the specific association with the drive task in the machines. *Mechatronic drive solutions* address this level. They are defined by the principal task that a drive performs in a machine rather than by the specific technical implementation, which depends on the individual application case. The tasks (e.g. conveying, lifting, positioning, winding) can be classified and described in a non-industry-specific manner. On the basis of these tasks, requirements can be defined for the configuration of the components (motor, inverter, gearbox) of the drive as well as for the software functions in the drive. What is more, once these drive tasks have been established, the components and software functions can be developed and prepared in such a way that the specific application can be implemented quickly and reliably.

This book offers a comprehensive description of these mechatronic drive solutions for highly automated production and logistics. It begins by describing their operating conditions. How are factories and logistics centres structured, what basic machine types do they have, and how are they automated (chapter 2)?

This is followed by a consideration of the basic structure of a drive and its components (chapter 3):

- What types of motor are there and what are their main features?
- How is an inverter constructed?
- What are the elements (gearbox, drive elements) that transfer the mechanical energy to the work machine?

Aspects of drive dimensioning and the reliability of drive systems are also covered in this chapter.

Then chapter 4 brings the two levels together. The entire range of applications is broken down into twelve mechatronic drive solutions. The many hundreds or thousands of drives in highly automated factories in the most disparate sectors of industry can be classified according to this system, while on the other side ten of these twelve drive solutions are to be found in the vast majority of factories. What tasks are performed by these drive solutions, what their operating conditions are and what a specific solution looks like are all covered in chapter 4.

Finally, chapter 5 addresses the engineering process. It looks at how a specific solution is developed and how ultra-efficient tools can be used for dimensioning, selection, software configuration and for commisioning and optimising the drive behaviour. It concludes by considering the entire life cycle and the associated costs.

The book concentrates on the fundamental issues, which are illustrated by means of specific usage examples. It is aimed at mechanical, electrical and mechatronic engineers working for machine or plant constructors or system integrators, who are faced with the task of designing and commissioning one of these highly automated machines or production plants. Engineers responsible for ensuring that the machines operate as efficiently as possible will gain a better understanding of how drive technology works in their plants. And engineers and students in the field of mechatronics and automation technology will learn how to find the optimal solution for drive tasks, quickly and reliably, to ensure that the machines and plants that are built execute their task in the best possible way.

2 Industrial production and automation

Dr. Edwin Kiel

This chapter looks at the application areas for controlled drives in highly automated production facilities. It starts by describing the typical characteristics of such production facilities. What is being produced, in what quantities and which production sequences are being deployed? The route from factory to end user (which usually passes via large-scale highly automated logistics centres) is also described.

The description of the typical features of these production facilities is followed by an outline of the production methods and types of machine used in them. The types of machine already provide an indication of the types of drive tasks to be solved. This chapter concludes with an overview of the automation structure in machines and production facilities. The automation provides the drives with the information they need to execute their task in the machine.

2.1 How production and logistics systems are structured

This chapter starts by describing the products for which highly automated production facilities are used, what their typical production rates are like and the nature of their production processes. It deals with mass production facilities for durable goods and consumer goods, since it is these which are characterised by high levels of automation and need a correspondingly high number of drives. In contrast, production facilities for investment goods (e.g. for machines) usually have fewer automated sequences and significantly less repetition, with the result that universal machines such as machine tools are used and assembly is in most cases carried out manually.

Highly automated production facilities replace manual resources with mechanised and automated sequences. The primary effect of this has been a general increase in prosperity, since people no longer have to work all day just to meet their basic needs (food, clothing, accommodation) and can instead dedicate their time to other activities.

2.1.1 Goods produced

Goods produced in highly automated facilities can be found in all aspects of daily life. An overview of these is provided in Table 2-1.

Table 2-1. Mass-produced goods

Sector	Products	Facilities
Food	Foodstuffs, beverages	Food producers, beverage filling plants, mills, sugar factories, dairies
Clothing	Textiles, items of clothing	Textile plants
Home and domestic	Building materials, furniture, household appliances, interiors	Building material plants (for stone, cement, floor coverings, flagstones, windows), furniture factories, plants producing household appliances
Health	Personal hygiene, medication	Pharmaceutical plants
Automotive	Vehicles	Vehicle manufacturers and suppliers
Communication	Telephones, computers	Telephone factories, mobile phone factories, components for computers
Entertainment, culture	Books, magazines, CDs, DVDs	Printing houses, bookbinders, newspaper printers, CD pressing plants

Mass production is the rule for products common to daily life; exceptions exist only:

- If the materials are not entirely suitable for automated processing (e.g. limp materials such as textiles or meat).
- If demand is low and individuality is prized (handcrafted goods, luxury goods).
- If long storage or transport times are not possible because the goods have to remain fresh (bread and baked goods, fruit and vegetables).

Most durable goods and consumer goods common to daily life and even goods designed for long-term use such as cars, furniture and household appliances are manufactured in highly automated facilities.

2.1.2 Production quantities

In order to determine the efficient size of highly automated production facilities, the rule of thumb is that increasing production rates will increase cost-effectiveness. Limits are only reached when the production rate becomes so high that it ceases to be efficient or, due to concentration on a very small number of sites, transport costs become unreasonable. This rule of thumb explains why machine performance will continue to increase until either physical limits are reached or production rates cease to be efficient.

How is the economically feasible limit of a production rate determined? There are a number of contributing factors:

- How big is the achievable market?
- What is the situation as regards market saturation for this product; in other words, what is the maximum number of people (as a percentage of the total population) who will buy this product?
- What is the competition like for this product; in other words, how many other factories does the facility need to compete with on the market?
- How high is the market potential resulting from typical regular demand for the product?

All four of these factors will be considered in more detail below and examples of each will be presented in order that conclusions can be drawn.

Size of the achievable market. The size of the achievable market is determined by the number of people to whom a facility can supply its products. Limits can be political in nature (trade barriers); however, these have been and continue to be breached gradually in the course of political unification processes for the most developed regions of the world. Population figures for the most industrialised regions of the world appear in Table 2-2.

Table 2-2. Size of achievable markets

Region	Inhabitants
European Union (EU)	Approx. 500 million people
NAFTA (USA, Canada, Mexico)	Approx. 430 million people
Japan	Approx. 120 million people

Transportation routes can impose further limits. For all products which do not age quickly (and/or items which can be made non-perishable such as packaged foodstuffs) and for which, therefore, actual transport time is of no relevance, the distance from the product facility to the consumer is not a primary concern.

The situation is quite the opposite in cases where a product's freshness and/or sell-by date are of the essence. Daily newspapers are an example. They have to be with the consumer by five o'clock every morning but go to print as late as possible in order to ensure that the stories appearing from the previous day are topical. In real terms, this means that a press printing a local newspaper is only able to supply a radius of approximately 100 km. However, within such a radius, the increasing productivity of printing machines has contributed to the joining together of regional newspapers which were once independent, with the result that one region usually only has one daily newspaper (large cities are an exception to this rule).

Where products such as bread and baked goods are concerned, the need for freshness means that although highly automated production is possible in technical terms, it is not widely established.

It can generally be assumed that a facility producing non-perishable goods is able to reach a sales market of at least 100 million people.

Market saturation. Many new products take only a short time to achieve very high levels of market penetration. In recent decades, the time needed for new products in the consumer electronics sector to penetrate the market has dropped to just a few years.

Every new product reaches a point at which market penetration can no longer be significantly increased. For example, in Germany there are now 60 cars for every 100 people, in the USA this figure is 75 for every 100.

Generally speaking, the following market saturation rates can be assumed:

- mobile phones: Approx. 90 to 120% (many people have more than one mobile phone)
- consumer electronics appliances: Approx. 80 to 100%
- cars: 50 to 70%
- household appliances: 30 to 70% (e.g. 1 to 2 fridges per household)
- computers: Approx. 35% (70% of all households)
- books: Approaching 100%
- cigarettes: 20 to 30%
- non-alcoholic drinks: 100%
- alcoholic drinks: Approx. 50%

If factories are producing products which have already achieved maximum market penetration, this market saturation must be taken into account when determining the economically feasible production rate. This issue is critical if market penetration continues to rise. In this case, initial purchase demand is greater than subsequent replacement purchase demand. This introduces the risk that factories will be built too large and subsequently be overdimensioned.

Competition. The next factor determining the economically feasible production rate is the competitive environment. Tastes and interests vary from one person to the next and some of us have a conscious desire to be different from those around us. Accordingly, all products are not going to appeal to all people. Furthermore, there will be different customer needs within the same product group (e.g. cars) - one person will want a budget car, the second a fast car and the third a large family car - and this generates demand for different products. Ultimately, our economic system is based on competition. Competitors compete for customers and want to use differentiation to gain a competitive advantage.

This effect can be described with the average share a product can achieve in a product group. This value can vary dramatically.

- Where cars are concerned, the most popular model sold in Germany has a market share of 6%. An average market share for cars is approximately 2% for each product range. All product ranges with smaller market shares are either import with their primary market located elsewhere or exclusive models such as luxury vehicles.
- For many foodstuffs, it can be assumed that a product will likewise be able to achieve a share in the region of 2 to 4%. In the supermarket, it is usual to find between 5 and 10 brands with several products competing; only in the case of discount stores is variety restricted.
- Only products of lesser interest to consumers and which usually do not have a very high market potential are able to achieve higher market shares. As there are only 2 to 3 suppliers of beef broth, adhesive tape or vacuum cleaner bags on the market for example, each accordingly has a market share of 30 to 50%.

An average market share of 1 to 2%, served by one factory, can be assumed for many products. A supplier whose market share is higher will usually distribute production across several sites. However, it is precisely products with low market potential and few competitors which are often supplied by just a small number of factories (2 to 4) on a market amounting to 100 million people.

Regular demand. Whilst the three factors discussed thus far have a variance of approximately 1 to 1.5 orders of magnitude, the situation is very different where demand is concerned. In this context, variance is much higher [Bu06a]:

- Cars have an average age of approximately 8 years; in other words, a car will last for approximately 15 years. Accordingly, the number of new vehicles registered per year in Germany is approximately 3.5 million and the total number of cars in the country is approximately 48 million.
- Other household goods designed for long-term use also have a period of use lasting between 10 and 20 years (household appliances such as washing machines, furniture).
- Consumer electronics appliances are used for between 5 and 10 years.
- Computers are used for between 3 and 5 years.
- On average, an individual will purchase 12 books per year.

- Foodstuffs such as tubs of margarine need to be purchased approximately once a month.
- Daily newspapers appear 6 days a week.
- 3 bottles of mineral water a day are sufficient to meet the demand for fluids.
- Smokers often smoke in excess of 20 cigarettes every day.

The demand rates quoted range from $7*10^3$ to $5*10^{-2}$ per year; in other words, they span 5 orders of magnitude.

Economically feasible production rate. By combining the four factors we can now calculate the following production rates (Fig. 2-1).

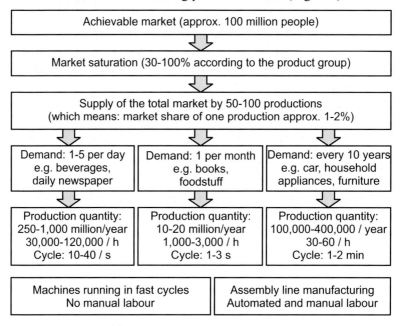

Fig. 2-1. Production volumes for mass production

The results of this calculation are in many cases identical to the production rate of corresponding machines and production lines. For example, car production lines generally work at a production rate of between 200,000 and 400,000 units per year. Books are produced at maximum rates of between 2,000 and 3,000 items every hour. Beverage filling plants fill approximately 30,000 units every hour. Cigarette machines have the highest production rate, producing more than 0.5 million units every hour.

In every one of these cases, production and machinery were optimised until the economically feasible production rate was reached. This chapter can be summarised as follows:

- Clock-cycle rates in the region of a minute work for products people will purchase every few years, e.g. cars and household appliances. This results in production rates of several hundred thousand units per year.
- Clock-cycle rates in the region of an hour work for products people will purchase every few weeks, e.g. books, tubs of margarine, etc.
- For products purchased on a daily basis (e.g. daily newspapers), clock-cycle rates in excess of 10 per second (or 40,000 per hour) are required.
- The highest production rates are needed and work for products consumed in high volumes on a daily basis (drinks, cigarettes).

In most cases, multiplying production rate by revenue per unit results in a turnover of between 10 and 50 million euros. Highly automated production facilities usually have a per-head turnover of between 200,000 and 500,000 euros per employee; in other words, such facilities will have between 50 and 200 employees. These are minimum values for multi-shift operation in order to be able to map the required functions and qualifications to individuals.

This calculation does not apply to automotive plants, where annual production rates amount to several billion euros and several thousand staff will work on one production line. Hence, automotive plants are the largest and most complex example of mass production and have therefore naturally been used as a model for other production environments. Technologies which have proved successful in automotive plants can also be used in other highly automated production environments.

2.1.3   Production processes

No finished good on the market is homogeneous and is made up of a single material. Even a basic foodstuff like sugar is packed in a bag and stacked on a pallet before it leaves the factory.

From a system perspective, production is a sequence of work steps during which parts are changed or combined with other parts (by means of joining or assembly, for example) to create assemblies.

We will now look at the typical workflow for various industries before considering a number of basic principles for mass production. These basic principles are the starting point for production techniques and processes in machines to be executed by drives.

Mechanical engineering. Machines are usually only manufactured in small quantities (annual volumes of 50 are common and more than 1,000 would be a major exception). The production sequence for machines is described in detail in the reference literature [Wi05a] and is characterised by the following basic elements (Fig. 2-2):

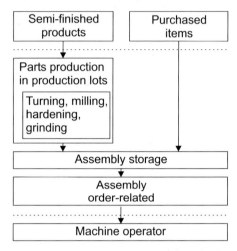

Fig. 2-2. Production sequence in mechanical engineering

- Machine-specific components are produced in parts production (drawing parts as opposed to standard parts).
- The work steps in parts production may require several machines (e.g. turning or milling for soft gear finishing, hardening, grinding for hard gear finishing at the end of the production sequence).
- The material used most frequently is metal (steel or aluminium).
- Production techniques can be split into primary forming (e.g. casting), forming (e.g. forging, pressing) and cutting (metal-cutting processes such as turning, milling, grinding). There are also techniques for surface treatment and joining.
- Wherever possible, parts are manufactured and put into temporary storage on an order-related basis.
- Parts production is followed by assembly, during which the component parts are combined to form units and ultimately a machine. This process is usually manual.
- Parts production and assembly are separate production areas.

This typical mechanical engineering production sequence differs from mass production on a number of counts:

- Production quantities are significantly lower.
- There is a tendency to use universal machines which are able to produce a variety of parts (e.g. machine tools).
- The flow of goods to the end product is not continuous and homogeneous (it is common for intermediate products to be put into temporary storage rather than being forwarded for further processing immediately).

- There is a clear organisational split between parts production and assembly.

It is for these reasons that the production principles of mechanical engineering are not applied universally in mass production and other basic principles are common.

Automotive production. Although many of the component parts which make up a vehicle are produced using similar techniques to those found in mechanical engineering (e.g. the combustion engine, the gearbox), the sequences are very different (Fig. 2-3):

- The flow of material through the various stages of production (body-in-white, paint shop, final assembly) is continuous and there is no opportunity for prolonged intermediate buffering.
- Production equipment is designed specifically for the model in production. In some cases, the same production equipment can be used to produce a mixture of a number of different models.
- Wherever possible, parts and units are forwarded to final assembly without prolonged intermediate buffering (just-in-time delivery).
- The entire automotive production workflow, and even the workflow associated with complex units such as combustion engines or gearboxes, is distributed across several hundred workstations with the same clock cycle (assembly line production). The machines in these workstations are customised for the corresponding work step.

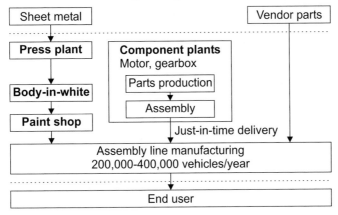

Fig. 2-3. Production sequence in the automotive industry

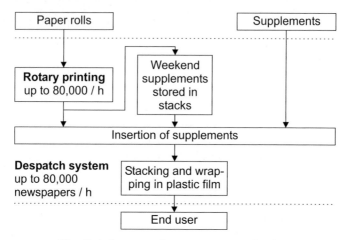

Fig. 2-4. Sequence in newspaper production

Daily newspaper. The next example is the production of a daily newspaper (Fig. 2-4). Production rates can reach up to 80,000 copies per hour. A newspaper printer working at this rate can print 400,000 copies during the print run which starts at the press deadline (9 o'clock in the evening) and finishes at the end of delivery (2 o'clock in the morning), thereby meeting an average demand for more than a million people.

The production process comprises two stages:

- Rotary printing, the process used to produce the sections of a newspaper. The raw materials are rolls of newsprint. The product is the printed, cut and folded newsprint which appears in a newspaper section.
- Newspaper processing, the process by means of which the books and supplements are folded together and then stacked; these newspaper stacks are then wrapped in plastic film to provide protection against moisture and soiling.

Not all parts of a newspaper are printed immediately. For example, parts of the Saturday edition (weekend supplements) are printed in advance during the week outside of the main print run and then stored temporarily in stacks or rolls. Advertising inserts are also usually printed in other locations for a number of newspapers and then forwarded to the production process via an appropriate form of intermediate storage (stacks, rolls).

These are the basic principles of newspaper printing:

- The raw material is stored in large quantities on rolls; rolls can also be used for intermediate storage.
- The flow of material is continuous; rotary printing and the despatch system work at the same production rate.

- The end product is combined in manageable despatch units (newspaper stacks) which can be forwarded directly to the next stage of transportation (HGV bound for newspaper retailers and their paper deliverers).
- There is no manual intervention in the production process. The first time human hands come into contact with the product is when the newspaper stacks are loaded and unloaded during transportation. Paper deliverers do not come into contact with the newspaper until they push it through the letterbox.

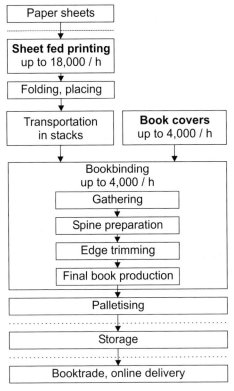

Fig. 2-5. Sequence for book production

Books. Although the production of books is in principle an example of mass production, it differs significantly from other products (Fig. 2-5). Books are manufactured with very high variance (individual print runs can vary greatly between 500 and up to 50,000 copies). The production process can be divided into the following stages:

- letterpress printing,
- bookbinding

Printed sheets are usually printed on a sheet-fed offset printing machine. The raw materials are stacks of unprinted paper. The end products are printed sheets (in a variety of colours, double-sided, the printed paper is sometimes also surface-treated). Today's production rates can reach up to 18,000 sheets per hour. A book is made up of between 10 and 30 sheets depending on the number of pages it contains; accordingly, between 600 and 1,800 books can be produced every hour.

These printed sheets are then folded and stored in stacks ready for further processing by the bookbindery.

In addition to the folded printed sheets, the bookbindery needs other base materials such as book covers. There are special machines for these.

The bookbindery starts by putting the printed sheets together. The spine is then added (using an adhesive or by stapling or stitching), the edges are made using a three-knife trimmer and finally the book goes into final production, where its cover is added. Bookbinding machines can run at speeds of up to 4,000 books per hour.

When books reach the end of the production sequence, they are stocked on pallets (they might be wrapped in plastic film prior to this). They are then stored on the pallets or despatched to distribution centres from where they are sent to booksellers or (in the case of online purchases) direct to consumers.

Printing and binding hardly ever take place on the same site; it is rare for the two to be linked in the same place. This is because variance rates (in particular the number of printed sheets per book) are too high. However, in larger-scale operations, there are links between the various work steps involved in the bookbinding process and all machines work at the same production rate. Sometimes intermediate times are built in due to longer transportation routes (cooling screws) to allow for cooling or conditioning.

Book production too shares the basic principles of mass production:

- High production rate.
- Highly specialised machinery is needed to complete the work steps (and this machinery has to be able to manage the variance at which books are produced, e.g. in terms of paper quality, number of pages, format, cover).
- Automatic transportation of materials.
- The end product is ready for forwarding on pallets (stacks, sometimes wrapped in plastic film).

Beverage filling. Due to high levels of consumption, beverages are produced at very high production rates. The overall production process (Fig. 2-6) can be divided into the following stages:

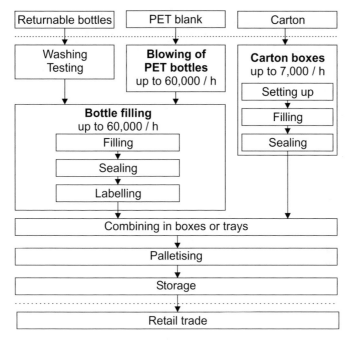

Fig. 2-6. Sequence for beverage filling

- The beverage is first produced in large containers. In the case of wine, we are familiar with the barrels which have since given way to stainless steel tanks. Beer is also brewed in large brewing tanks; the production process (fermentation followed by storage) takes approximately 5 to 7 weeks. Fruit juices are also produced initially in tanks from water and juice concentrate produced in the country of origin.
- Glass bottles are bought in from glassworks or, in the case of recycled bottles, cleaned and then tested for sound condition.
- PET bottles are blown from blanks on the actual beverage filling site. The machines used for this purpose can produce up to 60,000 bottles an hour. Within the machine, the total production time (2 seconds) is distributed across between 8 and 40 blowing stations arranged on a carousel.
- Bottle-filling is a three-stage process. First the bottles are filled, then they are sealed and finally they are labelled. Different machines linked via conveyors are used for each stage. Here too, the maximum production rate is 60,000 bottles an hour.
- There are machines which can take the base material for beverage cartons supplied in the form of stacks or on rolls, manufacture the cartons, fill them and then seal them. These work at production rates of up to 7,000 cartons an hour.

- Once the bottles or beverage cartons have been filled, they are either packed in crates (reusable packaging) or put on trays. Trays are folded boxes which can hold a number of bottles and are covered with plastic film.
- These crates or trays are then stacked on pallets which are transported to storage via conveyor belts. They then move from storage to HGVs bound for distribution centres or consumer markets.

Here too, the basic principles are the same:

- The flow of material is continuous.
- Specialist machinery which is linked in a chain and works at the same production rate is used.
- The product inside the beverage carton is produced in high volumes and then dosed into individual containers.
- On leaving production, the containers are grouped in trading units (crates or trays) and stacked on pallets which are then put directly into storage prior to forwarding for sale.
- There is no manual intervention; in the majority of cases, human hands do not come into contact with the containers during the bottle-filling process.

Basic rules for mass production. The following basic rules can be derived from the examples:

- The raw materials are supplied or produced in large quantities. In the case of two-dimensional materials (sheet, paper, textiles, film) these are rolls or stacks; in the case of liquids these are containers and tanks.
- Some production processes are continuous with subsequent separation; others are intermittent.
- Linking requires the same production rate throughout the entire production sequence.
- Automatic transport between machines is achieved using conveyors. Intermediate storage is avoided where possible (examples of exceptions are printed sheets in letterpress printing and weekend supplements).
- The production sequence includes grouping into units and stacking on pallets for storage and transportation.
- There is no manual intervention in the production sequence (except in the case of non-automated assembly processes); many products are not touched by human hand until the last stage of the process (handover to the consumer).

Now that we have considered a number of different production processes, we need to look at how goods produced in such large quantities are subsequently stored and distributed. For this purpose, we will now consider the systems used for the distribution of goods.

2.1.4   Distribution of goods

The concentration of production on a small number of sites, from where products are supplied to millions of people, means that the demand for the movement of goods (and accordingly logistics) is high. It is the job of logistics to make sure that consumers get the goods they need at the right time, in the right place, in the right amount at as low a cost as possible.

As only a very small number of products are produced to order (the major exception to this rule is probably the car, whereby production is triggered following configuration and placing of an order by a customer), intermediate storage is essential whilst en route to the consumer. Every storage phase is a compromise for the following mutually-contradictory requests:

- Minimum storage (capital commitment) or in other words maximum turnover of stock.
- Maximum possible production lots in order to minimise non-productive time and set-up time in the event of production switching over to a different product.
- Optimum transport routes and transport units.
- Minimum possible number of stock turnover and picking processes.

Taking these issues into consideration, distribution logistics aims to create as efficient a system as possible to transport finished goods from production to the consumer. The systems commonly used to achieve this today are described below.

A central element of all logistics systems for the distribution of goods is a distribution centre (also known as a hub) where goods are received from all directions, distributed and then despatched (again in all directions). Usually, a distribution centre of this type will also have a storage facility. In the case of distribution-only centres (postal or parcel distribution centres, freight carrier hubs), incoming goods are forwarded as quickly as possible.

Logistics in the retail trade. If goods reach consumers via the retail trade, the overall goods flow usually looks like this (Fig. 2-7):

- Production facilities supply goods to the distribution centres of retail chains or purchasing organisations. Depending on the scope of the product range, retail chains in Germany will have one or more of these distribution centres. Producing companies supply goods in complete pallets (loading units) either directly from production or from a finished goods store.

- Goods are despatched from distribution centres and reach sales outlets within a matter of days. Scanner tills at the point of sale enable stock information to be accessed at any time so that reorders can be placed before goods sell out. For retail outlets, deliveries are usually put together (picked) on the basis of trading units, i.e. not complete pallets but the next smallest packaging units. Containers with wheels are frequently used to transport goods. In recent years, some of these picking processes have become automated.
- In total, the product is stored in three places: in the producing company's finished goods store, in the distribution centre and in the retail outlet. There are fewer storage points if the product is supplied directly to the retail outlet by the producing company. However, this only makes sense for products with high turnover rates (e.g. beverages). Having said that, complete pallets can also be transported directly to distribution centres from production without first being put into storage. This means that producing companies have to be extremely flexible in respect of their production techniques since these have to be in line with the requirements of their customers.

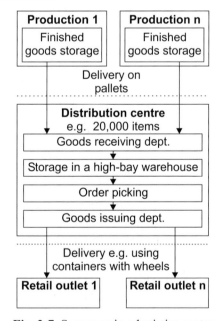

Fig. 2-7. Sequence in a logistics centre

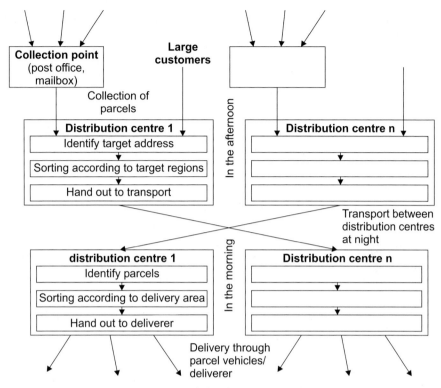

Fig. 2-8. Logistics for parcel services

Logistics for parcel services. In the case of parcel services, the following logistics sequence is common (Fig. 2-8):

- During the course of an afternoon, the parcels are either collected by the parcel service or, in the case of large customers, delivered directly to the parcel distribution centre.
- That evening, they are sorted according to destination parcel distribution centre.
- Once sorted and grouped, the consignments travel via HGV overnight to the destination parcel distribution centre.
- Following their arrival, the next morning they are distributed on the routes taken by the various delivery lorries, ensuring that they reach their destination before lunch.

This system supports next-day delivery. Express services covering huge distances use aeroplanes instead of lorries. However, this requires an additional central transition point to which all parcel distribution centres make deliveries from all directions. Parcels are then sorted according to the destination to which the aeroplane will return following sorting and loading.

In a very small number of cases, the delivery times cited here are not fast enough. Drugs and medicines, for example, have to be delivered to pharmacies or patients within a matter of hours. Local distribution systems distributing goods several times a day cater for this need.

Online shopping generally relies on a combination of goods distribution centres and parcel service.

Containers are usually used for intercontinental trade. Goods are dropped off in and collected from container terminals and transportation times with container ships are longer. However, goods are ultimately moved in accordance with the same basic principle (collect, move, distribute) and with a high level of automation.

Distribution centres have to perform the following functions for logistics purposes:

- goods acceptance (identification, storage),
- storage, often in high-bay warehouses for pallets or containers,
- Picking of delivery orders (removal of some goods from a pallet and grouping in a delivery),
- sorting in parcel distribution centres,
- despatch (the consignment might need to be packaged before being transferred to a means of transport for forwarding).

None of these processes would be possible without a linked EDP system capable of locating goods at all times and planning their onward journey or without automated materials handling system plus related drives under the control of this EDP system. If human resources are deployed (e.g. for unpacking or picking of delivery orders), tracking of the individual processes (often using bar code readers) is used in an attempt to avoid every possible error source. The manual crossing off of goods from pick lists as orders are made out is a technique which is seldom employed.

A modern distribution centre for retail chain stores approximately 20,000 products in 40,000 bays and completes in excess of 100,000 storage and retrieval sequences every day. The majority of the storage and retrieval sequences are completed automatically by corresponding conveyor systems and material handling systems. A logistics centre of this type will have several thousand drives and employ 200 people.

Now that we have seen how modern mass production and distribution systems work, we need to look at how the machinery they use is structured in order to enable these production rates and the high level of automation to be achieved.

2.2 Machines in production and logistics

The previous chapter described which products are produced using highly automated facilities and cited typical production rates. It also described how these products reach the consumer. Production rates and automation now both have a direct effect on production techniques. As a starting point, the route taken by an end product to reach a consumer can be divided into two groups:

- the production steps used to make the product,
- the transport tasks used to move the product

In highly automated production facilities, both tasks are performed using a multitude of machines which in turn contain numerous drives controlled by means of automation. In the centre of this chapter we will look at the associations between machine types and drive function.

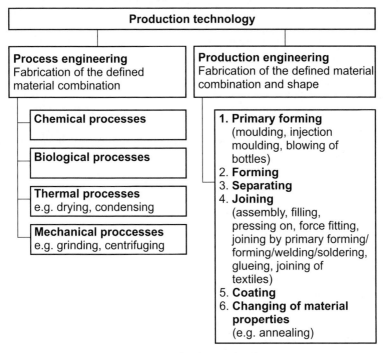

Fig. 2-9. Production technology

Table 2-3. Materials for durable goods and consumer goods

Organic materials	Plants used in foodstuffs	Cereals Fruit, vegetables
	Plants used in non-foodstuffs	Textile fibres (cotton, hemp), Wood Pulp (paper)
	Animals used in foodstuffs	Meat, milk
	Animals used in non-foodstuffs	Wool
	Mineral oil products	Plastics
Inorganic materials	Metals	Iron, aluminium, non-ferrous metals
	Minerals	Ceramics, cement
	Glass	
	Water	

2.2.1   Production equipment

Generally speaking, production technology can be divided into (Fig. 2-9):

- Process engineering (chemical, biological, thermal, mechanical), where products are given a defined material composition but not yet a defined shape.
- Production engineering, where products are given a defined shape.

All durable goods and consumer goods reach consumers as individual units; in other words, they are supplied with a defined material composition and shape. Accordingly, their production always involves manufacturing processes, at the latest when a product is packaged, even if the products themselves are initially produced using process engineering (e.g. beverages) [St00].

Materials for durable goods and consumer goods. A list of the materials used to produce durable goods and consumer goods appears below. Unlike in the case of machines, which are primarily made from metal, a multiplicity of different materials are used.

The list (Table 2-3) shows that metal processing is rare, whilst many products are based on renewable organic materials (this is true of both edible goods and non-edible goods such as textiles, wood and paper). The use of inorganic materials such as glass and minerals is also common.

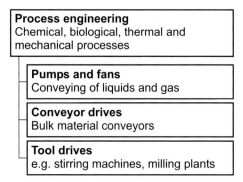

Fig. 2-10. Drives in process engineering

Process engineering. Process engineering takes in the technical and economic performance of all processes which change the composition, properties and nature of materials. These processes can be divided into four groups:

- chemical processes based on chemical reactions, e.g. oxidisation, polymerisation,
- biological processes, e.g. fermentation,
- thermal processes such as vaporising, distilling, absorption, extraction, crystallisation, drying,
- mechanical processes such as breaking up, milling, separation, filtering, centrifugation, agitating, mixing.

Process engineering processes are usually used to produce raw materials in high quantities. Typical processes are:

- the smelting of iron ore in furnaces,
- the manufacture of pulp in paper production,
- the manufacture of plastics in chemical plants,
- glass melting in glassworks,
- the condensing and centrifugation of sugar in sugar factories,
- the fermentation and storage of beer in breweries,
- most food production processes prior to products being packaged.

The concept of economies of scale also applies to process engineering plants, but size restrictions still apply due to physical limits (the process has to remain manageable) and the volumes that can be handled by the plant concerned. In the case of many basic materials (e.g. flour, sugar, milk products) there is a strong concentration on a small number of operating companies and it is then often the case that raw materials have to be transported to these sites from a wide catchment area.

A multitude of drives are used in process engineering plants (Fig. 2-10):

- drives for pumps and fans moving liquids and gases,
- drives for belt conveyors for bulk solids,
- drives for mechanical processes such as milling plants, centrifuges, stirring machines.

The drives in process engineering plants are more powerful than in production plants and do not perform highly dynamic tasks. They are often used directly for process control, whereby the process is controlled directly by means of conveying velocity (e.g. the air flow for thermal processes such as drying). Due to the power of the drives, their energy efficiency is of major importance.

Production engineering. Production engineering takes in all processes by means of which work pieces with a defined shape are produced and changed. Production techniques are listed according to process in Table 2-4 (DIN 8580).

Table 2-4. Overview of production techniques

Create cohesion	Maintain cohesion	Reduce cohesion	Increase cohesion	
1. Primary forming	2. Forming	3. Separating	4. Joining	5. Coating
	6. Changing material properties (e.g. hardening)			

Of these production techniques, joining is very important, since it can be used to link multiple components. DIN 8593 defines the following specific joining techniques:

- assembly,
- filling,
- pressing on, force fitting,
- joining by primary forming,
- joining by forming,
- joining by welding,
- joining by soldering,
- joining by gluing,
- textile joining.

Durable goods and consumer goods are produced in large batch or mass production. This explains why the uses of production techniques which can be mechanised and automated are common. As the use of metal as a process material is rare (in most cases metal only continues to be used in the production of vehicles and white goods), metal-cutting machining is nowhere near as important as it is in mechanical engineering.

There are usually many stages involved in the production of a product. The information earlier in this book about production quantities makes it clear that products are produced in very large quantities with short cycle times. The total production time is much longer than the production cycle. This means that the production process has to be broken down into stages which can run in parallel. In principle, this can be achieved in two ways:

- Several machines are used in parallel to execute all (or many) of the steps involved in the production process.
- The various tasks are completed for each work piece one after the other by a machine process specialising in the corresponding work step (division of work, assembly line principle).

In the majority of cases it is more efficient to adopt the second approach in respect of the division of work, since this means that overall production will involve less redundancy in terms of tools and techniques, etc.

A different technique can be adopted in respect of production techniques for large batch and mass production:

- division of the overall process into individual steps which can be mechanised and automated,
- use of techniques for the rapid duplication of geometries (primary forming, forming, coating),
- mechanisation of simple steps in the production process (forming, separating, joining).

Duplication of geometries. One of the most important production techniques for mass production is based on the duplication of geometries, where the work piece is given a specific shape. This technique, which involves either primary forming or forming, is used in a wide variety of applications:

- casting of metals,
- injection moulding of plastics (injection moulding machines),
- blow moulding of glass bottles,
- compression moulding (e.g. punching),
- blow moulding of plastic bottles (in principle this is a form of transfer moulding, where a plastic blank is softened by heat and pressure and then forced into a mould),
- printing (in principle this is a coating technique, where the coating material, in this case the printing ink, is applied selectively to the item to be coated, e.g. paper webs, in accordance with a template),
- structuring processes in semiconductor manufacturing.

A feature common to all of these techniques is the very frequent use of a mould (or template) and the fact that the work piece is given the essential features of its shape in just a single work step. In turn, special manufacturing processes are needed to manufacture the moulds (or templates) themselves. In the case of metal forming processes (casting, injection moulding, transfer moulding) machine tools are used for this purpose; however, coating techniques are preferred for print templates. In most cases the moulds or templates are generated automatically directly from the data on the basis of which the work piece is constructed (geometry data). Eroding machines and 3D milling machines are used for metal moulds and laser or electron beam equipment for print templates and masks for semiconductor manufacturing. Most duplication processes are intermittent processes.

Mechanisation of individual manufacturing steps. Mass production features numerous manufacturing steps which need to be completed within a short cycle time. These include:

- forming processes, e.g. the bending of cardboard for a package,
- separation processes such as cutting (e.g. edge cutting of books) or punching,
- joining processes such as filling, welding (e.g. of a film cover onto a plastic package) or gluing.

In addition to the individual manufacturing steps, the format of the manufacturing process is also determined by how these steps can be completed. Three basis types can be identified:

- In complete processing a number of manufacturing steps is completed one after the other on a single machine.
- In continuous production the manufacturing steps are completed with continuous material infeed.
- In intermittent production or assembly line production the individual manufacturing steps are completed one after the other in a fixed time cycle, thereby eliminating the need for buffers between workstations.

The characteristic features of these three manufacturing sequences are described in more detail below.

Complete processing. In complete processing a work piece undergoes a number of work steps one after the other. Examples of complete processing include:

- metal-cutting machining of a work piece in a machining centre with more than one tool,
- insertion of PCB components in electronics production with multiple component parts,
- injection moulding of a plastic component.

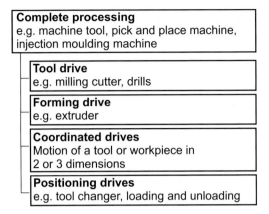

Fig. 2-11. Drives in complete processing

Complete processing is always the technique of choice if time and equipment are required to fix and align the work piece whilst the tools can be changed quite quickly. Even if precision drive engineering (Fig. 2-11) and control is needed to position the tool, this manufacturing principle is advantageous.

In pick-and-place machines for electronic components, revolver pick-and-place heads pick up a number of component parts from feeders before placing them on the printed circuit board. It is common for the total pick-and-place time to be split between several pick-and-place heads. This technique enables pick-and-place rates of up to 60,000 components per hour to be achieved. A functional principle with a separate pick-and-place station for each component is not economically feasible in this context.

In mass production, complete processing always hits problems when non-productive times due to tool changes become too long, when high clock-pulse rates are required and when it is simply too complicated to combine the various activities in a single workstation (e.g. because machining the material significantly changes its shape, as is the case when making cardboard boxes).

Continuous production. Processing a material continuously is described as continuous production. The movement of the material is uniform during processing. Typical examples of continuous production are:

- the rolling of sheet in metal manufacturing,
- wire drawing,
- continuous extrusion processes for plastic pipes or film,
- the manufacture of paper,
- the production of sheet glass,
- stranding processes in the production of ropes and cables

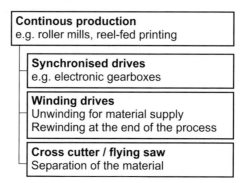

Fig. 2-12. Drives in continuous production

- many processes in textile production: spinning, coiling, finishing operations such as dyeing and dressing,
- reel-fed printing.

The advantage of continuous processes is that the material does not constantly have to be accelerated or decelerated. Constant speed and mass inertias result in a production process which is highly stable and of excellent quality.

Continuous production processes require a number of drives (Fig. 2-12), to feed in the material (synchronised drives). Rolls are used for the feed. If material is being stretched (e.g. in wire drawing applications, metal rolls), the feed rate will vary from drive to drive.

On reaching the end of production, the material has to be either separated or taken up on rolls as a continuous web. Separation, for which drive functions such as cross cutters or flying saws are used, marks the transition from continuous to intermittent production.

Continuous production is commonly used to produce primary materials or semi-finished goods. Only a small number of end products go to market wound continuously on rolls. Even toilet paper is perforated and packaged prior to arriving at its point of sale.

Intermittent production. Intermittent production or assembly line production, in which a sequence of discrete machining steps is combined to form one continuous sequence, is the production principle most commonly used in mass production processes. For this purpose, the overall operating sequence is divided into numerous work steps completed on individual workstations. The work piece passes through these workstations one after the other. As the production cycle is identical for all stations there is no need for intermediate buffering.

One advantage of intermittent production techniques where work is divided is that each workstation only needs to support or make provision for the processes, tools and materials used on it. A disadvantage of intermittent production techniques is that all work steps have to have the same processing time. This means that operating sequences have to be precisely planned. Optimisation strategies to increase productivity always have to start by optimising the point at which bottlenecks occur, since this determines the overall operating speed. A further disadvantage is that just one individual fault is enough to bring the entire sequence to a standstill. The sequence does not usually make provision for redundancy. Therefore, the individual processing steps have to be designed so that they can be completed with a high level of reliability. Faults have to be rectified very quickly.

There are two basic types of intermittent production:

- assembly line production,
- machines running in fast cycles

Assembly line production. Assembly line production moves the product under production through the entire production process. The most prominent example of this technique is automotive production, where, in the early days, revolving chains were used. The individual work steps would be completed one after the other on workstations arranged one behind the other, usually by means of manual labour. Workers had to complete their work steps within the cycle time. Today, instead of being forwarded to the next workstation as part of a continuous motion some of the goods to be produced "jump" to the next station. The individual production steps can be automated or completed manually. Typical cycle times for this type of production technique vary between 20 seconds and 5 minutes. Significantly longer production cycles (between 4 and 20 minutes) are used for products manufactured in smaller quantities but with division of work in assembly line production (e.g. series machinery, commercial vehicles, industrial trucks, agricultural machinery).

Machines running in fast cycles. Machines with a central drive are designed so that one production step is completed per revolution of that central drive. This central operating cycle is distributed throughout the machine via a line shaft. Distributed movements for the execution of the individual processing functions are then derived from the movement of the line shaft. Once the processing step has been completed, the work piece is forwarded to the next workstation. A typical operating cycle for a machine of this type is in the region of one second. At this processing rate the individual processing functions can no longer be completed manually.

Typical examples of machines based on this principle are the form, fill and seal machines found in packaging technology [Bl+03]. Fig. 2-13 shows an example of the operating sequence in a machine of this type for beverage cups. All production steps are completed at the same cycle rate and the work piece is forwarded from one workstation to the next.

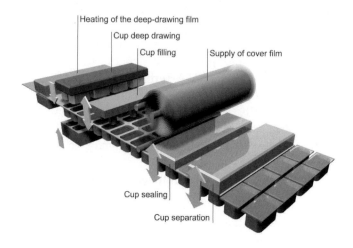

Heating of the deep-drawing film

Cup deep drawing

Cup filling

Supply of cover film

Cup sealing

Cup separation

Fig. 2-13. Production sequence for a packaging machine running in fast cycles

If a work step cannot be split, the processing sequence has to be made parallel. An example is the blowing of PET bottles. The production time is approximately 2 s, whilst the production cycle might be 0.125 s. This means that a total of 16 moulds and blowing stations are used. A suitable mechanism such as a carousel principle takes up the blank in the mould and, once 16 production cycles have been completed, forwards the finished bottle for despatch. Similar principles are applied when prolonged idle times have to be managed (e.g. cooling times, heat treatment for shrinking foil, drying times). In such applications, the work pieces are taken up by a conveyor system which moves at a correspondingly low speed. Capacity is dimensioned in such a way as to ensure that the required number of work pieces can be taken up corresponding to the cycle time and the process time. On completion of the cycle, the work piece bridges the required time. To save space, processes such as these can also be implemented using loop accumulators.

Rotary indexing tables are often used for intermittent processes during which the work piece has to be positioned very precisely. All machining stations are freely accessible from the outside and material can even be fed in from the outside. Furthermore, rotary indexing tables have a very high repeat accuracy as regards work piece positioning from one station to the next, since this depends only on the mechanical tolerance of the work piece holders on the table and the indexing of the individual positions (which can be achieved either by means of precise angle measurement or using index holes with a mechanical hold-down).

Machines with line shaft. The basic design of machines running in fast cycles often hails from a time during which they were implemented with a mechanical line shaft and individual motions derived from the line shaft for machining processes. Accordingly, the individual machining steps are more like simple functions, since this is all that can be achieved with corresponding mechanisms and gearboxes. In this context, cams, which facilitate user-defined distributed motion sequences, are a common feature.

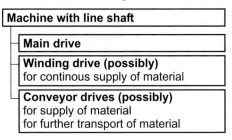

Fig. 2-14. Drives in machines with line shaft

In intermittent production machines, line shafts perform two functions:

- they synchronise the individual machining steps in time,
- they distribute energy to the individual machining stations.

A disadvantage of all machines with a line shaft is that in the event of a change of product involving a change of format, the individual machining steps have to be adapted and readjusted. The transmission of drive power via long shafts also poses problems.

For this reason, the use of distributed servo drives running in synchronism as electrical line shafts is advantageous (Fig. 2-14). The machining steps can then be adapted by means of a software modification or by making changes to software parameters. No mechanical setting work is required. The use of servo drives involves the use of mechanisms which convert the rotary movement of the drive into the machining sequence. Since these designs previously entailed direct coupling, they are designed in such a way that the drive is able to support highly constant motion. This also means that when servo drives are used, requirements in respect of dynamic response are low, and this in turn has major advantages where energy is concerned.

Machines with synchronised single drives. Machines running in fast cycles with motion control distributed across single drives require the following drive functions (Fig. 2-15):

- synchronised drives to move the material,
- winding drives to feed the material in from a roll,

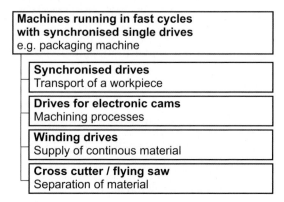

Fig. 2-15. Drives in production machines with single drives

- drives for cross cutters and flying saws to separate the material whilst it is in motion,
- cam drives to perform other machining processes whilst the material is in motion.

All these movements synchronise with the central machine cycle, which is distributed electronically to the single drives.

2.2.2   Material flow systems

From the description of highly automated production techniques and processes and goods distribution systems it is clear that the demand for automated material flow and logistics systems is high. This chapter intends to describe the basic structures of automated in-house logistics systems, since these need a large number of controlled drives [Ma06a].

This chapter does not deal with:

- simple manual means of transport which do not have their own drives (carts, containers with wheels and pallet trucks, for example),
- manually driving material handling equipment such as industrial trucks (fork lifts) which often features electric drives but is not automated,
- all long-distance means of transport (commercial vehicles, freight trains, ships).

The overall in-house logistics system can be divided into the following task areas:

- procurement logistics to process goods from suppliers before making them available for further processing or distribution within the organisation,

- production logistics to ensure that primary goods are made available to individual machining stations in the right amounts and at the right time,
- distribution logistics to get finished goods to customers,
- disposal logistics to ensure the right and proper disposal or reuse (recycling) of waste products.

The production techniques and processes and the logistics systems described in this book produce or distribute goods almost exclusively in the form of individual units. Bulk solids are only relevant to production in their role as raw materials. It is for this reason that we only need to consider transportation and storage systems for goods in the form of individual units.

Although direct forwarding is not unheard of, goods in the form of individual units are often forwarded in or on transport containers (containers, pallets). Mass-produced items in particular are put into containers in large quantities (forming trading units) for the purpose of storage and forwarding before being put onto pallets in loading units. Goods in the form of individual units then travel all over the world in standardised containers.

There are numerous features which can be said to characterise in-house means of transport:

- transport area (line, floor, room),
- transport direction (horizontal, on a slope, vertical),
- manoeuvrability (stationary, guided, free),
- principle of operation (continuous, discontinuous),
- transport level (floor level, overhead, below floor level).

The combination of these features means that a small number of means of transport are commonly used; these shall be described in more detail below. Particular emphasis has been placed on describing the drive tasks which have to be implemented.

Conveying belts and other linear conveyors. Conveying belts are one of the most frequently used means of in-house transport. They move goods along a line. In principle, conveying belts are continuous conveyors, but sections of belt are often switched on and off according to requirements. A belt on which goods to be conveyed have been placed can also be stopped so that a defined end position is reached. Accordingly, conveying belts can become non-continuous conveyors.

Belt conveyors often feature a rubber track on which the goods to be conveyed are moved. Wire mesh belt conveyors and steel belt conveyors are also used.

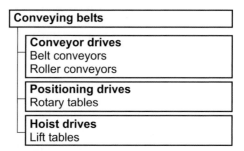

Fig. 2-16. Drives in conveying belts

Roller conveyors or roller tracks are often used to transport pallets and containers. In such applications, roller conveyors have drives. In most cases, a number of rollers are set in motion by a single drive, with power transmission between rollers via rubber belts, V-belts or chains. Roller tracks use gravitational force or their mass inertia to move the goods being conveyed subsequent to previous acceleration.

Conveying belts and roller conveyors are often combined and linked via rotary tables. Pallets or containers can also be fed to the side by a roller conveyor using chain conveyors. In this case, the pallet is stopped in a defined position on the roller conveyor; the chains are lifted and take up the pallet for transfer to a right-angled conveyor.

In goods and postal distribution centres, automated sorters which move the goods being conveyed from one belt to another, thereby determining the direction of onward transport, are common. Pushers can be used to achieve the same aim for parcels and suitcases. The use of tilt tray sorters is another option.

Linear material handling equipment moves the goods being conveyed at an average speed of between 0.3 and 2 m/s.

Comprehensive transport systems for the (usually automatic) conveying of pallets are quite common upstream of finished goods stores in producing companies and distribution stores in goods distribution centres in particular. A drive is located every 1 to 2 m. The conveying drives are supplemented by drives controlling the direction of transport (sorters, pushers, rotary tables, cross conveyors) so that the individual route of an item being conveyed can be defined. The automation system controls the sequence; photoelectric barriers and in some cases identification systems (bar code, RFID systems) track the route taken by goods in order to ensure that the image in the automation system and the actual location of the relevant goods match.

Hoists. Hoists have to be used to overcome differences in height. They usually take the form of non-continuous conveyors. Typical hoists in automated transport are:

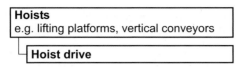

Fig. 2-17. Drives in hoists

- Lifting platforms, often set up as scissor-type lift tables. A pantograph mechanism converts a linear motion into a lift motion to lift the surface on which the goods being conveyed have been placed. The work surface is often a roller conveyor to enable pallets to be taken up. As well as hydraulic cylinders, toothed belts working in conjunction with geared motors can be used to generate linear motion.
- Vertical conveyors are in principle goods lifts on which the lift motion is generated hydraulically or by means of ropes, chains or toothed belts.

Since drives in hoists work against gravitational force (Fig. 2-17), specific conditions must be complied with in respect of the design of the equipment and its safety.

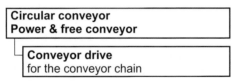

Fig. 2-18. Drives in circular conveyors and power & free conveyors

Circular conveyors, Power & free conveyors. Circular conveyors have a central drive (Fig. 2-18) which drives a chain as a traction mechanism in a rail. Carriages run on the rail and the goods to be conveyed are attached to hangers. In most cases these rails are mounted on the ceiling. Loading and unloading stations are used to put goods into and then take them back out of the transport system. Power & free conveyor rails are split into two sections: one for the traction chain and the other to guide the carriages. A driver connects the carriage to the chain for onward transport.

The use of circular conveyors in mass production is wide and varied. Possible areas of application include:

- automotive production,
- production of white goods,
- paint finishing,
- cooling and drying rooms.

Conveying speeds are between 0.2 and 0.4 m/s. A single drive can drive circular conveyors up to a length of 500 m; the use of a second enables the transport distance to be increased to up to 2,000 m.

In the case of longer transport distances and low goods frequency, transition to a monorail overhead conveyor on which each item is driven individually, is recommended.

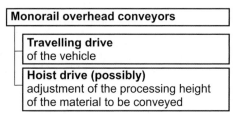

Fig. 2-19. Drives in monorail overhead conveyors

Monorail overhead conveyors. Monorail overhead conveyors are a railbound means of transport whose rails are fixed to the factory ceiling or to supports. The carriage, which has its own drive with a friction wheel, moves along the rail. The goods to be conveyed are attached via hangers. Branching elements in the rail system facilitate route flexibility.

Energy has to be transmitted to the travelling drive. This can be achieved with contacts or via busbars, or contactlessly by means of inductive energy transfer. Each vehicle has to have a vehicle control system to evaluate the signals from the master control, determine the position on the route, activate the drive and, using an appropriate distance sensor system, make sure that the vehicles do not collide.

The trolleys travel at between 10 and 100 m/min. Travelling drives with gearboxes customised in line with the design characteristics of the monorail overhead conveyor are used (Fig. 2-19).

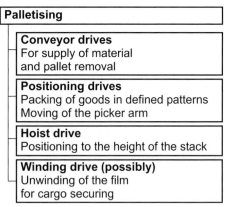

Fig. 2-20. Drives of a Palletiser

Palletising. Palletising describes the process in which goods in the form of individual units are grouped to form trading units or loading units. This process is usually fully automated (Fig. 2-20) and comprises the following functions:

- The goods to be palletised are usually fed in from upstream production on a conveying belt.
- They have to be taken up and then set down in layers via a multi-axis handling system. Compliance with packing patterns is vital. The handling system takes the form of either a robot or a special multi-axis gantry system.
- The individual layers are stacked on top of each other.
- In many cases, the process concludes with a cargo securing being fixed to the pallet. This can take the form of a layer of film being wrapped all the way around the consignment (heat shrink or compression wraps). This method of cargo securing requires an unwinder for the wrap and, in some cases, a rotary table so that the pallet can be wrapped.

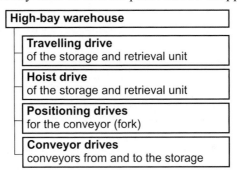

Fig. 2-21. Drives in a high-bay warehouse

High-bay warehouse. The technique which has proved most popular for the storage of goods in the form of individual units in recent years is the high-bay warehouse, in which containers and pallets are stored on several levels one above the other. The conveying equipment (storage and retrieval unit) accesses the storage bays via the aisles. The advantage of high-bay warehouses is the high rate of storage space utilisation, along with the ability to access individual storage bays via three dimensions (select aisle, approach level and bay), thereby accelerating access times. Today, virtually every large producing company and most logistics centres have high-bay warehouses.

If individual items do not have to be removed (picked) from pallets or containers (in accordance with the "man-to-goods" principle) and storage and retrieval functions solely on the basis of complete pallets or containers, high-bay warehouses are absolutely ideal for complete automation. This is achieved using storage and retrieval units with the following functions and drives (Fig. 2-21):

- A travelling drive which moves the vehicle in the bay aisle. Usually rail-bound. The drive can be installed stationary or on the vehicle. Toothed belts, chains, rack and pinions or friction wheels working in conjunction with the rails are used for power transmission.
- A hoist drive which sets the height of the goods being conveyed. Toothed belts or chains are used here also.
- A positioning drive which retracts the conveyor (fork) out in order that the pallet can be taken up.
- Other drives might also be used to take up and fix the goods being conveyed.

The primary contribution made by storage and retrieval units has been the resulting increase in speed. Today, travelling speeds of up to 4 m/s and hoisting speeds of up to 2 m/s can be achieved. Acceleration rates of up to 2 m/s^2 are possible. Storage heights can reach 18 m. A storage or retrieve operation takes on average 5 to 30 s to complete, with the result that a single storage and retrieval unit can complete up to a million operations per year.

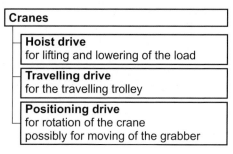

Fig. 2-22. Drives in cranes

Cranes. Cranes have traditionally been associated with production. Every building site has one, every container terminal has a long line of container cranes and many factories too have overhead cranes mounted on their ceilings.

Cranes have a hoist drive and – dependent on design – additional drives for horizontal movement (travelling trolley, part-turn actuator, travelling drive). Grabbers are moved using positioning drives (Fig. 2-22).

Industrial trucks. Industrial trucks (e.g. fork lifts) are the solution of choice for transport on factory floors. They are usually operated manually. Driverless industrial trucks (which might or might not run on rails) facilitate increased automation. Energy is either stored on the vehicle (battery, fuel with combustion engine for energy conversion) or fed in via sliding contacts or inductively (and therefore contactlessly). Navigation is either forcibly guided (on rails) or relies on markers on the travel way (optical or magnetic markers or inductive transponders are used). Alternatively, laser navigation can even be used. Driverless industrial trucks need to support comprehensive safety engineering (bumper, emergency stop pushbutton, safety edges and laser scanners to the sides). They also have to be supplied with control information via a host system and a communication system. The travelling speed is usually 1.1 m/s, with a maximum of up to 1.67 m/s (6 km/h) being permitted.

Driverless trucks have a travelling drive and a steering drive. They also have to have drives for load transfer if this is achieved automatically (Fig. 2-23).

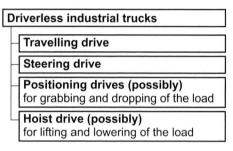

Fig. 2-23. Drives in industrial trucks

Making out of delivery orders. In goods distribution centres, outbound deliveries are usually made out as delivery orders. For this purpose, individual units have to be taken out of storage and put together to form a consignment. In many cases, the processes involved in the making out of delivery orders are still completed manually and two distinct basic principles can be identified:

- Man-to-goods: Here, the person making out the delivery order goes to the goods storage location (either manually or on a picking vehicle) and removes the required quantity from the container.
- Goods-to-man: Here, a transport system moves the container or pallet from the storage bay to a picking station, where the person making out the order removes the required quantity.

In order that delivery orders can be made out efficiently, powerful software packages are used in today's warehouses to minimise both the time and effort needed to plan sequences and the distances needing to be travelled or transport movements required.

The process of making out delivery orders can to some extent be automated. The biggest obstacle preventing this is primarily the fact that the goods to be picked come in so many different formats, necessitating the use of suitable sensor technology to detect and plan motion sequences. This explains why the successful introduction of automatic systems for the making out of delivery orders has only been possible where individual goods have uniform or at least similar formats (examples include boxes for tablets and music or video software on CDs and DVDs). In such circumstances, goods can for example be stacked and then pushed along a belt conveyor which forwards them to the picking container. This type of automated system can work at rates of up to 10,000 units per hour.

Recently, a new system which automates the workflow described below has been introduced for the automatic making out of delivery orders in trading units bound for retail markets:

- Incoming goods are initially stored on pallets in a high-bay warehouse.
- The individual trading units are then depalletised automatically and stored individually on carriers in a second high-bay warehouse.
- From this second high-bay warehouse, picking machines pack the trading units onto containers with wheels. The containers are then despatched to consumer markets complete with the packed consignments.

In this system, the stacking sequences must be planned very precisely. Furthermore, material handling must be as gentle as possible, since the goods are usually only wrapped in simple cardboard packaging. The picking machines, which are ultimately specialist multi-axis handling units fitted with positioning drives to facilitate the making out of delivery orders, can achieve picking rates of up to 1,000 units per hour.

Automated making out of delivery orders is gaining in significance as it provides a means of reducing the amount of human intervention required and enables warehouses to work around the clock. The driving force behind such systems is provided by powerful software for planning the processes involved in the making out of delivery orders as well as flexible handling equipment to execute the motion sequences.

As this part of the chapter describing the various types of productions and material flow systems and outlining the drives used in those systems draws to a close, we now need to look at how the automation facilities within these systems are structured.

2.3 Structure of automation systems

Today, the production and distribution of durable goods and consumer goods would be inconceivable without automation and, accordingly, the efficient use of information technology [Za00]. Ultimately, we have now reached a stage where each individual step in the process to produce and distribute goods is made ready, controlled and assisted by computers. The actual physical process in the real world is always mapped in a database image in the world of information technology and, naturally, the two are constantly compared.

All processes in information technology can be allocated to one of the following three groups:

- the planning of (or making ready for) actions (e.g. production orders, orders to make out delivery orders for shipment),
- the direct control of these actions,
- the monitoring and supervision of the results of these actions.

In this context, each action is usually already a complex process relying on the interaction of numerous items of equipment (provision of material, control of machinery for the individual processing steps, transport between workstations, control of the large number of routes taken during storing and retrieving).

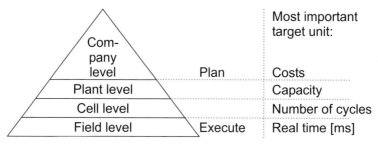

Fig. 2-24. The automation pyramid

Automation pyramid. In order to illustrate how automation and information technology tasks gradually become more and more specialised within the context of production and logistics, they can be presented in the form of a pyramid (Fig. 2-24). Enterprise resource planning is located at the top, representing the following tasks:

- Planning for individual production orders (what will be produced and/or delivered when and in what quantities, what is required in terms of procurement/upstream production, stock planning), often based on orders placed remotely (purchase orders sent via fax, online and even electronic data transmission in business-to-business transactions).
- Link with commercial EDP facilities for the crediting of finished goods, debiting of material consumption, accounts entries for resources used (working hours, machine running times), recording of quality data.

In this context a factory is part of overall enterprise resource planning, which today is usually managed using an ERP system.

The processes on this level do not have high real-time requirements. Planning and evaluation can to a large extent be achieved without a direct relationship to the points in time at which the associated processes will be completed. It is often the case that planning and evaluation processes are calculated in central ERP systems at longer time intervals (e.g. every 24 hours at night). In this context, all that is important is that these calculations are completed in good time, so that the necessary information is made available to production and logistics. This explains why standard IT systems typically used in office environments, rather than special real-time systems, are used on this level.

The situation on the next level, on which the actual manufacturing sequence is controlled, is quite different. This is where the individual cells for an interrelated area in the factory are controlled, which then have to control the individual actions required for the execution of a job. This is where the transition from planning to control level takes place. The computers used need to be real-time capable in order that they can track and intervene in the physical processes. In particular where the starting of operational or motional sequences is concerned, processes cannot be delayed. This is why special real-time systems which place specific requirements on software deterministic and computing time are used at cell and control level.

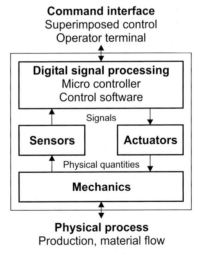

Fig. 2-25. Structure of a mechatronic system

Mechatronic system. A cell comprises controllers which intervene in the process directly via actuators and sensors. On this level, the structure resembles that of the basis system principle behind a mechatronic system (Fig. 2-25) [HeGePo98, Is99]:

- In the physical world, where the structure or simply the position of products undergo changes (in the context of production or transport), the process involved is usually mechanical and is controlled by the selective application of forces.

- Forces are applied via actuators in accordance with signals they receive from the control system.
- The resulting changes in the process are detected by the sensors. This is necessary since every action intended is subject to errors and might not always be completed as anticipated.
- The overall control loop is closed via signal processing. In the first instance, commands relating to the actions to be completed are received from the higher level. Correcting variables are then calculated from these commands and sent to the actuators as setpoint values. Sensors detect the actual value and, accordingly, the status of the process. If the signal processing detects a system deviation affecting the intended action, it will make corrections for the actuators by adjusting the setpoints accordingly. Signal processing is usually achieved using software running on digital computers. The software executes open-loop (applying setpoints to actuators) and closed-loop (direct corrections further to a setpoint/actual value comparison to determine the system deviation via a closed control loop) control functions.

The actual selection of actuators and sensors, as well as that of the control system, is determined essentially by production techniques and, accordingly, by the process to be automated. The basic concepts involved are described in more detail below.

2.3.1 Controls

Controls are the central element in automation systems [We01b]. Today, only software-driven digital controllers remain relevant. They are available in various types and designs:

- programmable logic controls (PLCs),
- numerical controls (CNC, computerised numerical control),
- motion control,
- special machine controls based on microcomputers (embedded controllers),
- Industrial PCs with soft PLC or soft CNC.

These terms are not part of a system and their assignment can be explained primarily in terms of historical background. Hence, we shall start by looking at what they have in common.

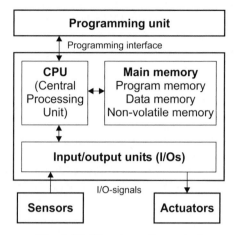

Fig. 2-26. Structure of a control

Structure of a control. All controls are computer systems whose hardware has the following basic components (Fig. 2-26):

- Central processing unit (CPU) running the software. This is usually a microcontroller, in other words a CPU integrated in a semiconductor with additional functions such as memory, I/O.
- Main memory for this CPU (not only working memory and non-volatile memory but also program memory and data memory).
- Current supply.
- Interface via which the control is programmed.
- I/O functions for the connection with sensors and actuators, operator control and monitoring equipment and the higher levels of the automation pyramid.

Depending on the requirements to be met by the control, the type and design of the individual elements will vary:

- The word width in the CPU can range from 8 bits for basic PLCs to 32 bits for CNCs, industrial PCs and complex PLC systems.
- Memory size can also vary from a few kByte to several hundred MByte.
- Non-volatile memory can take the form of flash memory (on embedded controls and flash disks on industrial PCs) or battery-backed RAM (on PLC systems) and hard disks (on CNCs and industrial PCs).
- The current supply is usually provided by feeding in 24 V DC or mains voltage; battery-backed standby power supplies are used in some cases.
- Operator keyboards and screens are controlled directly by industrial PCs (and often also by CNCs); in other cases operating units are located remotely and controlled via a communication system.

- Some interfaces with sensors and actuators (known as I/Os) are integrated directly in small PLC systems. In larger PLC systems, the I/O modules are connected remotely via communication systems.

Software is the critical part for the operation of controls. Software programs are designed to execute defined computing steps (arithmetic and logic operations) sequentially. The sequence can be modified on the basis of the results computed (comparisons followed by branching operations).

PLCs. Many descriptions of open-loop and closed-loop control functions hail from a period when they were implemented with parallel switching functions. Accordingly, relay controls are made up of individual switching elements whose input is determined by a logic operation (AND or OR functions) involving upstream switching elements, and whose output in turn controls downstream switching elements. This is supplemented by elements such as flags (1-bit memories or flip-flops) and timer relays, which allow a set period of time to elapse before activating their output contact.

Switching elements of this type can be used to build a sequence control for a discrete machine. Following a start signal (e.g. start pushbutton or start signal from an upstream machine), the first action is started. Once this has been completed (e.g. reported by a limit switch), the next action gets underway. In other words, every step is triggered by the linking of the results of upstream actions. However, this means that every step also has to activate a signal on reaching its completion.

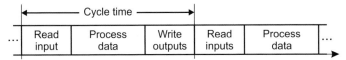

Fig. 2-27. PLC program execution

A programmable logic control executes this type of control in a microcomputer. The basic sequence is illustrated in Fig. 2-27.

- the input signals are read (process image),
- the individual logic operations are executed by the program (and the output values are calculated),
- at the end of the computing cycle, the output values calculated are switched to the outputs,
- the sequence then starts again from the beginning.

One of the key differences compared with relay controls is the fact that results are not updated continuously but only on completion of each cycle; in other words, the control uses a sampling technique. However, this does not have any major implications for the process being controlled provided that the signals do not change too fast for the sampling rate. Modern PLC systems are able to compute several thousand instructions per second, meaning that cycle times remain within the range of 1 and 20 ms even in the case complex control tasks. This is sufficient for many control tasks.

The advantage of a PLC compared with a relay control needs no further explanation. Configuring the function on the programming device's screen is much easier and more efficient than wiring and hardware requirements for complex control tasks are significantly reduced since only the inputs and outputs, but not the internal signals, have to be wired. Flexibility in the event of changes is high and the control software can easily be reused.

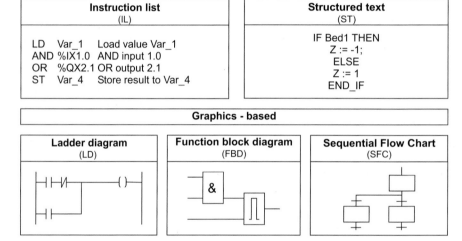

Fig. 2-28. PLC programming languages

PLCs are programmed using programming devices or programming systems running as application software on a PC. Various programming languages based on a variety of programming conventions are now standardised in IEC 61131-3 (Fig. 2-28):

- Function block programming (FB diagram) resembles a logic diagram with individual function blocks (e.g. logic elements, timer elements, counters) with the connections between the blocks represented by lines.
- Ladder diagram programming shares some of the characteristics associated with relay interconnection. AND operations are series connections between contacts and OR operations are parallel connections.
- Programming with an instruction list is equivalent to programming in the machine language of a computer (assembler programming). Values are read (loaded), processed according to the value saved in the accumu-

lator by means of logic commands (AND, OR) and then written (stored).

These three programming languages (two of which are graphics-based and one of which is text-based) are simple and intuitive; they can be used by skilled workers and are taught as part of vocational training.

Other programming languages standardised in IEC 61131-3 are higher-level programming languages (Structured Text, ST) or sequential flow charts (SFC).

A complete PLC program is made up of a number of program blocks, each of which contains instructions for a function. These program blocks can be used to build a complete program in a modular way; for functions which are to be implemented many times in the same way, program blocks are then called a corresponding number of times and wired specifically at their interface. This enables PLC programs to be standardised.

Where the implementation of an automation system is concerned, the advantage of a PLC lies in the fact that it is modular; in other words, in the fact that each of the logic operations for the output signals can be pro-grammed entirely independently. The relationships between machine func-tions must be considered. Downstream sequences are set up in accordance with the handshake principle: once one action is complete, the next one starts. This maps, for example, the passage of a product through consecu-tive conveying routes. PLC programs also provide an easy means of im-plementing actions for deviations from standard procedure. A correspond-ing logic operation can be used to indicate a status requiring special treatment. This will then trigger appropriate actions and possibly prevent an inappropriate action (for this, the status must only be compared with others via a NOT function). Once the special action is complete, the stan-dard sequence can be resumed. Today, PLC programs for automation sys-tems usually need up to 80% of their resources for special operations of this type (faults, setting-up operation, etc.).

PLCs are the workhorse of factory automation. Several million PLCs are produced world-wide every year. Large numbers of them can be found in almost every machine, production line and logistics system. PLCs are implemented as hardware systems in various sizes and performance classes (from compact controls to modular controls) and as soft PLCs in the form of software functions in industrial PCs.

CNC control. The task for of a CNC control is completely different. It has to control a number of axes (e.g. two-dimensional with X and Y axis or three-dimensional with X, Y and Z axis) to guide the cutting tool on a machine tool in such a way that the required work piece contour and sur-face characteristics are produced. The sequence is determined by the in-tended geometry of the work piece. The overall sequence is made up of

individual traversing motions to be made by the tool or work piece at a
defined speed and is described in NC blocks. In this context, an NC block
is the equivalent of a machining step describing relative motion in a multi-
dimensional space. DIN 66025, or "G-Code", is used to describe these
machining steps.

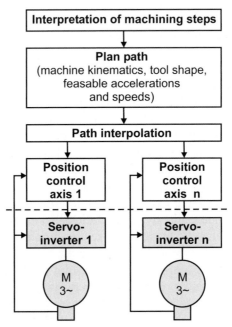

Fig. 2-29. Program execution in a CNC control

The CNC control executes the following functions in order to process this
machining sequence (Fig. 2-29):

- An interpreter processes the individual machining blocks and triggers
  internal program functions.
- Geometry data processing uses the individual motions to calculate characteristic curves for the individual axes of the machine tool. For this it
  takes into account the machine kinematics (transformations in the directions of movement of the servo axes), tool contours (tool offsets) and
  achievable speeds and acceleration rates.
- The interpolator uses sections of the path to calculate angular position
  setpoints for each axis in a fixed time base.
- Position setpoints are then converted into corresponding motion sequences by the electric drive with its drive control functions (position,
  speed and torque control).

Here too an internal program is executed cyclically in order to generate
output values for the subsequent actuator (servo drive) at fixed time inter-

vals. Since a number of drives which have to generate a precise 2D or 3D contour on the work piece are driven synchronously, the position setpoints have to be absolutely time-synchronous; in other words, jitter-free. This is the main difference between the temporal behaviour of a PLC and that of a motion control system, e.g. a CNC control:

- Due to its cyclic program execution, the sequence control (PLC) starts the next cycle as soon as the previous one is complete. Like this, no computing time is wasted, but a lack of constancy of the computing time will lead to a jitter in sampling times. The timing for program execution is not entirely deterministic.

- In contrast to sequence control, motion control has to be jitter-free and therefore deterministic. Every violation of real time will lead to a motion fault and be visible in the contour of the work piece. The maximum program execution time must be a known quantity and must not exceed the fixed repeat interval. Distributed tasks must be synchronised precisely in time.

Complex and computing-intensive algorithms for geometry data processing and interpolation mean that CNC computing power requirements are high. Furthermore, CNC controls usually have local operator terminals in order that the machining sequence can be visualised and programming can be carried out locally (workshop-based programming, data acceptance from CAD systems). Also, they have to have sufficient local memory for high volumes of data and the capacity for downloading this data from higher-level controls. This is why machine tool controls are often built with an industrial PC. A real-time operating system takes care of time-based requirements.

Robot controls. Robots are manipulating devices designed for universal use. There are various geometries which require the synchronous motion of between three and eight axes. Industrial robots are designed as kinematic chains requiring complex mathematical transformations.

Programming is another area where robots differ from machine tools. It involves a mixture of a sequence program and motion commands. Positions to be approached are often defined approaching and saving a position manually (known as the "teach-in" technique).

The internal structure of a robot control is very similar to that of a machine tool control; here too, the use of industrial PCs is common.

General approach to drive control. The following generally valid principle for controlling drives in an automation system can be derived from the description of the two principles of PLC and CNC control. It splits software functions into three levels as follows (Fig. 2-30):

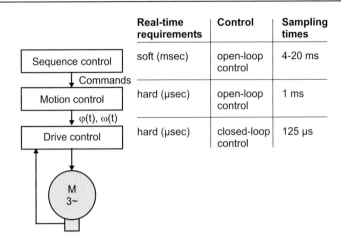

	Real-time requirements	Control	Sampling times
Sequence control	soft (msec)	open-loop control	4-20 ms
Motion control	hard (µsec)	open-loop control	1 ms
Drive control	hard (µsec)	closed-loop control	125 µs

Fig. 2-30. Software levels in controls and drives

- Sequence control, which is responsible for the sequence of the individual steps, is located at the highest software function level.
- Motion control is located below sequence control. It is charged with the task of calculating instantaneous setpoints for drive control. Motion control is dependent upon the task the drive has to execute in the machine. This level has to be time-deterministic where dynamic and precise motion sequences are concerned in order to execute motion profiles corresponding to the setpoint curve (i.e. without excessive contouring errors).
- Drive control, which is charged with the task of ensuring that the drive is compliant with its direct setpoint value specifications (usually a combination of angle, speed and torque variables), is located at the lowest level. Disturbance couplings are corrected directly here. Drive control is often implemented using a closed control loop.

The two upper levels comprise open-loop control functions, whilst a closed-loop function is executed at the lowest level. On the other hand, the two lower levels have to work with strict time determinism, whilst sequence control usually makes allowance for temporal jitter. The software functions for motion control and drive control are described in more detail in chapter 3.4.

Decentralised and central motion control. A number of drives are usually assigned to one control. Motion control can be implemented either decentrally in the drives or centrally in the control (Fig. 2-31) [KiKu06].

In the case of decentralised motion control, the control (usually a PLC for sequence control) sends commands to the axes. Furthermore, the communication system or other techniques can be used to synchronise multiple axes. This architecture is always to be recommended if a large number of axes are being driven by a single control and/or if the movements of the axes either do not need to be synchronised very closely (e.g. conveyor drives in a plant) or are dependent upon a master motion (e.g. synchronised movements in continuous production or on production machines running in fast cycles).

Central motion control should always be used if a common motion sequence for a number of axes has to be calculated on the basis of a multidimensional motion profile. Typical applications are coordinated drives in machine tools and robots.

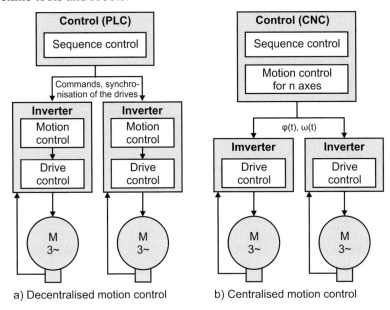

a) Decentralised motion control   b) Centralised motion control

Fig. 2-31. Decentralised and central motion control

Programming of controls. Controls are very universal devices whose actual function in a machine or plant is defined by software and software programming. This programming is carried out at the following levels (Fig. 2-32):

- Control manufacturers equip their systems with a basic set of functions. These will be infrastructure components (operating system, device drivers, communication system) and software functions which either interpret or directly execute (following compilation, i.e. building the directly executable software) the application program (which is dependent on how the machine is actually used for a production order).
- Machine manufacturers then program the application program so that it is able to implement the functions of the actual machine. In order to be able to do this, the program must take not only the machine's I/O (in other words, which sensors and actuators are there, what is the nature of

their function) but also its operating mode and geometry into account. Since most machines produce a range of products (rather than the same one all the time), possible settings must be available for the machine operator.

- The operator uses the machine. An actual production order usually requires settings to be made affecting the control program. On production machines, these settings are usually made using operator dialogues describing the goods to be produced (possible options might include selection from a catalogue of standard programs, selection of recipes, entry of formats). On a machine tool and a robot, on the other hand, the actual program can be created by the machine operator to match the contour of the work piece to be produced or alternatively it can be exported from CAD data. In such cases, the programming options available to the machine operator are extensive.

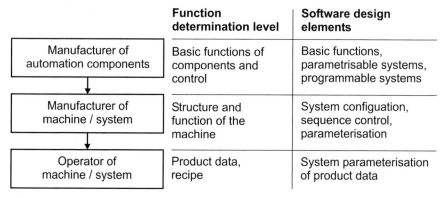

	Function determination level	Software design elements
Manufacturer of automation components	Basic functions of components and control	Basic functions, parametrisable systems, programmable systems
Manufacturer of machine / system	Structure and function of the machine	System configuation, sequence control, parameterisation
Operator of machine / system	Product data, recipe	System parameterisation of product data

Fig. 2-32. Programming levels for controls: Equipment manufacturer, machine manufacturer, machine operator

The various programming and parameterisation levels result in the use of a whole range of languages and procedures:

- PLC programming languages (either compliant with IEC 61131-3 or manufacturer-specific) suitable for programming sequences. Some of these are supplemented with specific function blocks for the motion control of drives (PLCopen function blocks).
- Geometry-based NC programming compliant with DIN 66025. Advanced developments of this technique are able to process geometry data (e.g. Step NC).
- Parameter setting dialogues for the selection of programs and the entry of values for actual production orders on production machines.

Controls then have to use one of these options in their software to convert the defined function into the actual machining sequence and the cyclic control of actuators. The range of algorithms required to achieve this is very wide and extremely specific for each of the different machine types.

2.3.2 Actuators

Actuators convert control output signals into forces to control the process; accordingly, they are the "muscles of the automation system". On the one hand they are actuated by the control by means of electrical signals and on the other hand they apply a suitable principle to affect the process. This book concentrates on the control by means of motion.

The signals from the control can be split into two very general groups:

- Switching data exhibiting one of two statuses (ON or OFF, open or closed) at a given point in time is output via digital signals. In industrial control technology, 24 V DC signals as defined standard IEC 61131-2 are usually used for this purpose.
- Continuous signals which can be transmitted either as analog signals (e.g. 0...10 V or ±10 V) or as data words (8, 16 or 32 bits in width) via a communication system.

Electromagnets. In the vast majority of cases, digital switching data is converted using electromagnets, which are common to many applications:

- Use in valves for controlling e.g. the flow of gases or liquids (i.e. fluids). Electromagnetic pneumatic valves mark the transition to pneumatic drive technology and accordingly hydraulic valves to hydraulic drive technology.
- Use in electrical contactors for the purpose of switching electrical energy. Frequently used as a means of switching electric motors on and off; in such cases, there is no scope for further control of their motion.
- Use in electromagnetic brakes and clutches which control either braking forces or power flows.

Drives take over process control from this point on. We need to start by looking at the typical characteristics of fluid-based (pneumatic and hydraulic) drive principles.

Pneumatic drives. On pneumatic drives compressed air moves piston rods in cylinders, usually between two fixed limit stops. The motion itself is influenced only by the opening and closing of valves by the control, the actual sequence is defined by the pressure, the selection of the cylinder and the counteracting forces. Typical operating pressures are in the region of 5 bar (50 N/cm^2). As the medium is compressible, end limit stops have to be used.

The use of pneumatic drives in machines for simple motion is very common. This is due to the relatively low cost of pneumatic elements compared with electric drives and the lack of a need for further conversion into linear motion. In principle, it ought to be possible to use continuously controlled proportional valves and a system of position transducers to run pneumatic drives with position control, but this would counteract the cost advantage previously gained over electric drives. Pneumatic drives rely on a compressed air supply; whilst many industrial plants already have this facility in place, an additional infrastructure would be required for smaller machines.

Pneumatic suction cups controlled by valves are also commonly used to fix work pieces ready for subsequent motion. Here too control is achieved by means of digital switching data.

Hydraulic drives. Whilst low costs are the most attractive feature of pneumatic drives, high force densities are the major draw of hydraulic drives. The operating principle of hydraulic drives is similar to that of pneumatic drives (both are fluid drives), the only difference being that oil is used instead of air. However, although many of the basic principles (linear motion via pistons in cylinders, motion between two fixed limit stops, control based on switching data via valves, the need for an infrastructure with a pressure accumulator) are identical, the operating pressure of 100...300 bar (1,000 to 3,000 N/cm^2) is significantly higher. As the medium's percentage compressibility is low, it is possible to use proportional valves on hydraulic drives, by means of which pressure and therefore force and motion can be very precisely controlled. Disadvantages include low efficiency and problems involving leaks.

In addition to linear hydraulic cylinders, there are rotary hydraulic motors which can be used for example for travelling drives.

Switched electric motors. Alongside fluid drives, which are controlled via magnetic valves, electric motors can also run in switched mode. On small drives with low voltages, this can be achieved directly using the control (voltages of 5 V, 12 V, 24 V or 48 V). On larger drives, however, contactors set in motion by means of electromagnets are used; this involves switching the mains voltage (400 or 480 V). If electric motors are simply switched on and off, the motion sequence will be dependent upon the characteristic of the motor and the counteracting forces. The speed of the motor cannot be controlled.

Use of power electronics. Common to all drives running in switched mode is the fact that the motion sequence is not controlled directly by the control's software. Accordingly, these drives are not mechatronic systems. Switching data simply starts processes, it does not control them. The acceleration process of a machine is defined by the features of the mains, the features (speed/torque characteristic) of the electric motor and the features (load torque and moment of inertia) of the processing machine. With direct motor start-up it is difficult to take any influence on the acceleration process. The use of controlled drives ensures the benefit of improved starting properties.

Switched motion sequences are often not ideal for reasons associated with power. For example, starting up a motor induces significant loss, since the counter voltage induced is proportional to the speed and therefore does not start to drop until the motor accelerates. Until the steady-state operating point is reached, the voltage drop between the input voltage and the induced counter voltage combined with the current for torque generation represents a high power loss. Accordingly, the power loss in the electrical circuit on start-up is the direct equivalent of the kinetic energy when the motor accelerates. Fig. 2-33 contains a simple diagram illustrating power loss on start-up for motors with direct mains switching and motors with speed-controlled acceleration.

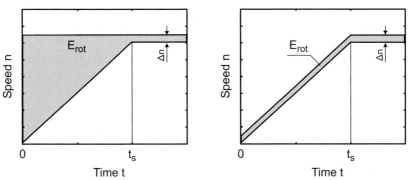

Fig. 2-33. Schematic diagram for uncontrolled and controlled motor start-up

The use of electromagnets also has negative connotations where power is concerned. For example, to achieve motion, a higher current must be applied than would be needed to hold the armature. If magnets are simply activated and deactivated, the current for holding will be too high, in turn generating unnecessarily high losses.

This set of circumstances explains the increasing use of power electronics to control power conversion as appropriate for the prevailing operating point. The use of power electronics also makes it possible to use signal processing to control power conversion. This then gives rise to mechatronic systems, of which controlled drives are also a component. The operating mode of controlled electric drives is described in detail in chapter 3.4.

2.3.3 Sensors

Sensors measure the status of process quantities and report this information to the control. As there are different types of actuator, so there are different types of sensor:

- Which process quantities are detected (position, speed, force, pressure, temperature but also the identification of goods)?
- What is the nature of the measuring technique?
- Which signal does the sensor supply to the control (switching data in the form of a digital signal or a continuous measured value)?
- What are the nature of the measuring range and the accuracy of the detected signal?

As our focus is on motion, sensors for process factors such as pressure and temperature as well as the flow of fluids and the fill level of containers will not be discussed in further detail here. For more information, please refer to [GuLi99].

Sensors can be subdivided into further categories on the basis of their function:

- They have a direct influence on the control sequence and as such are an integral component of the mechatronic system. Proximity switches, photoelectric barriers and sensors built into drives for the detection of positions and speeds are members of this group.
- Sensors in the second group have a less direct influence on the control sequence, since the information they acquire is sent to the cell level and used to plan its subsequent progression. Examples of sensors belonging to this group include identification sensors (bar code and RFID scanners) and image processing systems. This group of sensors will not be discussed in any further detail since they do not usually interact directly with drives.
- Sensors which check quality during the production sequence and output an error message immediately if the required levels of quality are not reached represent a mixture of these two groups. They have a direct influence on the sequence because, as a result of their intervention, it will either be interrupted or the faulty part will be taken out of production automatically without the overall sequence being brought to a complete standstill. In this case the sensor initiates a specific action of the sequence. In production processes running in fast cycles in particular, these types of sensor are often used to implement early warning routines for faulty products, thereby safeguarding overall quality. In such applications, the sensitivity of these sensors must be set very carefully in order to prevent the risk of pseudo fault detection (when a fault has not ac-

tually occurred but is detected) and/or actual faults slipping through the net (when a fault has occurred but is not detected).

Switching data. The acquisition of switching data is of significant importance in automation systems. There are two main sensor principles:

- Optical sensors are used to interrupt a light beam. This leads to photo electric sensors either directly in the form of retro-flective sensors (because the transmitter and receiver can then be installed in the same housing) or as through-beam photo electric sensors with transmitters and receivers in separate housings. Photo electric sensors are able to work with all materials which are not transparent. A disadvantage is the risk of soiling, common to many industrial applications. Here, preventive diagnostics procedures (weakening of the received signal) can help to avoid faults.
- On inductive proximity switches a change in a magnetic field is induced when a metallic object approaches, and switching data is detected. Inductive proximity switches are highly dependent on magnetically conductive materials and are therefore ideal for use on moving machine parts.

Both sensors are contactless and as such are not subject to wear. For this reason they have the edge over microswitches, on which an electric contact is activated via a pushbutton.

The acquisition of switching data is commonly used in automated production sequences and material flows. On belt conveyors, for example, the movement of the goods being conveyed (e.g. the individual packets, containers and pallets) is tracked using photo electric sensors. The control will then use the information provided to decide what action to take next.

Position sensors. Proximity switches only generate one item of switching data which can then initiate a subsequent control action. If a continuous flow of data is required, as is the case for example for the control of a motion sequence, position sensors will be needed. Position sensors can be divided into two groups:

- sensors built into motors,
- sensors mounted outside motors on the mechanical system.

The advantage of building sensors into the motor is that they can be used both for drive control and to determine the actual position of the material as it moves. A disadvantage is that inaccuracies due to the mechanical transmission between motor and machine cannot be detected. Sometimes a software-based solution is used to compensate this (e.g. leadscrew pitch compensation). The angular sensors used in the motor are described in detail in chapter 3.3.6.

If the measurement facilities supported by the motor are not accurate enough, an additional load-side encoder will need to be used. The following options are compatible:

- Encoders which support rotary measurement; they are either read serially via a communication system or send incremental signals to the control or drive.
- Linear encoders which are either optical or magnetic in design. Optical linear measuring systems are used extensively on high-precision machine tools in particular.
- Laser distance sensors which are used for example in storage and retrieval units to determine the position of the vehicle.
- Measuring tapes affixed to rails and belts which are scanned by a sensor on a vehicle (e.g. in a monorail overhead conveyor system) and are able to determine the actual position to within 1 mm.

Image processing. Some automation systems already use sensors with image processing (e.g. on robots). These can then detect the position and level of work pieces and derive information from this for the motion sequence. The use of image-processing sensors is bound to increase in the future, as they offer more flexibility and are becoming less expensive as the price of microelectronics falls.

Fig. 2-34. Operator terminals

2.3.4   Visualisation and operation

The automation of production processes and logistics systems has meant that the primary task of personnel is now to monitor processes and rectify faults (although automation quality is excellent, there are still some situations where human assistance is required). The better an automation system works, the less often manual intervention will be required. Having said that, there is more for service personnel to oversee. Furthermore, interruptions in the operation of high-performance production and manufacturing systems can quickly lead to major downtimes. It is for this reason that it needs to be possible to carry out repairs as quickly as possible (e.g. by replacing devices in just a few steps). Car manufacturing plants have the most complex requirements due to their high production value and the fact that their production sequences do not usually support redundancy. One fault will bring the entire production line to a halt. This is why eliminating faults quickly is so important.

In order to be able to analyse the instantaneous status of a system quickly, operators and service personnel need information. Visualisation systems are used for this purpose (Fig. 2-34). These systems are usually also used to operate the system. Accordingly, as well as the components described above, an automation system will also have stations for visualisation and operation, via which personnel can gain an overview of the status of the system and intervene manually if necessary. This is located in the following locations:

- The overall production process is monitored in control centres. Here, large screens and overview panels providing a general picture of the situation and fingertip access to specific details within the overall process. These control centres use communication systems to access data at cell or control level and even individual actuators and sensors.
- Control centres are supplemented by operator terminals installed on the machine or system which support direct local monitoring, fault diagnosis and troubleshooting. For this purpose, the operator terminals are designed with an increased class of protection such as IP 54.

Operator terminals are available in a variety of types and designs ranging from simple text-based displays to extensive PC-based SCADA systems. A trend towards graphics-based displays is becoming evident. These are a cost-effective option thanks to inexpensive LCD displays and enable operators to assimilate the displayed information more quickly. Operator terminals are usually connected to controls and drives via a communication system from which they read the information to be displayed. Corresponding software tools supporting both text-based and graphics-based visualisation are used to program the display. For productive machine operation, intuitive control schematics are very important as they can be processed more quickly and easily by operating and service personnel.

2.3.5   Communication systems

Communication systems are incredibly important, if not vital, to today's automation systems. Therefore, it is easy to see how vertical integration from management level right down to control level will necessarily involve the exchange of high volumes of data:

- Data for planning: Data associated with production orders (which product, which work steps, recipes and control data for workstation configuration).
- Control data: Completion messages from production steps, finished piece numbers, required times, scrap, quality data.
- Protocol data: Data about the actual production or manufacturing step which needs to be recorded (obligation to produce supporting documents, traceability of the production technique/process, what was used, how the machine was set, which actual production step was carried out, the nature of the measurement data acquired).
- Diagnostic data: Actual data from production enables the operation to be optimised, in turn helping to increase productivity.

This vertical communication is supplemented by networking at control level and with the actuators and sensors. Since the majority of the components of the overall system are now controlled by software, a communication system can be used to integrate them into an overall network which can then be used for a variety of purposes:

- Adaptation of components to the production order at hand (parameterisation).
- Component diagnostics (detection and reading out of warnings and error messages).
- Acquisition of additional information about the process by components supporting needs beyond those associated with the automation system. An example of this is condition monitoring, which provides information for process diagnostics and preventive maintenance.

As electric drives in particular are very flexible in terms of adaptation and acquire all sorts of information, they are ideal for electronic communication at field level [Po03].

Traditionally, the various components of a control system were interconnected directly by means of wiring, bringing advantages in the form of direct signal transmission and support for direct diagnostics but disadvantages in terms of high wiring costs, in particular if a large number of signals need to be exchanged between two components. Many simple sensors and actuators are linked to control I/O directly by means of parallel wiring.

Replacing direct parallel wiring with a communication system enables the signals to be transmitted to be packaged in a data telegram which is then sent via serial transmission. The communication tasks to be performed can be divided into two groups:

- In process data communication data has to be transmitted continuously. Although communication systems are not able to simulate the direct and virtually instantaneous transmission achieved with parallel wiring due to their serial mode of operation, they do try to achieve this as much as possible. Since all control systems implemented in software work with sampling, it is sufficient if all data is transferred within one software sampling period. Cyclic transmission mechanisms without acknowledgement are usually used for process data communication.

- In service data transmission data is transmitted on an event-controlled basis or as and when required. Typical applications are the reading and writing of parameters, the execution of commands and the transmission of larger data volumes (e.g. data files, programs). For service data transmission, a time slot is usually reserved within cyclic data transmission and communication mechanisms with acknowledgement are used for the actual transmission process.

For systems running with time synchronisation (this is typical of motion controls and drives), the communication system can also be used to synchronise the systems.

Components are usually standardised in order to ensure that they can interact via a communication system. Standardisation provides an assurance that devices supplied by different manufacturers can exchange data without needing to be adapted specifically for this purpose.

In general terms, communication systems are characterised by the following features:

- baud rate,
- maximum number of nodes,
- transmission medium (cable type, optical fibres),
- topology (bus, line, star, tree),
- communication relationships supported (master/slave, publisher/subscriber),
- optional service data transmission,
- time synchronicity including jitter.

The fieldbus systems introduced in the 1990s were based on baud rates of 500 kbit/s to 16 Mbit/s. In the office world, Ethernet (introduced during the mid-1990s) has proved the most popular solution, supporting a baud rate of 100 Mbit/s (Fast Ethernet) and even 1,000 Mbit/s (Gigabit Ethernet). The widespread use of PCs means that components for Ethernet are relatively inexpensive. Furthermore, Ethernet's mechanisms have been designed to network office environments across distances of up to 100 m, lending it interference immunity and durability. Other communication systems from the PC environment (USB, IEEE1394 "Fire Wire") are difficult to use in industrial environments as they do not support simple electrical isolation.

It was against this background that Ethernet was also introduced in industrial communication [PrSt03]. One of the first obstacles to be overcome is Ethernet's inability to rank nodes on the basis of priority and the lack of time-based control for communication. This can lead to collisions (two nodes wishing to use the same transmission medium at the same time) which have to be resolved. The fact that the mechanisms used are not real-time capable puts a further obstacle in the way of using Ethernet for process control tasks.

Table 2-4. Overview of industrial communication systems

The system	Max. baud rate	Topology	Max. no. of nodes	Communication relationship
AS-i	167 kbit/s	Bus	124	Master/slave
CAN	1 Mbit/s	Bus	127	Publisher/subscriber
PROFIBUS	12 Mbit/s	Bus	125	Master/slave
DeviceNet	0.5 Mbit/s	Bus	64	Master/slave
INTERBUS	2 Mbit/s	Line	512	Master/slave
SERCOS	16 Mbit/s	Ring	255	Master/slave
PROFINET	100 Mbit/s	Line, star	Unlimited	Master/slave
EtherCAT	100 Mbit/s	Line	65.535	Master/slave
Powerlink	100 (1,000) Mbit/s	Line, star, tree	254	Publisher/subscriber

A number of initiatives and system standards have enabled progress to be made on the way to overcoming these restrictions. Almost all fieldbus organisations have also defined systems which use Ethernet as their communication medium and are real-time capable.

Table 2-4 provides an overview of the world's most popular industrial communication systems and their main features.

2.3.6 Safety engineering

Stefan Witte

Automated machines are expected to provide an assurance of personal safety. Whilst twenty years ago, accidents at work involving machinery causing serious injuries were more frequent, the use of safety devices prescribed by standards has brought about a significant reduction in the number of such incidents.

Safety engineering in automated systems is an important element in this respect. This usually involves the use of special sensors to detect the actual position of machines. Safety controls set the corresponding machine part to a status in which it will not pose a risk to anyone in the vicinity (e.g. the drives are shut down). In order for such actions to be effective, safety technology and automation must work in close collaboration.

Accordingly, safety engineering and automation are increasingly growing together, exchanging information and using common sensors and actuators. Because safety components have to meet more complex technical requirements than automation components, this directly influences the design of the individual components [Wr07].

The movement of machine parts and tools generated by drives is one of the primary sources of danger in machinery. Automation often makes it impossible for machine operators to anticipate what the drives are actually going to do. Therefore, it is absolutely vital that appropriate safety measures are put in place to ensure that when operators are in danger zones, the drives behave in such a way that they do not pose a risk. This is the task of safety devices.

Machinery Directive. Machinery Directive 98/37/EC (the revised form 2006/42/EC is due to come into force on 29/12/2009) provides the legal framework mechanical engineers must comply with to ensure the safety of their machinery. The Machinery Directive replaces previously applicable sectoral safety regulations and specifies that standards to assess and reduce risks to an acceptable level (by taking appropriate safety precautions) must be applied.

Risk analysis. The required safety level is determined on a machine-specific basis by carrying out a risk analysis (Fig. 2-35). Such analyses must be carried out as follows:

- definition of the machine's limits and application as directed,
- identification of hazards and their associated hazardous situations,
- evaluation of risk for each hazard and hazardous situation identified including maloperation and foreseeable misuse,
- risk assessment and making of decisions to minimise risk.

A risk graph is used to assess aspects of the severity of an injury, the frequency or duration of a risk, the possibility of the risk being avoided and the probability of an undesirable event occurring. The machine is then categorised on the basis of safety level (category in accordance with EN 954-1 or performance level (PL) in accordance with EN ISO 13849-1 and SIL in accordance with EN 61508). The following categories are most frequently applied for machines with moving machine parts located in areas which can be accessed by persons.

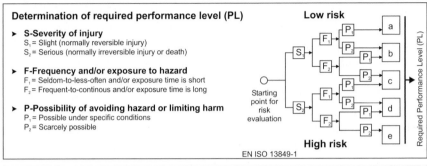

Determination of required performance level (PL)

➤ **S-Severity of injury**
S_1 = Slight (normally reversible injury)
S_2 = Serious (normally irreversible injury or death)

➤ **F-Frequency and/or exposure to hazard**
F_1 = Seldom-to-less-often and/or exposure time is short
F_2 = Frequent-to-continous and/or exposure time is long

➤ **P-Possibility of avoiding hazard or limiting harm**
P_1 = Possible under specific conditions
P_2 = Scarcely possible

Starting point for risk evaluation

Low risk

High risk

Required Performance Level (PL)

EN ISO 13849-1

Risk assessment and determination of the required Safety Integrity Levels (SIL)

Consequences and severity	Se.	Frequency and duration	Fr.	Probability of hazardous event	Pr.	Avoidance	Av.	Class 3-4	5-7	8-10	11-13	14-15
Death, losing an eye or arm	4	Up to 1 hour	5	Often	5			SIL2	SIL2	SIL2	SIL3	SIL3
Permanent, losing fingers	3	1 h ... 1 day	5	Likely	4				AM	SIL1	SIL2	SIL3
Reversible, medical attention	2	1 day ... 2 wks	4	Possible	3	Impossible	5			AM	SIL1	SIL2
Reversible, first aid	1	2 wks ... 1 year	3	Rarely	2	Possible	3			AM	SIL1	SIL2
		> 1 year	2	Negligible	1	Likely	1					

EN IEC 62061

Fig. 2-35. Risk graph and probability of occurrence in accordance with EN ISO 13849-1 and EN IEC 62061

- Category 3 or PL d: This risk class must always be applied if provision has been made to ensure that moving parts in automated machinery cannot be accessed (covers, barriers such as protective fences, safety doors, light curtains) and these barriers must only be opened for the purpose of work relating to machine set-up, the elimination of faults or maintenance. In this case, when the barriers are opened the drives are shut down and thereby set to a safe status. In some cases the work being carried out will require that it is possible for the drives to be put into restricted operation (e.g. limited speed, limited increment), but this must not be allowed to pose a risk to personnel.

- Category 4 or PL e: This risk class must be applied in the event of permanent human presence in the operating range of machines posing a high level of risk. The most well-known example are presses, into which work pieces are inserted by hand and where extensive safety devices have to be provided.

The safety level identified must be implemented universally in the machine in order to reduce the risk posed to the people present in its danger zone (commissioners, fitters and operators) to an acceptable level. This also includes indirect risks (in the case of machines which only pose a risk if damaged).

The degree of safety to be applied determines the regulations in accordance with which the safety devices need to be designed and built. Generally speaking, safety devices have a very low residual error rate, since the occurrence of an error must not be allowed to put the safety function at risk. In many cases, in particular where control systems for Categories 3 and 4 are concerned, this gives rise to two-channel redundant systems, ensuring that *a single* error cannot take the safety function out of service.

Compared with the regulations governing the design and manufacture of non-safety-relevant components, this is a significantly more complex requirement and incurs higher costs. This is why special safety components are required; the automation components alone are not able to assure the required safety level.

EN 954-1 continues to pursue a deterministic approach. Accordingly, it specifies presumed errors and possible means of elimination for components, which are then applied to analyse the safety of the circuit in which those components are used. For electronic circuits and in particular programmable systems, this standard is no longer sufficient. The approach of more recent standards such as EN 61508 and EN 62061 (sectoral) is based on failure probability. These standards also define regulations for the development methodology to be applied, since this can be an effective way to significantly reduce error rates. All in all, they provide a comprehensive framework for the development and use of electronic and programmable control systems for the implementation of safety-relevant machine controls [Ap04].

The following standards (Table 2-6) play an important role where the safety engineering of machines and control systems is concerned:

Table 2-6. Overview of standards for safety engineering

Machinery Directive 98/37/EC	The Machinery Directive imposes a legal obligation to assess the hazards posed by a machine and take appropriate action to reduce the risk to an acceptable level. The Official Journal of the European Communities lists standards associated with this directive. It is assumed that machines which comply with these harmonised standards will meet the requirements of the Machinery Directive. The Machinery Directive is binding for anyone importing machinery into or manufacturing, assembling or commissioning machinery in member countries of the European Union. All machines meeting the requirements of the Machinery Directive must bear the CE mark.

Revised form of the Machinery Directive 2006/42/EC	Will replace Machinery Directive 98/37/EC on 29 December 2009.
EN 1050/EN ISO 14121	Safety of machinery – Principles for risk assessment. Its content includes risk investigation and risk assessment techniques and procedures for hazard analysis.
EN 1037	Safety of machinery, prevention of unexpected start-up. Its content includes definitions and terms used for power disconnection and power dissipation equipment as well as measures (other than those for power disconnection and power dissipation) for the prevention of unexpected start-up.
EN 60204	Safety of machinery – Electrical equipment of machines – Part 1: General requirements. Its content includes a definition of stop categories. Stop categories define how a motor is brought to a standstill in the event of a safety function request. There are three different stop categories: 0, 1 and 2. Stop category 0 defines uncontrolled stopping, i.e. the motor is simply shut down and coasts or mechanical braking is applied. In stop category 1, the motor is brought to a controlled standstill and then the drive torque is cut off. In stop category 2 the motor is brought to a controlled standstill but motor control remains active even after the motor has come to a stop. The use of this stop category for shutdown in a hazardous situation is not permitted.
EN 954-1	Safety of machinery – Safety-related parts of control systems – Part 1: General principles for design. The standard defines a measure for achievable safety in the form of five control categories ranging from "proven-in-use" (Cat. B and 1) and the use of generally accepted safety principles (Cat. 2) to single fault security (Cat. 3) and the handling of multiple faults (Cat. 4). Machine structure is an important factor in this standard. For example, Categories 3 and 4 require that the safety function is set up with redundancy. The standard applies to electrical, electronic, programmable, mechanical, pneumatic and hydraulic controls and adopts a deterministic approach (error lists and means of eliminating errors). EN 954-1 is no longer applicable to programmable systems today. EN 954-1 was replaced by EN ISO 13849-1 in March 2007. A transition period will remain in place until 2009.
EN 61508	Functional safety of safety-related programmable electronic systems, PES. This standard (which is not harmonised) considers the entire life cycle of a device or machine. It defines probabilities for

	residual errors which could lead to a hazardous failure, on the basis of which it specifies a safety integrity level (SIL). The structure of a machine is an indirect factor in this respect. Although a redundant safety function usually has a less hazardous failure probability than a non-redundant structure, a non-redundant structure can achieve a higher SIL by means of test routines. Compliance with EN 61508 is not mandatory.
EN 62061	Sectoral standard of EN 61508 defining the "Functional safety of electrical/electronic/programmable electronic safety-related systems". It defines requirements and provides recommendations for the design, integration and validation of these control systems for machines. It does not define requirements for non-electrical (pneumatic, hydraulic, electromechanical) control elements in machines.
EN ISO 13849-1	Safety-related parts of control systems - Part 1, General principles for design. EN ISO 13849-1 combines the clear categorisation of the machine using risk graphs from EN 954-1 with the determination of failure probabilities from EN 61508 and defines a performance level based on potential risk. MTTF (mean time to failure), DC (diagnostic coverage, i.e. diagnostic scope for detecting possible errors), CCF (common cause failure), structure, the behaviour of the safety function under error conditions (e.g. fail-safe behaviour), safety-related software, systematic failures and the ability to execute a safety function under foreseeable ambient conditions are considered. The standard applies to electrical, electronic, programmable, mechanical, pneumatic and hydraulic controls.

Modern safety concepts for machines. In the past, safety technology was built as an addition to automation and usually comprised electromechanical components. Machines would respond to human presence in the danger zone by shutting down immediately (emergency off), generally leading to a major interruption in production.

However, requirements for high machinery availability are making it necessary for situations in which there is a need for personnel to carry out work in or on the machine (in other words, be present in the danger zone) to be coordinated with the automation of the machine in such a way as to facilitate faster restarting. In many cases, safely limited speed is required for machine setting-up operation. Accordingly, the safety technology and automation must be closely linked. The ultimate aim must be for the machine control to also take over the safety functions.

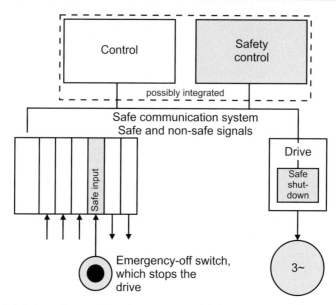

Fig. 2-36. A modern automation concept with integrated safety technology

This gives rise to concepts in which the safety technology becomes part of the overall automation (Fig. 2-36) [Be03, Gr03]:

- Safety sensors such as emergency-off switches and lightgrids are connected to safe inputs.
- These signals are transmitted to the control via a communication system which supports the transmission of both safety-relevant and non-safe signals.
- The machine control is either a safety control or is supplemented by an additional safety control. The control functions for the automation and the safety technology access the same signals and exchange data.
- Safe output data is then transmitted via the communication system. Safety-relevant components such as drives in particular evaluate these signals directly, thereby assuring that the appropriate safe status is set (e.g. protection against restart, safe standstill, safely limited speed).

The ultimate effect of an integrated safety system of this type is a significant increase in the overall productivity of a machine and plant which does not come at the cost of restricted safety. Modern safety standards ensure that electronic and programmable systems are able to assure a level of safety which is at least equivalent to that afforded by conventional concepts.

This chapter 2 opened by considering the products produced using automated production techniques and the actual techniques themselves. The prevailing production and material flow systems were presented and the drives used in them were described. The techniques used to automate these machines and systems were also discussed. We now need to look at how the drives used in large numbers in the types of production processes and logistics systems we have considered are structured.

3 The drive system and its components

Following the introduction of Production and Automation Systems, and hence the field of application of drives, in chapter 2, this chapter will explain the drive system and its individual components in greater detail. This will lay the foundation for then being able to deal with the individual drive solutions, which implement the drive functions in machines and which will be described in chapter 4 below.

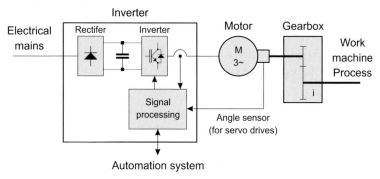

Fig. 3-1. Structure of a drive system

Controlled three-phase AC drives, which comprise a frequency or servo inverter and a three-phase AC motor, represent what is currently the most economical solution for a drive system for outputs above 100 W for use in automated production [Br99]. The biggest advantage is that these drives can be supplied directly from the mains without a transformer. The foundation for this technology has been established through the availability of capable power semiconductor switches (IGBTs, MOSFETs) and powerful microcontrollers.

Low voltage drives (small, permanently excited DC motors, stepping motors, EC motors) dominate the market for lower outputs. DC motors with thyristor DC speed controllers have been superseded by controlled AC drives that require less maintenance and are more cost effective. Hydraulic and pneumatic drives are generally not operated with closed-loop control.

The advantage of controlled three-phase AC drives is that their behaviour can be controlled by software very well and that their operating ranges and their dynamic responses are fully adequate for a large number of machine applications.

The components of the drive system will be presented in detail below (Fig. 3-1):

- At the heart is the *motor*, responsible for the conversion of electrical energy into mechanical energy (chapter 3.3).
- It is controlled by an *inverter* which controls the conversion of electrical energy in such a way that the motor executes the desired motion. A digital signal processor runs software functions for drive control and motion control to this end (chapter. 3.4). A distinction is made between *frequency inverters* which control the motor without measuring the speed, and *servo inverters* which apply precise angular and speed control to the motor by evaluating the output from an angle sensor.
- The operating point of the motor is often not optimally matched to the operating point of the machine. *Gearboxes* then convert the speed of rotation and the torque (chapter 3.5).
- Additional *driving elements* then further convert the motion at the output shaft of the motor or gearbox until it reaches the actual work process (chapter 3.6).

Before these components are described, we will first explain how a drive task is described and then dimensioned (chapter 3.1). Drive dimensioning is the precondition for the selection of the drive components. The operating conditions for drive systems will also be presented, as these are also an important marginal condition (chapter 3.2).

Following a presentation of the individual components, we will then go into the general aspects of the entire system (chapter 3.7). Initially these optimisation points for the entire drive train relate to matching the individual components to each other. This chapter will be rounded off by considering reliability which will describe how, on the one hand, the reliability desired in use can be achieved by the manufacturer of the drive components and how, on the other hand, it is possible to ensure in specific applications that the drives possess the appropriate level of reliability, being an essential element for the overall productivity of a machine (chapter 3.8).

The content of this chapter thus provides the basic information relating to the drive components allowing individual drive solutions to be described in detail in chapter 4 below.

3.1   Dimensioning of drives

Johann Peter Vogt, Dr. Carsten Fräger

Below we will present how a drive task is described in order to derive from the description the requirements on the individual drive components.

Positioning tasks. Many drives, in particular those used in materials handling and conveying technology, have the task of moving a work piece or material to a defined location. Processing in many manufacturing processes is also based on bringing tool and work piece into a defined position in relation to each other. In continuous or intermittent production processes, this relative positioning must be achieved while the work piece is being moved. Thus all these drive tasks are ultimately positioning tasks. The sequence of movements and the forces and outputs necessary for its execution are decisive for the description and dimensioning of a positioning task. In many cases, a positioning task is a dynamic task in which the instantaneous value for the power required varies significantly.

Drives for work machines. Other drive tasks involve supplying mechanical energy to a work machine which executes a process. In pumps, the work machine delivers a liquid, in mills materials are made smaller by the application of energy. Machining processes require energy for the cutting forces. While in the case of positioning tasks, movement with a change of location is in the foreground, in these drive tasks, the provision of mechanical energy to the work machine is the central aspect. Drives for working processes of this nature generally fulfil static tasks. In these, the instantaneous value of the output does not vary so much as in the positioning tasks and the drives are generally operated for a longer time at a constant speed. On the other hand, there can be significant output variations as a consequence of the processes that are not as predictable as in movement tasks. For this reason, good matching to the process is necessary.

Overall, the drive task is described by means of the physical parameters of the machine and the movement profile. Compared to a design of drives running at constant speed, the design of dynamically operated drives must go into greater detail, in order to be sure to pick up the influences of the changes in the speed of rotation and the torque [GaSc96].

3.1.1 Linear and rotary motion

The starting point for any type of motion is Newton's axiom that a force F acting on a mass m will accelerate this mass [CzHe04]:

$$F = m \cdot a \tag{3.1}$$

Where F is the sum of the forces acting on the body. These are composed of the following forces in drive applications:

- The force exerted by the drive F_A,
- the force due to weight F_G in the case of a motion against the earth's gravity,
- friction forces F_{fr} counteracting the movement,
- the counterforce F_L resulting from the mechanical process,
- further forces (flow resistance, Coriolis force) which will be ignored here.

Integrating the acceleration a gives us the velocity v, and a further integration gives us the position s:

$$v(t) = \int a(t)\, dt \tag{3.2}$$

$$s(t) = \int v(t)\, dt \tag{3.3}$$

A mass must first be accelerated by a force if it is to be set in motion. An opposing force must be applied to bring the mass to the rest position.

Near the earth's surface, a body of mass m is accelerated at the acceleration due to gravity g. The force due to weight F_G results as follows:

$$F_G = m \cdot g \tag{3.4}$$

The equations illustrated here are valid for a linear motion. Electric drives and many drive applications frequently operate with rotary motion. In the description of a rotary motion the variables F, m, a, v and s are replaced by torque τ, the moment of inertia J, the angular acceleration α, the angular velocity ω and the angle φ:

$$\tau = J \cdot \alpha \tag{3.5}$$

$$\omega(t) = \int \alpha(t)\, dt \tag{3.6}$$

$$\varphi(t) = \int \omega(t)\, dt \tag{3.7}$$

A speed of rotation n, often expressed in the unit rpm, is converted to an angular velocity ω using the following equation:

$$\omega = 2\pi \cdot \frac{n}{60} \tag{3.8}$$

Applications with linear motion are, for instance, belt conveyors, trucks and hoists. Rotary applications include winders, cross cutters and mills. In many drive applications, a linear motion generated by the rotation of an electric motor is required. The rotary motion is converted into a linear motion in the drive train to achieve this goal.

Rotation and linear motion can be related to one another by the following equation:

$$s = r \cdot \varphi \qquad (3.9)$$

where r is the radius at which a rotary motion is converted into a linear motion.

3.1.2 Work, power and energy

Work W is the product of force F and position s:

$$W = F \cdot s \qquad (3.10)$$

Power P is the work performed per unit of time. From this results, by differentiation of equation (3.10):

$$P = \frac{dW}{dt} = F \cdot \frac{ds}{dt} = F \cdot v \qquad (3.11)$$

This equation is fundamental to drive dimensioning as it describes what power must be applied to generate a movement.

The power for a rotary motion is calculated as follows:

$$P = \tau \cdot \omega \qquad (3.12)$$

Energy is the capability of performing work. According to the law of the conservation of energy, no energy is lost during work, but is only converted into other forms of energy. If a body of mass m is accelerated to a velocity v, it stores kinetic energy E_{kin}:

$$E_{kin} = \frac{1}{2} \cdot m \cdot v^2 \qquad (3.13)$$

The kinetic energy of a rotating body with moment of inertia J is calculated as:

$$E_{kin} = \frac{1}{2} \cdot J \cdot \omega^2 \qquad (3.14)$$

If a body of mass m is moved against gravity to a height Δh, it will store potential energy E_{pot}:

$$E_{pot} = F_G \cdot \Delta h = m \cdot g \cdot \Delta h \qquad (3.15)$$

Potential energy may also be stored in a pre-stressed spring.

The movement of a payload stores kinetic energy in it. This is released again on braking. The task of a drive is precisely this: to supply energy during acceleration and to remove it again for braking. During acceleration this is described as operation in motor mode, in which the power of the drive is transferred to the mechanics, while operation in generator mode prevails during braking. Here the power flows from the mechanics into the drive. Fig. 3-2 shows the four resulting drive quadrants.

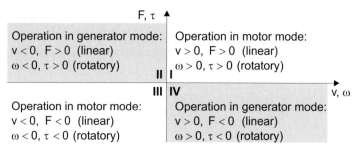

Fig. 3-2. Drive quadrants

3.1.3   Mass inertia

The equivalent to the mass m in a rotary motion is the mass inertia J. In general, the mass inertia for any body may be calculated as follows:

$$J = \iiint_V \rho(\vec{r}) \cdot r^2 dV \qquad (3.16)$$

Where ρ is the specific mass of the material. In practical application, rotationally symmetric pieces in which the above equation is significantly simplified are often used. The following results for a cylinder with total mass m and radius r:

$$J = \frac{1}{2} \cdot m \cdot r^2 \qquad (3.17)$$

Masses in linear motion can be converted into rotating inertia using the law of the conservation of energy.

$$\frac{1}{2} \cdot m \cdot v^2 = \frac{1}{2} \cdot J \cdot \omega^2 \rightarrow m \cdot v^2 = \left(\frac{v}{r}\right)^2 \cdot J \rightarrow J = m \cdot r^2 \ (3.18)$$

Using this equation it is possible to determine the equivalent moment of inertia caused by a mass in linear motion in a rotary drive.

In order to analyse the motion sequence of a drive, all the mass inertia components in the entire drive train must be determined and added. In addition to the mass inertia of the motor and the load in motion, there is the mass inertia of the other mechanical transmission elements. The influence of any gearbox used must be taken into consideration. This will be described in the next chapter.

3.1.4   Use of a gearbox and load matching

In many applications, the speeds of rotation of the application and of the motor are not optimally matched. In these cases a gearbox is used. The gearbox converts the speed of rotation of the application n_2 to the speed of rotation optimised for a motor n_1.

$$i = \frac{n_1}{n_2} = \frac{\omega_1}{\omega_2} \qquad (3.19)$$

The reduction of the mass inertia by a gearbox may also be calculated from the law of the conservation of energy:

$$\frac{1}{2} \cdot J_1 \cdot \omega_1^2 = \frac{1}{2} \cdot J_2 \cdot \omega_2^2 \rightarrow J_1 = \frac{J_2}{i^2} \qquad (3.20)$$

Load matching. Using a gearbox makes it possible not only to match the speeds of rotation, but also the mass inertia in the entire drive train. Decisive here for good controllability is the ratio of the mass inertia of the load J_{Load} with the gearbox ratio i converted at the motor end, and the mass inertia of the geared motor J_{Motor}. The load-matching factor k_J clearly affects the control behaviour. It may be calculated as:

$$k_J = \frac{J_{Load}}{i^2 \cdot J_{Motor}} \qquad (3.21)$$

In the case of small values of k_J, a large proportion of the torque is needed to accelerate the motor. If the motor is made even larger, it may quickly become the case that the required acceleration behaviour can no longer be achieved. Increasing the gearbox ratio within the framework of what is possible (taking into consideration the maximum speed of the motor and gearbox) or the use of lower inertia motors generally leads to better results here.

In the case of higher values of k_J, the load is dominant in the balance of the moments of inertia. In this case, a larger motor can lead to better control quality and system stability.

In the case of dynamically operated drives, the total drive is optimised at $k_J = 1$.

Fig. 3-3 shows the relationship between the moment of acceleration τ_{dyn}, which the motor must apply in comparison with the minimum τ_{dynref} at $k_J = 1$:

$$\frac{\tau_{dyn}}{\tau_{dynref}} = 1 + \frac{1}{k_J} \qquad (3.22)$$

The load-matching factor should be located in the mid range, where possible, in the case of drives with a high accelerating power.

Overall, the desirable load-matching factor k_J is located in the following range:

$$k_J = 0,5...10 \qquad (3.23)$$

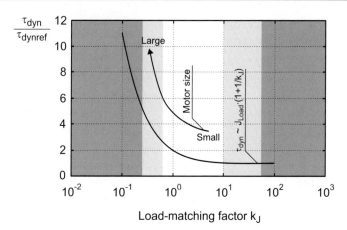

Fig. 3-3. Load-matching factor k_J

$k_J = 1$ provides a good control behaviour. The optimum gearbox ratio i_{opt} to achieve this is determined as:

$$i_{opt} = \sqrt{\frac{J_{Last}}{J_{Motor}}} \qquad (3.24)$$

Winding drives typically have a variable moment of inertia dependent on the winding diameter. As winding drives run at steady state, the moment of inertia of the load is significantly higher than that of the motor. The optimal load matching illustrated here is not applicable for these applications.

3.1.5 Friction

Friction occurs in any mechanical system. It always acts against the movement of the body. If a body is to be moved along a path, friction forces will arise. These may be differentiated according to their causes. The most important types for straight-line movement are static friction, sliding friction and rolling friction. The friction force F_{fr} is proportional to the normal force F_N, by which the body is pressed against the friction surface. This relationship is described by the coefficient of friction μ.

$$F_{fr} = \mu \cdot F_N \qquad (3.25)$$

In all horizontal movements, the normal force is equal to the force due to weight. In these cases, the friction force is calculated in accordance with the equation:

$$F_{fr} = \mu \cdot g \cdot m \qquad (3.26)$$

Static friction occurs when a force is applied to a movable body all the time this body remains in a resting position. Once the body begins to move, static friction is reduced and sliding friction occurs. Static friction and sliding friction are also known as Coulomb's friction. The coefficient of friction depends on the material pairing and any lubricant present (water, oil). Typical coefficients of static friction μ_0 lie in the range from 0.15 to 0.8, while the coefficients of sliding friction μ lie between 0.1 and 0.6.

The principle of rolling is frequently used in technology. Rolling friction is significantly less than sliding friction. Typical coefficients of rolling friction μ' lie in the range of 0.002 to 0.04.

Friction plays an important role in movement processes. Thus a proportion of the power loss within the drive components is due to friction forces. It is therefore important to use components that are as low loss as possible.

In the case of horizontal movement with low dynamics, the stationary power of the drive system is generally determined by friction.

In some applications, the transfer of forces is based on the principle of friction (friction wheels, synchronised drives with rollers, cables on driving pulleys, vehicles on rails). Here the coefficient of static friction limits the maximum transferable forces and hence the maximum possible acceleration.

Friction may, however, be ignored in many dynamic applications with high accelerations or in hoist applications since the biggest part of the drive power is used for building up kinetic or potential energy.

3.1.6 Process forces

In addition to the forces described above arising from the motion sequence, there are other forces and torques that arise from the process of the work machine in drive applications. The variation of the torque with the speed of rotation here depends on the process and can vary in different ways.

- Some processes in forming have a torque that is constant over the speed of rotation,
- in the case of calendars, the torque increases linearly with the speed of rotation,
- in pumps and fans, the torque increases with the square of the speed,
- in winding processes with a constant web tension, the torque decreases in inverse proportion to the speed of rotation.

If drives are driving work processes, the variation of the torque over the speed of rotation must be determined and taken into consideration for the analysis and dimensioning of the drive.

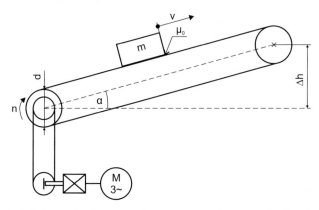

Fig. 3-4. Schematic diagram of a conveyor drive on an inclined plane

3.1.7   Speed and torque of a motion sequence

The development of the speed of rotation and the drive torque required can be determined for a motion sequence on the basis of the relationships illustrated. This is then decisive for the dimensioning of the drive, in particularly if it is operated very dynamically [GeSc96].

This will be explained using an example (fig. 3-4). A mass m is moved over a conveyed distance at an inclination of α. The force due to weight F_G is split by the inclined plane into a normal force F_N and a force acting parallel to the inclined plane F_F (Fig. 3-5):

$$F_N = m \cdot g \cdot \cos \alpha \qquad (3.27)$$

$$F_G = m \cdot g \cdot \sin \alpha \qquad (3.28)$$

Friction acts at two points. Thus there is static friction μ_0 between the mass m and the base which is moving the mass. The maximum acceleration a_{max} must be selected to be so small that the force of acceleration F_A does not exceed the friction force as otherwise the mass would slip on the substrate:

$$m \cdot a_{max} < \mu_0 \cdot m \cdot g \cdot \cos \alpha$$
$$\qquad (3.29)$$
$$a_{max} < \mu_0 \cdot g \cdot \cos \alpha$$

In the case of a coefficient of friction μ_0 of 0.1 and of small values of α, the acceleration a_{max} must hence remain below 1 m/s^2.

Furthermore there is friction acting counter to the movement of the conveyor. If a roller belt is used, then only rolling friction μ' is acting and this is substantially less than the sliding friction of a conveyor system such as a belt conveyor on a substrate.

$$F_{fr} = \mu' \cdot m \cdot g \cdot \cos \alpha \qquad (3.30)$$

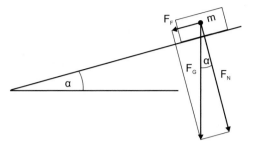

Fig. 3-5. Analysis of the force due to gravity F_G into normal force F_N and frictional force F_F

Thus the following forces are acting overall for the movement of mass m:

- The force of acceleration $F_A = m \cdot a$,
- the frictional force (acting parallel to the inclined plane) F_F,
- the friction force F_{fr}.

The maximum drive power $P_{A,max}$ is generated at the time at which the maximum acceleration a and the maximum speed v occur. It is calculated as follows:

$$P_{A,max} = m \cdot \left(a + g \cdot \left(\mu' \cdot \cos \alpha + \sin \alpha\right)\right) \cdot v \qquad (3.31)$$

The following motion sequence is now assumed for the variation of speed of rotation and torque over time:

- The movement occurs over a distance of 5 m in 2 s, initially against gravity and then with gravity back to the starting position.
- The accelerating time up to the travelling speed and the braking time are 0.5 s.

The travelling speed v is then 0.445 m/s and the acceleration a 0.889 m/s². Because of the diameter d of the drive roll on the conveyer line of 350 mm, this leads to a speed of rotation of 2.543 rad/s (24.3 rpm) on the output end of the drive.

The total passage of a motion sequence may be divided up into sections with a constant torque pattern and a speed of rotation that is changing on a linear basis. This load cycle is thus defined by m times t_z, $z = 1...m$, $T = t_m$ with the associated speeds of rotation $n_{L,z}$ and torques $\tau_{L,z}$ (Fig. 3-6).

The variation of the torque over the time shows the dominance of the torque that needs to be applied in order to compensate for the force acting parallel to the inclined plane. During acceleration and braking, the acceleration or braking torque is superimposed. The difference between the torque when travelling in the positive and the negative directions results from the friction force which always counteracts the movement.

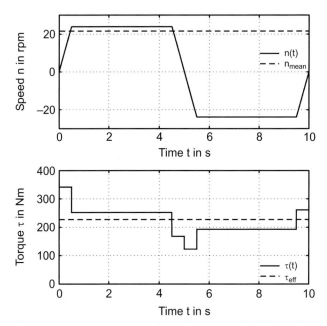

Fig. 3-6. Torque characteristic and speed characteristic of a motion sequence

The values required by the drive for the process then result from the torque characteristic τ_z:

- Maximum torque $\tau_{P,max}$:

$$\tau_{P,\mathrm{max}} = \mathrm{max}(\tau_z) \tag{3.32}$$

- R.m.s. torque τ_{eff}:

$$\tau_{eff} = \sqrt{\frac{1}{T}\sum_{z=1}^{m}\tau_z^2 \cdot \Delta t_z} \tag{3.33}$$

- Average speed n_{mean}:

$$n_{mean} = \overline{|n_{L,z}|} = \frac{1}{T}\sum_{z=1}^{m}|n_{L,z}| \cdot \Delta t_z \tag{3.34}$$

- Maximum speed n_{max}:

$$n_{\mathrm{max}} = \mathrm{max}(n_{L,z}) \tag{3.35}$$

These four variables are decisive for the selection of the components, in particular the motor, the gearbox and the inverter.

The motion sequence over time or the individual operating points may also be illustrated in two dimensions in an τ/n graph (Fig. 3-7). This also provides speed-of-rotation-dependent reference points for the torque demand, which for its part is an important input variable for the selection of the motor.

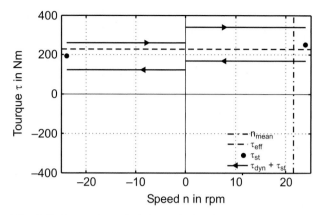

Fig. 3-7. Torque/speed characteristic of a motion sequence

The determination of the process power required is also decisive for the selection of the drive. This is determined with the following equations from the variables calculated:

$$P_{eff} = \tau_{eff} \cdot \frac{2\pi \cdot n_{mean}}{60} \tag{3.36}$$

$$P_{max} = \tau_{P,max} \cdot \frac{2\pi \cdot n_{max}}{60} \tag{3.37}$$

In the example above with a mass m of 500 kg and an inclination α of 15° and a coefficient of rolling friction μ' of 0.01, the following values result:

- $n_{max} = 24.1$ rpm,
- $n_{mean} = 22.1$ rpm,
- $\tau_{max} = 341$ Nm,
- $\tau_{eff} = 228$ Nm,
- $P_{eff} = 0.53$ kW,
- $P_{max} = 0.865$ kW.

3.1.8   Elastic coupling of the load

In many cases a drive does not act directly on the load, but its torque – or its force, after a conversion into a linear movement – is transferred via driving elements. These driving elements do not have infinite stiffness, but have elasticity and damping. Fig. 3-8 shows the relationships in principle.

This spring-mass system constitutes a dual-mass oscillator having a resonant frequency. The relationships will be illustrated for a system with linear motion, they can also be directly transferred onto a rotary system.

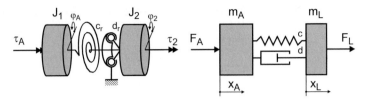

Fig. 3-8. Dual-mass oscillators with rotating and linear motion

The following equations apply for the movement of both masses m_A and m_L:

$$m_A \ddot{x}_A = F_A - c \cdot (x_A - x_L) - d \cdot (\dot{x}_A + \dot{x}_L) \qquad (3.38)$$

$$F_L = m_L \ddot{x}_L = c \cdot (x_A - x_L) + d \cdot (\dot{x}_A - \dot{x}_L) \qquad (3.39)$$

where c is the spring constant and d the damping constant of the elastic coupling of the two masses. This spring-mass system is excited by a change in force F_A to commence vibrations with the peak amplitude x_A-x_L. The resonant frequency ω_0 is calculated as follows:

$$\omega_0 = \sqrt{\frac{c}{m_A} + \frac{c}{m_L}} \qquad (3.40)$$

Typical resonant frequencies of mechanical transmission elements such as toothed belts and leadscrews lie in the range from 5 to 200 Hz, while gearboxes have significantly higher resonant frequencies [Dr01].

In the event of rapid variations in the drive force F_A, the system initially reacts by building up the spring pressure, which then generates vibrations. The resulting force acting on the mass m_L can rise up to twice the force F_A here. If, on the other hand force F_A is built up slowly, the movement x_L of the mass m_L follows the movement x_A of m_A. Fig. 3-9 shows the resulting force F_2 for the cases where F_A is built up very quickly or significantly more slowly by way of a ramp with the rise time $t_{a,max}$.

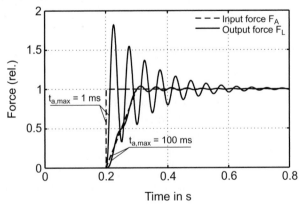

Fig. 3-9. Excitation of a dual-mass oscillator

The example is based on a resonant frequency of 20 Hz. These relationships demonstrate that it is useful to build up the acceleration and thus the torque in a drive without jerking, i.e. with a finite rate of change, to avoid these vibrations as far as is possible. The time in which the maximum acceleration is reached should be at least twice as long as the period of the resonant frequency for the dual-mass oscillator.

In practice, the resonant frequency depends on further variables in addition to the drive elements used (e.g. mass and position of the work piece or the material to be conveyed) meaning that it is necessary to determine and to take into consideration the minimum resonant frequency in all operating states.

In addition to the mechanical aspects that have been described in this chapter, there are further conditions of use that will be explained below.

3.2   Operating conditions of drives

Thorsten Gaubisch

Following the description of the dimensioning of drives from the point of view of the mechanical process, this chapter describes which further operating conditions drives encounter which have an influence on the design and selection of drive components. These operating conditions include the mains voltage and mains supply types as well as the ambient conditions on site.

Regulations, standards. Inverter drives that are operated on a low voltage supply between 50 and 1000 V_{AC} are subject to the Low-Voltage Directive (73/23/EEC or 2006/95/EC) and are therefore marked with the CE mark. Most devices further have UL approval (UL 508C) to assist in their use on the American market. Many operating conditions that drives must satisfy in order to ensure safe operation are already specified by these standards. The requirements of the Low-Voltage Directive on drive systems are laid down in IEC/EN 61800-5-1. This standard is the basis for the Declaration of Conformity for drive systems.

3.2.1   Mains voltages and mains supply types

Mains voltages. There are various mains voltages to be found around the world (Table 3-1).

Table 3-1. Mains voltages and frequencies

Mains voltage			Selected regions or states
Single-phase	Three-phase	Frequency	
110 V	190 V	50 Hz	Northern Japan
110 V	190 V	60 Hz	Southern Japan, Taiwan
120 V	200-240 V / 480 V	60 Hz	USA
120 V	200-240 V / 600 V	60 Hz	Canada
127 / 220 V	220 / 380 / 440 V	60 Hz	Brazil
220 V	380 V	50 Hz	China, Hong Kong, Philippines Argentina some nations in Eastern Europe
230 V	400 V	50 Hz	Europe, including the Russian Federation large parts of Asia including Turkey, India, Indonesia, etc. South Africa
240 V	415 V	50 Hz	Australia, Malaysia

Special voltages of 500 to 690 V are used sporadically in various regions or large industrial facilities

Asynchronous motors operated at the mains adopt their speed to the mains frequency. The mains frequency is of no importance in the case of inverters. The mains voltage is rectified and transformed into the desired motor voltage and frequency by the inverter.

Inverters have a DC bus designed for a mains voltage of 230 V or 400–480 V. The following permissible mains input voltages and mains frequencies result:

- 230 V equipment (single-phase): 160 V_{AC}–264 V_{AC}; 45 Hz–65 Hz,
- 230 V equipment (three-phase): 100 V_{AC}–264 V_{AC}; 45 Hz–65 Hz,
- 400 V/480 V equipment (three-phase): 320 V_{AC}–550 V_{AC}; 45 Hz–65 Hz.

If a number of single-phase 230 V devices are operated on a three-phase supply (L1, L2, L3 and N), they should be distributed as uniformly as possible over the three phases in order to achieve a symmetrical current load on the phases. As the mains current of an inverter is non-sinusoidal (as in the case of an ohmic resistance loading, for instance) the current in the neutral conductor is not zero, even if the three phases of the mains are loaded symmetrically, but is up to twice the phase current. This behaviour must be taken into consideration for the dimensioning of the wiring (cross-section) and load capacity of the neutral conductor.

If a number of single-phase devices are to be connected only between one phase and the neutral conductor (e.g. L1 and N) the total currents are distributed over the phase and the neutral conductor. This can cause an unbalanced load on the feeding three-phase supply.

Mains supply types. There is a wide variety of mains supply types around the world differing by the way in which the low voltage transformer in the mains supply is connected and by the nature of the earthing or occasionally by the insulation against earth. The usual mains supply types are TN, TT and IT systems.

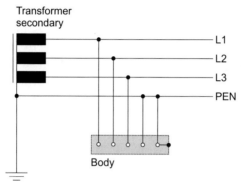

Fig. 3-10. TN-C system

TN system. The TN system (French Terre Neutre – earth neutral) is the most frequent supply form in low voltage systems. The star point of the transformer is connected directly to earth in this case. The electrically conductive bodies of the connected consumers are connected to the star point of the transformer by way of the protective conductor (PE). The protective conductor (PE) and the neutral conductor (N) are laid as a common conductor, depending on the cable cross-section. This configuration is known as a TN-C system (Fig. 3-10, French Terre Neutre Combiné (combined earth neutral)). This mains configuration is only permitted if the cable cross-sections are adequate as high current flows through the neutral conductor in the event of uneven loading of the phase conductors.

If they are laid as two independent conductors, this is known as a TN-S system (Fig. 3-11, French Terre Neutre Separé (separated earth neutral)). As a rule, the neutral conductor is split into two cables for N and PE at the point from which the cable cross-section falls below the minimum.

Inverters are exclusively suited for TN systems (including variants).

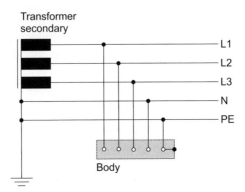

Fig. 3-11. TN-S system

TT system. In facilities in which the operation is inherently hazardous (especially building sites), a TT system (French Terre Terre (earth earth)) is generally used (Fig. 3-12) as it is forbidden to make a direct connection between the body and the operating power circuit (PEN) here. The secondary side of the transformer is connected in a star connection. The star point is earthed and is run as a separate neutral conductor (N). The neutral conductor has no protective function. The consumer device must be earthed separately. This will provide the protective earth for the body. Connection to TT systems is also possible for inverters without restriction.

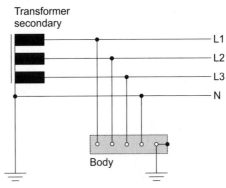

Fig. 3-12. TT system

IT system. The IT system (French Isolé Terre (isolated earth)) is used on installations required to provide high protection against fire and protection of persons. Also those which must initially continue to run in the event of an earth fault or short circuit to frame in order to minimise consequential damage in the process. In order to achieve enhanced reliability in the event of insulation faults, the star point of the transformer is insulated in the IT system. The body of the consumer is earthed directly via a protective earth (Fig. 3-13). The insulation against earth of all phase conductors must be constantly measured by an insulation monitoring device in order to be able to detect a fault. Any insulation fault determined must be repaired immediately. The protective measure will no longer work in the event of a second fault case. The current will then flow via earth between the two damaged phase conductors.

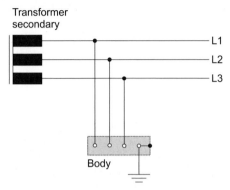

Fig. 3-13. IT system

IT systems are used in the following areas, for example:

- public supply system in France and Belgium,
- the chemical industry,
- mining,
- isolated power supplies (own generators),
- hospitals,
- ships' systems,
- crane, materials handling and lift installations,
- the paper and pulp industries.

No EMC emission level is defined for IT systems, as there is no reference of the mains to the PE (see chapter 3.4.8 relating to electromagnetic compatibility).

As a rule, the operation of inverter drives is initially only permissible on TT systems, TN systems or systems with an earthed neutral. Most drive manufacturers can provide special hardware variants that allow operation on IT systems. Some of the devices that are designed for TT or TN systems have a hardware switch which removes the connection of the Y capacitors in the device to the PE. This then permits operation on an IT system.

Earth-leakage circuit breaker. The additional installation of residual current devices (RCD, earth-leakage circuit breakers) in TT and TN systems is specified for the protection of persons in certain applications. These devices have differential current detection which trigger from a specified level of fault currents to PE and switch off the mains. The drives may trigger the RCD unnecessarily because of the discharge currents caused by an inverter, depending on the installation design (e.g. number of devices, motor cable lengths). The RCD is triggered when the inverter is switched on. In some cases it is necessary to use RCDs with higher trigger current ratings. It will then be necessary to check that the complete safety concept is still adequate.

In order to avoid this situation there are some device designs with particularly low discharge currents. The use of additional isolating transformers is also sometimes necessary.

3.2.2 Ambient conditions

The ambient conditions affecting drives include the climatic conditions such as temperature, air humidity, site altitude and atmosphere. These are especially relevant for the electronic components in the inverter for which the appropriate limit values are specified in the technical specification.

In addition to this, particular industrial sectors can require special operating conditions (e.g. explosion protection) to be met by the drive systems. This requirement rather affects geared motors as these can be installed in the danger area.

Climatic conditions. In use drives are required to guarantee reliable operation in the widest variety of climatic ambient conditions. These conditions are specified in the standard EN 60664.

The temperature range has a decisive role to play here. Most inverters permit use at an ambient temperature from -10 to +40°C. In this temperature range they are able to provide their rated power for the application. At temperatures above this, further use at reduced power may be possible (known as derating) by the reduction of the rated output current by typically 2.5% / °C. The maximum ambient temperature is generally around +55°C. The reason for the necessity of this current derating is in the cooling power of the heatsink. At ambient temperatures between +40°C and +55°C the heat generated inside the device can no longer be dissipated adequately and this can lead to an inverter to turn off. In addition, excessive heating can cause the maximum permissible temperature for some components to be exceeded and the ageing process for the electrolytic capacitors used to be accelerated. Current derating offers the capability of using the drives at ambient temperatures of up to +55°C.

The site altitude of the devices also has a role to play here as the cooling performance declines at altitude due to the reduced air pressure. Basically, drives may be used at their rated power at a site altitude of 0 to 1000 m. The output power of the units at site altitudes of 1,000 to 4,000 m has to be limited by reducing the rated output current by 5% per 1,000 m. This compensates for the reduced cooling capacity of the air at low air pressure and avoids overheating of the inverter.

Drives may also be used in an environment which does not exceed an average relative humidity of 85% and does not lead to condensation in the inverter. Otherwise dust deposited on the PCBs can be compounded into an electrically conducting fluid which can lead to the electrical destruction of the device.

Application conditions for special industrial sectors. The drives used for applications in special industrial sectors (foodstuffs, chemicals or wood processing industries) must be adapted to the particular operating conditions on site.

Explosion protection (ex-protection). The fundamental requirements for an explosion are the presence of a flammable material (gas, dust, liquid), an adequate volume of oxygen and a source of ignition. Thirteen different sources have now been identified, of which just a half are of electrical origin. Hot surfaces, sparks of mechanical origin or ultrasound are important in addition to sparks, electric arcs or static charges. Explosion protection is not restricted just to the field of combustible gases or vaporised liquids. A significant number of explosions can now instead be traced back to inadequately protected atmospheres carrying dust.

In most applications only the motor and the gearbox are in use in the potentially explosive atmosphere itself. These must be ATEX compliant (French Atmosphère explosible (explosive atmosphere) in accordance with European directives 94/9/EC (Product Directive, implemented in German law in the Regulation regarding protection against explosions 11th GPSGV) and 99/92/EC (Operating directive). A specially approved monitoring device must be used for motor temperature monitoring to guarantee secure shutdown of the motor in the event of overtemperature. This special monitor is not implemented in conventional inverters. The associated inverters are usually located outside this explosion-protected zone as it is not feasible to implement the ATEX conformity for the inverter.

Foodstuffs industry. Special geared motors are available tailored to the particular requirements of the foodstuffs industry. These requirements include the hygiene regulations and suitability for use in extended temperature ranges, absolute oil seal tightness and no sensitivity to water and cleaning materials.

Aggressive gases. The drives used in environments exposed to aggressive gases must be protected against these gases. An aggressive environment of this nature can damage the electronic assemblies in the inverter so much that the installation can fail. Measures such as sealing or pressure encapsulating the control cabinets or special surface coatings on motors and gear-boxes can contribute to an improvement in operational safety.

Vibration and shock resistance. Inverter drives are also exposed to different levels of vibration according to IEC 68227. Control cabinet devices are required to withstand lesser levels of vibration than is the case with devices mounted on motors, for instance. Control cabinet devices are usually designed for accelerations up to 0.7 g. Decentralised drives withstand up to 2 g.

Further standards that apply for the operating conditions for inverter drives are listed below (Table 3-2).

Table 3-2. Standards for the operating conditions

Operating condition	Standard
Climatic conditions	Class 3K3 in accordance with EN 60721
Soiling	Degree of soiling 2 in accordance with EN 60664 Part 1
Degree of protection	Device-dependent IP 20 to IP 65 in accordance with EN 60529

3.3 Motors

Dr. Carsten Fräger, Dr. Edwin Kiel

The motor is the heart of the drive. It converts electrical energy into mechanical energy. This book only considers AC motors, as this type of motor clearly dominates in the power range from 100 W to 100 kW for directly mains-operated drives. Three-phase AC drive technology has now displaced DC drives. For drives with smaller power ratings, other motor designs are also used (permanent-field DC motors, EC motors, stepping motors), but these are not covered in this book.

The three-phase AC motors which are used can be divided initially into two groups:

- *Standard three-phase AC motors*, which are also suitable for applications with constant speeds and can be directly driven from the mains. The speed of these motors can be varied by using frequency inverters.
- *Servo motors*, which are optimised for high dynamic performance and are only used in closed loop control with servo inverters.

In the following, we start by looking at the operating principle of three-phase AC motors. Then the standard motors and servo motors are described in more detail. Afterwards we also take a look at linear motors, which are used in applications where rotary servo motors reach their limits.

In order to use motors in speed or position-controlled applications, sensors are required to determine the angular position. The most important types of these sensors are presented below. At the end of the chapter we also take a look at brakes, which are required for any application in which a motor needs to be held at standstill.

The required motor properties for the application determine the selection of the suitable motor type. Table 3-3 shows a comparison of the features of standard three-phase AC motors and servo motors. Three-phase AC motors are used predominantly in applications requiring a robust drive with constant or slow-changing speeds. Servo motors are predestined for applications requiring fast speed variations, compact installation volume and high accuracy.

Table 3-3. Comparison between different types of electric motor

	Standard three-phase AC motor	Asynchronous servo motor	Synchronous servo motor
Dynamic performance	Average	High	Very high
Inertia	Average	Low	Very low
Overload capacity	Average	Very high	Very high
Power density	Average	High	Very high
Field weakening	Average	High	Low
Efficiency	Average to high	Average to high	High to very high

3.3.1 How three-phase AC motors work

A three-phase AC motor converts the energy P_{el} provided by the electric voltage supply into mechanical energy P_{mech}. During the conversion process, power loss $P_{V,mot}$ occurs. The power equation is as follows:

$$P_{el} = P_{mech} + P_{V,mot}$$

$$P_{el} = \sqrt{3} \cdot V \cdot I \cdot \cos \varphi \qquad (3.41)$$

$$P_{mech} = \tau \cdot 2\pi \cdot n$$

To understand how torque is generated in a three-phase AC motor, let us first draw an analogy to a DC motor. In general, torque is generated in an electric motor as a result of the fact that a force is generated when a current flows through a conductor which is positioned in a magnetic field. In the process, the resulting force is proportional to the magnitude of the magnetic field and the current. Accordingly, the following points contribute to the generation of torque:

- A *magnetic field* must be present. In a DC motor, this is done by means of the field winding located in the stator. The magnetisation is normally kept at a constant value.
- A *current* needs to be established perpendicular to this field. This is done with the aid of the armature winding located on the rotor. The torque is defined by the magnitude of the armature current.
- This current flow must remain perpendicular to the magnetic field even when the rotor is turning. This is achieved with the aid of the mechanical *commutator* of the DC motor, which reverses the polarity of the current flowing in the individual conductors of the rotor according to the instantaneous rotor position.
- Following the law of electromagnetic induction, a voltage is then generated as a result of the motion (i.e. rotation) of the rotor with the armature winding in the magnetic field. This voltage opposes the current flow, whereby the *induced voltage* is proportional to the magnetic field and the speed of rotation.

These four points all apply to synchronous and asynchronous motors as well, although the solutions are different.

In order to describe the behaviour of a three-phase AC system, it makes sense to introduce phasors or complex state variables. The total degree of freedom of a neutral conductor-free three-phase AC current system is two, because the sum of the three voltages or currents is equal to zero. As a result, the two dimensions of a phasor (length, phase angle) or a complex number (real and imaginary part) fit very well with the two degrees of freedom of the three-phase system. The conversion between the three variables of the three-phase system and the complex value is performed according to the following equation:

$$\underline{v}_S = v_{S1} + v_{S2} \cdot e^{j120°} + v_{S3} \cdot e^{j240°} \qquad (3.42)$$

The currents in the windings of a three-phase AC motor control both the angle and the magnitude of the magnetic field. In a three-phase AC motor the three windings are magnetically offset by 120° to each other. If a three-phase current vector is introduced, a magnetic field is generated with a magnitude which depends upon the amplitude of the current vector and an angle which depends upon the electrical phase angle of this vector.

Fig. 3-14 clarifies this effect [GiHaVo03]. An inverter, which is described in chapter 3.4, can freely adjust the length and angle of this phasor.

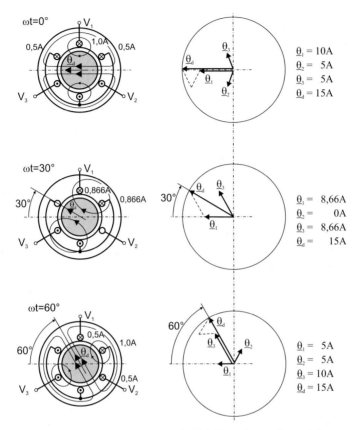

Fig. 3-14. Creation of the magnetic field in a three-phase AC motor

By means of a rotation through the instantaneous electric phase angle, it is possible to transform the current vector into a d/q coordinate system, which has the frequency zero and in which the i_q component creates the torque. The frequency with which it needs to rotate in order to do this depends on the operating principle of the motor. Initially we shall take a look at a synchronous motor, as this has a more simple operating principle.

In the case of the permanent-field synchronous motor, the magnetic field is generated with the aid of permanent magnets attached to the rotor of the motor. With field-oriented current injection, the orientation of the current vector is controlled by the inverter so that it is perpendicular to the instantaneous rotor position of the magnetisation. By determining the magnitude and orientation of the current, the inverter takes on the task of generating the current flow required for torque generation and orientation for the commutation of the motor. In order to satisfy this, the rotor position and therefore the orientation of the magnetic field need to be known. The encoder of the motor serves for this purpose. The permanent-field synchronous motor induces an alternating voltage in the three phases which is proportional to the speed of rotation.

Accordingly, the permanent-field synchronous motor can be described with the following equations:

$$\underline{v}_S = R_S \underline{i}_S + L_S \frac{d \, \underline{i}_S}{dt} + j \, \omega \, \Phi_F \, e^{j\varphi} \tag{3.43}$$

$$J \frac{d \, \omega}{dt} = \Phi_F \, i_{Sq} - \tau_L \tag{3.44}$$

$$\frac{d \, \varphi}{dt} = \omega \tag{3.45}$$

Equation (3.43) describes the electric circuit with the potential differences on the stator resistance R_S and the stator inductance L_S of the motor, as well as the voltage induced by the magnetisation Φ_F and motor speed ω. Equation (3.44) describes the generation of torque and (3.45) the integration of the speed and the angle.

The asynchronous motor requires slip to generate torque, i.e. a difference between the frequency of the three-phase stator system and the angular speed of the rotor. This means that, with a constant speed, the stator frequency needs to be varied according to the required torque. Despite this complex relationship, which cannot be described in full detail at this point (please refer instead to the additional literature [Le97, Sch98]), an analogy to the DC motor can also be found on the asynchronous motor.

To do this, the orientation is again chosen in such a way that the torque is generated as the product of the magnetisation and a torque-forming current i_q (the field orientation) [Bl72]. The magnetisation is then generated by means of the field-forming current component i_d, which is arranged perpendicular to i_q. During the build-up of magnetisation via i_d, a low pass with the rotor time constant t_R acts (further analogy to DC motors: the build-up of magnetisation via the field winding has a time constant which results from the resistance and the inductance of the field winding). From the instantaneous value of i_q and the magnetisation, the slip frequency ω_2 is then determined. Once the mechanical speed ω has been added, this defines the frequency of the current vector in the field coordinate ω_{mR}. Also in the case of an asynchronous motor, an induced voltage proportional to the frequency is generated.

The asynchronous motor is thus described by the following equations:

$$\underline{v}_S = R_S \, \underline{i}_S + \sigma \, L_S \frac{d \, \underline{i}_S}{dt}$$
$$+ (1 - \sigma) \, L_S \, (j \, \omega_{mR} \, i_{mR} + \frac{d \, i_{mR}}{dt}) \, e^{j\rho} \tag{3.46}$$

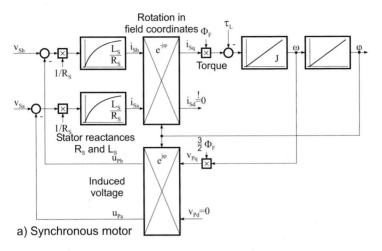

a) Synchronous motor

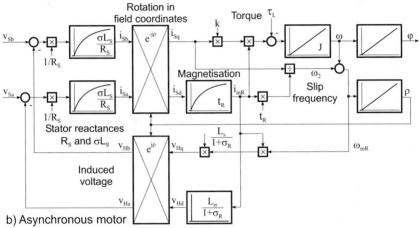

b) Asynchronous motor

Fig. 3-15. Block diagrams for the synchronous and asynchronous motor

$$J \frac{d\omega}{dt} = \frac{2}{3} \frac{L_\sigma}{1+\sigma_R} \cdot i_{mR} \cdot i_{Sq} - \tau_L \tag{3.47}$$

$$\frac{d\varphi}{dt} = \omega \tag{3.48}$$

$$i_{Sd} = i_{mR} + t_R \frac{d\,i_{mR}}{dt} \tag{3.49}$$

$$\frac{d\rho}{dt} = \omega_{mR} = \omega + \omega_2 = \omega + \frac{i_{Sq}}{t_R i_{mR}} \tag{3.50}$$

The structure of the first three equations corresponds to that of the synchronous motor. Equation (3.49) describes how the current component i_d creates the magnetisation, while equation (3.50) describes the calculation of the slip frequency ω_2 and the resulting frequency ω_{mR} of the current vector $\underline{i}_{d,q}$.

Fig. 3-15 presents the way in which the two motor types work in the form of block diagrams. The structural similarities to the generation of torque are evident. The differences primarily lie in the generation of the magnetisation and the formation of the slip frequency on asynchronous motors.

In comparison to a synchronous motor, in which only the pole position needs to be measured in order to determine the orientation of the current in the field coordinates, the orientation of the current vector in field coordinates is not measured on an asynchronous motor. Instead, it has to be determined with the aid of models and observers. The models in turn depend on motor data, which need to be known. Despite this, based on this description of how it works, field-oriented control has established itself as the most important technology for the fast control of asynchronous motors [Le91].

Now that the fundamental principles of how three-phase AC motors work have been presented, we go on to describe the specific motor designs in more detail below.

Table 3-4. Comparison of DC motors and three-phase AC motors

	DC motor	Synchronous motor	Asynchronous motor
Magnetic field generation	Field winding	Permanent magnets in the rotor	Current component i_d
Torque-generating current	Armature current	Current component i_q	Current component i_q
Commutation of the current	Mechanical commutator	Rotation of the current vector by the inverter	Rotation of the current vector by the inverter
Induced voltage	DC voltage	AC voltage	AC voltage
Synchronicity of the electric rotating field to the mechanical speed of rotation		Yes	No, slip speed proportional to the torque (approx. 3–5% for rated torque)

3.3.2   Standard three-phase AC motors

The standard three-phase AC motor is an asynchronous motor which converts electric power into mechanical power. In doing so, it either works directly from a three-phase system or is controlled by a frequency inverter. Characteristics of standard three-phase AC motors include:

- standardised main dimensions → standard motor (the electrical data are standardised in the various parts of IEC 60034, the mechanical data in IEC 60072),
- very robust,
- can be operated from the mains or via an inverter (provided the motor has the correct design),
- moderate to good efficiency,
- average power density.

The main dimensions, such as the shaft diameter, flange diameter and foot height are standardised. The electrical connections are generally made via terminals in the terminal box. With the aid of a variety of built-on accessories, such as brakes, encoders or fans, the motor can be adapted to the particular drive requirements.

Asynchronous motors are characterised by the following electromechanical variables:

- rated power P_N,
- rated speed n_N, rated slip s_N,
- rated torque τ_N,
- rated current I_N,
- rated voltage U_N,
- rated frequency f_N,
- rated power factor $\cos \varphi_N$,
- rated efficiency η_N,
- starting torque τ_d, starting current I_d,
- stalling torque τ_B, stalling slip s_B,
- pull-up torque τ_S.

The motor speed is set via the frequency of the electric voltage. With a fixed frequency f, the speed n decreases with the load (so-called slip s). The following equations describe the relationship for the case in which the stator winding resistance is ignored. This is acceptable for larger motors and for *illustration purposes*.

$$\text{Speed at no load } n_0 = \frac{f}{p} \qquad (3.51)$$

$$\text{Speed with load } n = (1 - s) \cdot n_0 \qquad (3.52)$$

$$\text{Torque } \tau = 0...\tau_N \rightarrow s \approx s_N \cdot \frac{\tau}{\tau_N} \qquad (3.53)$$

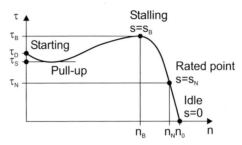

Fig. 3-16. Speed-torque-characteristic for an asynchronous motor with fixed-frequency mains operation

$$s = \left(\frac{\tau}{\tau_B} - \sqrt{\left(\frac{\tau_B}{\tau} \right)^2 - 1} \right) \cdot s_B \tag{3.54}$$

$$\frac{\tau}{\tau_B} = \frac{2}{\dfrac{s}{s_B} + \dfrac{s_B}{s}} \tag{3.55}$$

p is the number of motor pole pairs, which defines the ratio of the mains frequency and the mechanical motor speed. The use of motors with four poles ($p = 2$) is most widespread, but two-pole ($p = 1$) and six-pole ($p = 3$) motors are also used. Higher numbers of pole pairs are used for certain applications.

The speed-torque-characteristic at constant frequency is shown in Fig. 3-16. Starting from the no-load speed at $\tau = 0$, the torque increases to the stalling point. If this point is exceeded on account of an excessive load torque, then the motor will stall. The speed reduces to zero, because the required torque is not brought about. Continuous operation with high slip values is not permissible for the standard motor, as this would cause a significant build-up of heat which could destroy the motor. Corresponding measures to protect against overtemperatures must be used, such as thermal sensors in the windings or monitoring of the motor current.

The magnitude and phase of the motor current of an asynchronous motor can be determined from the current locus diagram (Fig. 3-17). The magnetic field in the motor is generated by the electric supply. To do this, a magnetising reactive current of around 30 to 70% of the rated current flows. The magnetising current corresponds approximately to the no-load current of the motor.

With the application of a load, an active current is also required in order to achieve a power flow. At the rated point, the value for cos φ_N usually lies in the region from 0.7 to 0.85. Fig. 3-18 shows the single-phase equivalent circuit diagram for an asynchronous motor.

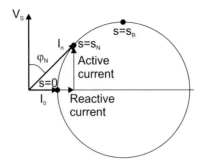

Fig. 3-17. Current locus diagram for a current-displacement-free asynchronous motor

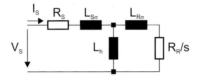

Fig. 3-18. Single-phase equivalent circuit diagram for an asynchronous motor

The motor can also work without problem in generator mode. The lower half of the current locus diagram describes this mode.

Direct start-up from the mains. During a direct start-up from the mains, a high starting torque is created in the region of 2-5 times the rated torque τ_N. At the same time a high current flows corresponding to 3-8 times the rated current I_N. High starting losses are generated in both the stator and in the rotor.

Star/delta start-up from the mains. If the starting current of the direct start-up is too high, it can be reduced to 58% by initially connecting the motor in a star layout for the start-up, and by then switching to a delta layout after successful start-up. However, this is only possible if the starting torque, which is reduced to 33% in a star layout, is sufficient for the application.

Soft-starter. A soft-starter also reduces the starting current of motors. They do this by reducing the voltage with the aid of a thyristor actuator. With this arrangement it is not possible to control the speed. In higher power ranges, soft-starter offer lower costs than frequency inverters.

Operation with a frequency inverter. For variable speed drives, motors are controlled by frequency inverters. Thanks to the variable frequency and voltage of the inverter, the three-phase AC motor can be operated throughout a wide speed range. Fig. 3-19 shows the operating ranges in motor mode and generator mode. In the range up to approximately the rated speed the motor operates with a full magnetic field, as a result of which a high torque can be delivered.

Above the rated speed, the magnetic field is reduced as the maximum output voltage of the inverter has been reached, and the motor is operated in the so-called field weakening range with reduced torque.

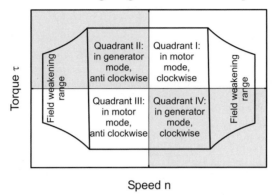

Fig. 3-19. Operating ranges of standard three-phase AC motors with inverter

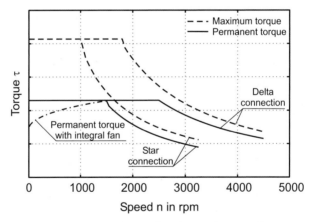

Fig. 3-20. Speed-torque-characteristics for four-pole asynchronous motors with inverters in star and delta connection

In order to ensure problem-free operation with a frequency inverter, the motor needs to be equipped with reinforced insulation. In the case of motors with integral fans it should be noted that, in the range below the rated speed, due to the reduced cooling effect of the integral fan, only a significantly lower torque is available than the rated torque in continuous operation. This restriction can be compensated for by using a blower (Fig. 3-20).

Table 3-5. Operating modes for an asynchronous motor

	Direct on the mains	Star-delta start-up	Soft-starter	Frequency inverter	Frequency inverter with 87 Hz operation
Speed adjustment	No	No	No	Yes	Yes
Max. starting torque	$3\text{--}5 \cdot \tau_N$	$1\text{--}1.5 \cdot \tau_N$	$1\text{--}2 \cdot \tau_N$	$2 \cdot \tau_N$	$2 \cdot \tau_N$
Starting current	$3\text{--}5 \cdot I_N$	$1\text{--}1.5 \cdot I_N$	$1\text{--}1.5 \cdot I_N$	$1\text{--}1.8 \cdot I_N$	$1.7\text{--}3 \cdot I_N$
Starting losses	High	Average	Average	Low	Low
Continuous power	P_N	P_N	P_N	P_N	$1.7 \cdot P_N$

87 Hz operation with an inverter. A frequency inverter can adjust the ratio of the output voltage to the output frequency via the V/f characteristic. This can be used to also operate a three-phase AC motor from a mains voltage of 400 V in a delta connection. This expands the speed range for three-phase AC motors. Fig. 3-20 shows the characteristics for star and delta connection for a four-pole motor. In a delta connection, approximately 1.7 times higher speeds are attained at the same torque. The rated frequency increases in a delta connection to approximately 87 Hz. Naturally, the current for this 1.7 times increase in power also increases by a factor of 1.7. Overall, 87 Hz operation can result in more cost-effective solutions for many applications in which a frequency inverter is used.

A general overview of the different operating mode of an asynchronous motor is shown in Table 3-5.

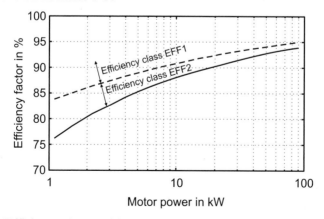

Fig. 3-21. Efficiency classes of four-pole standard three-phase AC motors according to CEMEP

Efficiency of standard three-phase AC motors. The efficiency of standard three-phase AC motors particularly at lower power ranges of up to approx. 1 kW is only 75% (Fig. 3-21). It increases to higher power levels until it reaches more than 90% at power levels above 22 kW [Do03]. Through design measures and the use of alternative materials (copper instead of aluminium in the rotor), the efficiency can also be increased to more than 80% on smaller motors [KiBoDo03]. This increase in efficiency does however also go hand in hand with increased costs for the motor.

Current discussions about energy efficiency (chapter 5.2.5) have created the result that the use of more efficient motors is increasing and that the introduction of a minimum efficiency for motors is being considered. Motor manufacturers often offer different standard three-phase AC motors which offer different efficiencies for the same power rating.

Fig. 3-22. Synchronous and asynchronous servo motors ranging from 0.25 to 60 kW

3.3.3   Asynchronous and synchronous servo motors

Servo motors are optimised for high dynamic performance. Characteristic features of servo motor operation include dynamic speed and torque variations, operation at standstill in order to hold positions and short-term operation under high overload [GrHaWi06].

Servo motors are characterised by the following features:

- slim design,
- high power and torque density,
- low inertia, high dynamic performance,
- high overload capacity,
- high efficiency,
- well suited for operation with a servo inverter.

In comparison to standard three-phase AC motors, these motors can be built into much more compact designs and have a significantly lower moment of inertia, which means that they accelerate much more quickly and can make the power available for the machine in a very small space.

In contrast to three-phase AC motors without speed feedback, a servo motor not only has three power connections and one earth connection, but also at least six connections for the position sensor and two connections for temperature monitoring. As the large number of connections can easily lead to wiring errors which delay commissioning unnecessarily, servo motors are equipped with plug connectors for error-free and quick connection to the servo inverter.

As servo motors are always operated with an inverter, their winding data (voltage, frequency) and therefore their speed are not geared to standardised mains frequencies. Quite the opposite, by breaking away from standardised mains frequencies there is a great deal of freedom in the way these motors can be operated.

Servo motors are offered both in asynchronous technology and synchronous technology (Fig. 3-22). The choice of motor principle depends on the relevant requirements for the drive.

Both motor principles can also operate without problem in generator mode.

The shaft diameters and flange diameters follow those of standard three-phase AC motors. With the aid of different built-on accessories, such as a brake, or different types of angular position measuring system, the motor can be adapted for the particular application.

As servo motors are always operated with an inverter, they are equipped with reinforced insulation.

Asynchronous servo motors. With asynchronous servo motors, the magnetic field in the motor is generated by the electric supply. To do this, the servo inverter injects a magnetising current into the motor. With the magnetising current, the servo inverter has control over the magnetic field in the motor.

The same equations which describe the electromagnetic behaviour apply as for standard three-phase AC motors. The servo motors have lower values for winding resistance and leakage inductance than standard motors.

Through field weakening operation, asynchronous servo motors have a wide speed range with reduced torque above the rated speed. Due to their higher mass inertia than synchronous servo motors, they are particularly well suited for oscillating mechanisms which are less dynamic, e.g. for drives with long toothed belts.

In comparison to the characteristics of synchronous servo motors, asynchronous servo motors do not have a fixed speed limit. Due to the wide field weakening range of the motors, it is possible to run the motors at lower torques in a wide speed range far above the rated speed (Fig. 3-23).

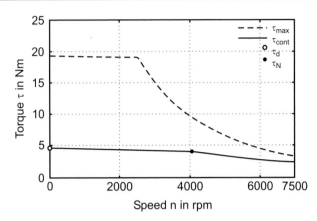

Fig. 3-23. Speed-torque-characteristics for an asynchronous servo motor

In lower power ranges asynchronous servo motors are designed as self-ventilated units. By using a blower it is possible to significantly increase the power density of a size from a rated power of 2 kW or higher.

In the power range from 10 to 500 kW, designs are also used in which an air flow is directed through the winding in the housing. The air flow is generated with the aid of powerful fans, which are arranged either in an axial direction or radially. The designs are similar to those of DC motors. Very high power densities can be achieved in this way. However, the degree of protection is reduced to the class IP 23.

Synchronous servo motors. Synchronous servo motors offer lower inertia than asynchronous servo motors with the same rated torque, and they are also more compact. The result is higher dynamic performance with large acceleration.

In a synchronous servo motor, the magnetic field is generated by means of high energy permanent magnets which are attached to the rotor. No magnetising reactive current is required. As there are none of the losses associated with generation of the magnetic field, synchronous servo motors offer excellent efficiency and can often be built with a smooth surface without any additional ventilation measures to a high degree of protection [Fr01].

Synchronous servo motors are predestined for highly dynamic applications and for operation in dusty environments.

Fig. 3-24 shows a cross-section through a synchronous servo motor with the active part for generating torque, a resolver for angle and speed measurement and a brake for holding the position in the deenergised state. It can be seen clearly that the active part for generating torque only takes up a small part of the overall volume. Generously dimensioned roller bearings, the holding brake and the resolver account for a large proportion of the motor's size.

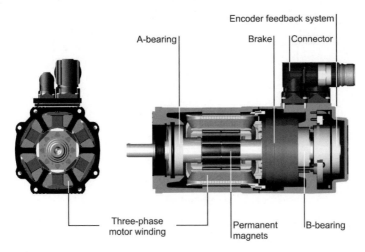

Fig. 3-24. Cross-section of a synchronous servo motor

Together with the servo inverter, servo motors make a speed-torque-range available in which the drive can be operated.

Below the maximum torque limit there is a characteristic for thermally permitted operation. The points on this S1 characteristic represent the operating points for continuous operation or for the operation equivalent to it. The area above the region of thermally permitted continuous operation is limited on the one hand by the available current. The current defines the maximum attainable torque (upper limit of the region). On the other hand, the maximum output voltage of the inverter limits the achievable motor speed (limitation on the right-hand side). The speed limit is dependent upon whether or not the control permits field weakening.

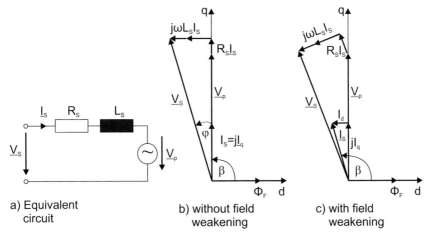

Fig. 3-25. Phasor diagram for the synchronous servo motor with and without field weakening

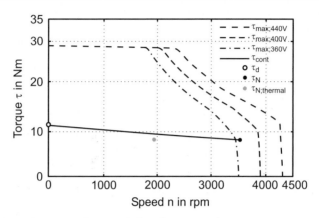

Fig. 3-26. Speed-torque characteristics for a synchronous servo motor without field weakening

Operation without field weakening. During operation without field weakening, only a current i_q is injected via the inverter in phase with the rotor voltage, so that – with the single-phase equivalent circuit diagram of the synchronous servo motor – we obtain the phasor diagram shown in Fig. 3-25 b). The result for a specific motor is shown in Fig. 3-26, which also shows the limiting characteristic for various mains voltages. The smaller the mains voltage, the smaller the attainable speed. The inverter limits the available voltage U_S according to the instantaneous mains voltage, so that with increasing speed the maximum current $I_{q\,max}$ which is possible in the motor and therefore also the maximum attainable torque is limited by the voltage.

In order to ensure reliable operation, the least advantageous limiting characteristic for the arising voltages needs to be taken into account. On weaker mains supplies this can lead to significant losses in power. By using a controlled mains supply in the inverter, its maximum output voltage becomes independent of the mains voltage, and as a result the working range of the servo motor does not need to be restricted (chapter 3.4.1).

Operation with field weakening. Naturally, in the case of a permanent-field synchronous motor, the rotor field is fixed by the magnets and cannot be adjusted during operation. This generates the torque together with the current I_q.

By injecting a current I_d, which is leading in relation to the rotor voltage V_P, in addition to the current I_q, the terminal voltage V_S is reduced at the same torque. The phasor diagram is shown in Fig. 3-25 c). If the current $I_d \neq 0$ then the voltage of the motor is reduced, or the associated speed is increased. As the torque is still generated in unchanged fashion by I_q, the limiting characteristic is shifted to higher speeds depending on the field weakening current I_d. The effect is shown in Fig. 3-27: the left-hand curve shows the limiting characteristic without field weakening, while the right-hand line shows the attainable limit with an optimum field-weakening current. The gain in speed as a result of field weakening is more than 800 rpm at torques of up to 20 Nm. This represents a significant increase in the drive power – without the need for increased power electronics [Fr03, KiHeKo04].

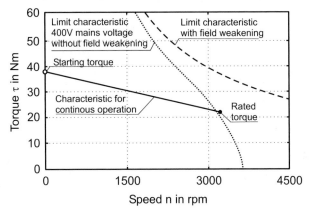

Fig. 3-27. Speed-torque characteristics for a synchronous servo motor with field weakening

The suitability of a synchronous servo motor for operation in the field weakening range depends to a large extent on the magnitude of its stator inductance. Hard motors have a very low leakage inductance and are therefore hardly suited to field weakening operation at all.

Among other things, field weakening operation of a synchronous servo motor offers the following advantages in terms of drive design:

- With the same inverter power it is possible to make increased shaft output power available.
- In the event of a voltage drop of the mains or the DC-bus voltage, the drive can continue to make its torque available without a speed drop, so that operation can be continued without interruption and operational reliability is increased.
- During generation of the speed setpoints, the current mains voltage or the influence of a higher power input from other consumers on the DC bus do not necessarily need to be taken into account, as in the short-term these influences on the speed limit are absorbed by the field weakening. This makes it possible to simplify the generation of speed setpoints in a complex control structure.
- In the case of drives with low torque requirements at high speeds, e.g. winding drives, the drive power no longer needs to be designed for the product of the highest torque and the highest speed or the highest cur-

rent and the highest induced voltage, and instead the highest power which is actually required determines the drive power that needs to be installed. As a result, this type of drive can be designed significantly more cost-effectively.

3.3.4   Linear motors and direct drives

With linear motors, the force required for the linear motion is generated directly by the interaction between the motor current and the magnetic field. Figuratively speaking, a linear motor is created as a result of the unwinding of a rotating motor. Typically, the magnetic field is generated with permanent magnets in the same way as in a synchronous servo motor. Designs with an iron-core active part in a flat-bed or U-shaped layout are used most commonly. Usually, the primary part has a 3-phase rotary current winding. The secondary part contains the permanent magnets (Fig. 3-28). As a rule, several secondary modules are attached to the fixed part of the machine, so that the overall length of the secondary part corresponds to the traverse path plus the length of the primary part.

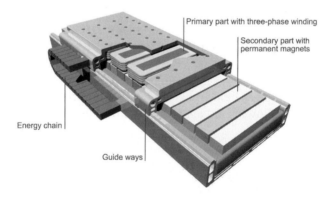

Fig. 3-28. Structure of a linear motor

The primary part is connected to and drives the moving part of the machine. To do this, the electric cable to the primary part is connected to the servo inverter via a flexible trailing cable (a so-called energy chain). For position detection a sensor is attached along the traverse path. The signals from this sensor are also transmitted via a flexible cable from the moving part of the machine to the fixed part.

Fundamentally it is also possible for the secondary part to move while the primary part remains stationary, but in practice the secondary part is often longer and cannot be housed inside the moving part for design reasons.

As their secondary part, asynchronous linear motors have an aluminium or copper rail as a short-circuit element. In production machines they can only be considered for small feed forces, as their efficiency is comparatively low and they have a large volume.

For the control of the linear motor, a linear encoder is required which must provide the signals for current, speed and position control. Here, optical and magnetic encoders are used, whereby the optical encoders offer higher resolution but are also more expensive. Similarly to a synchronous servo motor, the rotor position also needs to be known for the drive control of a linear motor. Linear encoders which supply absolute position information are expensive. Alternatively, the drive control can also perform a motor pole angle adjustment after power-up.

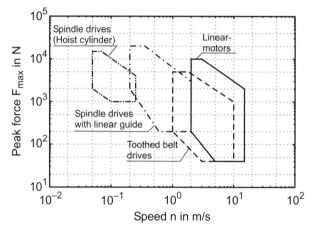

Fig. 3-29. Operating range of linear motors

Typical linear motor data:

- Peak force F_{max} = 100 N...10 kN,
- Continuous force F_{cont} = 50 N...5 kN,
- Speed v = 2...15 m/s,
- Acceleration up to a = 100 m/s².

Fig. 3-29 shows the operating ranges of linear motors in comparison to linear drives with mechanical conversion of the rotation of a servo motor into a linear motion (chapter 3.6.6).

The equations used to describe the operational performance of synchronous servo motors also apply to linear motors provided the rotary variables rotational speed n, torque τ and number of pole pairs p are replaced with the linear variables speed v, force F and pole pitch κ:

$$n \cdot p = \frac{v}{p \cdot \kappa} \tag{3.56}$$

$$\tau \cdot p = \frac{\pi \cdot F}{\kappa} \tag{3.57}$$

Linear motors also have a limiting characteristic which describes the permitted operating range. A general curve is shown in Fig. 3-30.

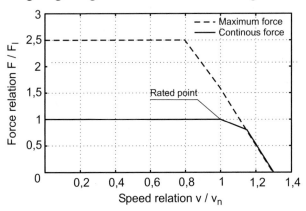

Fig. 3-30. Limiting characteristic for a synchronous linear motor

Table 3-6 shows the advantages of linear motors in comparison to linear drives with servo motors and a mechanical conversion. However, when using linear motors the restrictions detailed in Table 3-7 must also be taken into account.

Table 3-6. Advantages of linear drives with linear motors

Very good controllability, as the system has a much higher stiffness than conventional structures with a toothed belt drive and therefore also has higher resonant frequencies.	→ High controller gain
Very good acceleration values are possible, as due to the lack of resonances fewer vibrations are initiated.	→ High acceleration
Very high speeds, as critical rotational speeds or critical linear speeds on transmission elements, e.g. spindles, do not need to be taken into account.	→ High speeds
High accuracy, as there is no clearance.	→ High accuracy
High running smoothness or balance quality, as there are no mechanical transmission elements, particularly on ironless winding parts.	→ High running smoothness

Table 3-7. Limitations of linear drives with linear motors

The entire travel way is magnetic and therefore tends to pick up iron particles.	→ Sealing required
For perfect operation of the motor (air gap), the guide must be very accurate, even if the application does not require any accurate guides.	→ Complex and expensive guidance
The normal force load represents a significant load on the bearings – designs with low normal forces are more complex and expensive.	→ Reduced service life
Mechanical brakes for linear motors are complex and expensive.	→ Expensive, bulky, problems with vertically operated linear axes
The energy supply generally takes place via mobile energy chains, which take up space and are subject to wear (cable breakage, wear to chain parts).	→ Maintenance-intensive, reduced service life
The motor gives off its heat loss to the moving parts in the machine.	→ Interfering build-up of heat in the process, necessity for liquid cooling of the motor
The volume of the component is higher than a comparable mechanical solution, e.g. a motor with an active surface of approx. 10 x 25 cm^2 is required for a force of 2 kN; alternatively, it would also be possible to use a feed screw ($\varnothing$ 10 mm) and a nut ($\varnothing$ 24 x 10 mm).	→ Larger machine
Position feedback systems are very complex.	→ Expensive

Linear motors are used whenever conventional solutions cannot be implemented due to very high demands in terms of

- speed, acceleration,
- accuracy,
- limiting frequency or amplification factor,
- reliability and minimal maintenance requirements.

[ObSaPo01]. With requirements for machines increasing, the share of this type of machine is set to rise. Nonetheless, it can be assumed that, because of the cost disadvantages of linear motors, the conventional drives with rotary servo motor will provide by far the largest proportion of servo drives for quite a long time to come.

Torque motors. As well as linear direct drives, there are also slowly rotating rotary direct drives, which are known as torque motors. Although the size and therefore the costs of a motor are defined by its torque, and the maximum power can be attained by having the highest possible rated speed, in some cases the use of a gearbox for adaptation of the motor speed to the working process is not appropriate. In these cases direct drives with a high number of poles are used.

Torque motors are predominantly used as synchronous servo motors with a high number of pole pairs. Short designs with a large diameter dominate. The rated speed is adapted to the required speed via the winding. Rated speeds reach down as low as 50 rpm. Peak torque values up to more than 30,000 Nm are also available. Hollow shafts and large internal diameters are often used. Designs as motor kits without a housing to be integrated into the design of the machine are common. A liquid cooling system is used in many applications In order to provide the necessary cooling [HaSt03].

As the adaptation of a torque motor to the operating point of an application often necessitates a modification to the wiring, torque motors are generally not used as standard products, and instead they are offered as application-specific variants. Unlike standard servo motors with gearboxes, it is not quite as straightforward to define standard motor ranges which cover a broad application scope.

Similarly to linear motors, the use of torque motors is set to increase, but this will be limited to applications which have particularly high requirements in terms of the drive behaviour. The combination of a standard three-phase AC motor or servo motor with a gearbox will remain the dominant design for a drive system.

3.3.5 Operating limits of motors

For reliable operation of the electric motors, the permissible operating limits of the motor must not be exceeded. These include the following limits:

- maximum continuous torque, which is determined in accordance to thermal limits,
- maximum peak torque value, which is defined by the design of the magnetic circuit (e.g. iron saturation, demagnetisation of the permanent magnets on synchronous servo motors),
- maximum speed, defined by mechanical strengths and the bearings,
- maximum voltage, defined by the insulating system,
- field weakening limits,
- mechanical load capacity of the shaft,
- class of protection,

- suitability for hygiene areas in the food industry (stainless steel shaft, smooth surfaces),
- suitability for environments in which there is an increased risk of explosion.

The thermal limits of the motors are given with regard to the service life of the insulating materials and the bearing grease. Exceeding the temperatures shortens the service life, although. A short-term exceedance is generally not a problem.

No-load losses, which already heat the motor without any torque load being applied, and load-dependent losses are both created in the motor. The no-load losses P_{V0} are dependent upon the speed:

$$P_{V0} \sim n \dots n^2 \qquad (3.58)$$

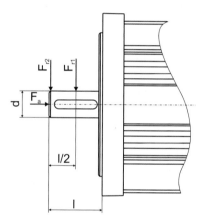

Fig. 3-31. Radial force and axial force on the motor shaft

The load-dependent losses P_{VL} increase approximately as a function of the square of the torque:

$$P_{VL} \sim \tau^2 \qquad (3.59)$$

The thermal limits are given by the rated power P_N or the rated torque M_N of the motor at the rated speed n_N and the holding torque τ_d which can be attained at standstill.

A thermal overload at the permissible ambient temperatures will be avoided if, in the speed-torque-diagram, the thermal operating point formed by the effective torque τ_{eff} and mean speed n_{mean} (chapter 3.1.7) is below the limit formed by the holding torque τ_d at standstill and the rated point n_N, τ_N.

The achievable speeds and torques are limited by the available voltage of the inverter and the limited inverter current. The interaction between inverter and motor is reflected in the speed-torque limiting characteristic. Operation is only possible if all of the required operating points are within the speed-torque limiting characteristic.

The mechanical structure and the magnetic properties, e.g. the properties of the permanent magnets on synchronous motors, limit the maximum permissible torque τ_{max}. This torque must not be exceeded even short-term. The same applies to the maximum speed n_{max}. With regard to the centrifugal forces and resonance effects on the rotor, this must also not be exceeded.

With regard to the load on the output shaft and the service life of the ball bearings of the motor, the radial and axial force loads on the shaft must not exceed the limits specified in the data sheet.

Fig. 3-31 shows the forces acting on the shaft. The forces arise when the torque is transferred onto the mechanical transmission elements, e.g. toothed belt, V-belt or gears.

Together with the torque, the forces place a load on the shaft. Accordingly, the combined effect of torque and force must not overload the shaft. Fig. 3-32 shows the limits of mechanical load capacity dependent on the shaft design (smooth shaft or shaft with keyway) for the combined load for a specific motor.

The forces are absorbed by the ball bearings of the motor. The service life of ball bearings depends upon the load and the speed. Normally, the service life L_{h10} is stated, and this figure indicates the service life which will be attained by 90% of the bearings. Fig. 3-33 shows the relationship between a load exerted by an axial force and a radial force as well as the number of revolutions attained by 90% of the bearings. Here, the inner load on the bearings resulting from the bearing pre-load in the motor is also taken into account. The number of revolutions u_{h10} together with the speed n determines the service life of the bearings.

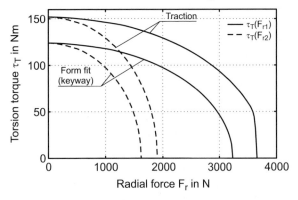

Fig. 3-32. Load capacity of the motor shaft for friction-fit and positive connections between the motor shaft and the transmission element

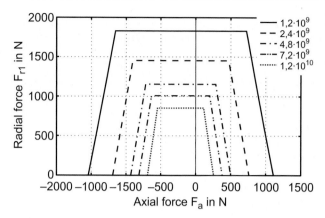

Fig. 3-33. Service life of the bearings as a function of radial and axial force in revolutions (5,000 to 50,000 operating hours and a speed of 4,000 rpm)

$$L_{h10} = \frac{u_{h10}}{n} \tag{3.60}$$

The stated relationships apply in the same way to linear motors, provided the force and speed of the motor are considered instead of torque and speed. Instead of the service life of the ball bearings, we look at the service life of the linear guide, which depends upon the speed and force load.

3.3.6   Angle and rotational speed sensors

Drives operated with closed-loop control are equipped with a measuring system for the speed and the angular position, so-called angle sensors or encoders, in order to achieve high speed and positioning accuracy. These position measuring systems or encoders close the control loop for speed and angle of rotation. They supply the actual speed and the actual angle to the controller for comparison with the setpoints. In addition, on drives with synchronous motors they also supply the rotor position information for the current control.

The speed is calculated by differentiation of the angle. This differentiation is part of the angle evaluation.

$$n = \frac{1}{2\pi} \frac{d\varphi}{dt} \tag{3.61}$$

The encoder is integrated in the motor. It is evaluated by the inverter.

Fig. 3-34. Encoder installed in a servo motor: a: Resolver, b: Incremental encoder, c: Sin-Cos absolute encoder

The requirements in terms of accuracy depend on the application in which the motor is going to be used. In order to be able to use the motors in a wide range of different applications, a number of different encoder systems are available. The arrangement of the encoder in a servo motor can be seen in Fig. 3-34. Optical encoders offer higher accuracy, but they also cause higher costs.

Table 3-8. Comparison of encoders

Encoder type	Principle	Accuracy	No. of resolved absolute revolutions
Resolver	Magnetic	±10 arcmin	1
Encoder	Optical	±10 arcmin	None
Sin-Cos single-turn absolute encoder	Optical	±2 arcmin	1
Sin-Cos multi-turn absolute encoder	Optical	±2 arcmin	Up to 4.096

Table 3-8 shows a comparison of the main features of the different encoders. This information alone is often enough to select the right encoder.

Details on the operating principles and features of encoders are contained in the sections below, where the encoders for rotary drives are described. The information can be transferred analogously to linear encoders.

Resolvers. Resolvers are magnetically operated angle sensors which comprise a rotary transformer and the actual resolver part with a single-phase rotor winding and a two-phase stator winding with the number of pole pairs p. The operating principle exploits the angle-dependent coupling between the windings in the rotor and in the stator. From the ratio of the induced voltages, the system can then calculate the angle and rotational speed.

Resolvers are electromechanical angle measuring devices without any electronic components. They are supplied and evaluated by an electronic circuit implemented in the servo inverter. This makes them extremely robust in comparison to other measuring devices, e.g. angle sensors with opto-electronic evaluation. Resolvers are made of the same materials as the servo motors: copper and iron (Fig. 3-35). As a result, they are significantly more robust than optical angle sensors. Resolvers can withstand external temperatures and vibrations and are resistant to voltage faults. The sensitive electronics are housed in the control cabinet, where they are protected from environmental influences.

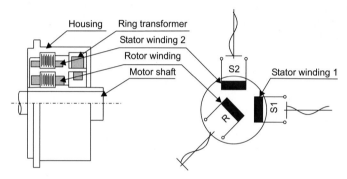

Fig. 3-35. Cross-section and operating principle of a resolver

Resolvers supply absolute angle information within a revolution, i.e. the angle information is not merely incremental. As a result, it is also possible to form the rotor position information for the current control of the synchronous servo motor directly from the resolver signals. This means that, as soon as the drive is powered up, the correct current can be set according to the angle, omitting the need for lengthy homing procedures or other measures for absolute angle measurement. The only requirement for this is that the number of motor pole pairs is an integer multiple of the number of resolver pole pairs. This makes it an attractive speed and angle measuring device for electric drives.

With the aid of the ring transformer, a carrier frequency voltage with the amplitude V_e is transmitted in the single-phase rotor. The carrier frequency voltage induces the voltages V_{cos} and V_{sin} in both phases of the stator winding. As the two phases are electrically offset by $90°$, the amplitudes of the voltages depend cosinusoidally or sinusoidally on the angle of rotation:

$$V_{cos} = V_e \cdot i_{tr} \cdot i_{rs} \cdot \cos(p \cdot \varphi) \qquad (3.62)$$

$$V_{sin} = V_e \cdot i_{tr} \cdot i_{rs} \cdot \sin(p \cdot \varphi) \qquad (3.63)$$

Here, i_{tr} is the transmission ratio of the ring transformer and i_{rs} the ratio between the rotor and stator windings. The angle of rotation can be calculated from the ratio of the voltages:

$$\varphi_{meas} = \frac{1}{p} \cdot \tan^{-1}\left(\frac{V_{sin}}{V_{cos}}\right) \tag{3.64}$$

The measured angle φ_{meas} deviates from the actual angle φ due to measuring inaccuracies, which amount to approximately ±10 angular minutes (arcmin). By measuring the error of the resolver and performing a compensation in the electronic evaluation, it is possible to reduce the error to ±4 arcmin. Accordingly, the accuracy is approx. 10 to 11 bits per revolution. By contrast, the resolution of the angle measurement can be increased to 14 to 16 bits through suitable methods.

Different methods are available for calculating φ_{meas}. As a rule, these methods form part of the signal processing of servo inverters. As the amplitudes V_{cos} and V_{sin} of the stator voltages are obtained through synchronous rectification or sampling with the carrier frequency, particular attention must be paid to injected faults with this carrier frequency.

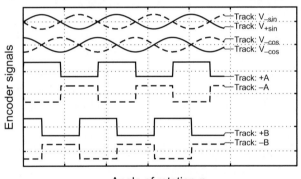

Fig. 3-36. Signals from different incremental encoders

Incremental encoder. With incremental encoders, an incremental disk is optically scanned. From the optical signal, the system generates square-wave signals or sinusoidal signals with preferably 2048 or 4096 pulses per revolution. Two signals offset by half a pulse length are generated, and these signals can be used for angle measurement and detection of the direction of rotation (Fig. 3-36).

With encoders with square-wave signals, the edges of the signals are used to count the angular motion. The direction of rotation can be detected by comparing the edge direction with the signal from the other track. This enables an angular resolution of up to four times the number of increments. For an incremental encoder with 2048 lines, this corresponds to 13 bits per revolution.

On encoders with sinusoidal and cosinusoidal signals, the ratio between the two signals is used to determine the angle within a period with the aid of the arc tangent function. As a result, incremental encoders with Sin-Cos evaluation deliver a higher resolution than incremental encoders with square-wave signals at the same number of increments [SchRoLe85, Ku05]. The angle within a period (z = no. of periods) is obtained from the sinusoidal and cosinusoidal voltages according to:

$$\varphi_{meas} = \frac{1}{z} \cdot \tan^{-1}\left(\frac{V_{+\sin} - V_{-\sin}}{V_{+\cos} - V_{-\cos}}\right) \quad (3.65)$$

With this evaluation, the resolution can be increased by a further 8 to 10 bits in comparison to a digital evaluation. Overall, this results in a measuring system which achieves an angular resolution of more than 20 bits per revolution. In addition to the high positioning accuracy, this high angular resolution also results in excellent smooth running performance at slow speeds.

The encoders only supply incremental angle signals, as a result of which the commutation information required for operation of synchronous motors is missing. Incremental encoders are therefore particularly suitable for asynchronous servo motors.

Incremental encoders with additional commutation signals are used in some cases for synchronous motors. In addition to the incremental signals, these encoders supply additional pulses which correspond to the number of pole pairs of the motors. For 3-phase synchronous motors, these are the three signals U, V or W, or for improved noise immunity and transmission security the six differential signals $+U$, $-U$, $+V$, $-V$, $+W$, $-W$. These signals are used by the servo inverter to inject the motor current into the three phases of the motor. This configuration has the advantage that it is inexpensive to implement the evaluation in the servo inverter. However, the large number of leads is a disadvantage. This results in the need for expensive cables and connectors.

Absolute encoders. In many applications, the absolute angular position of the drive needs to be known immediately after power-up of the plant. This means that either the angle within a single revolution (single-turn) needs to be known immediately, or the particular revolution within a defined range, e.g. 4096 revolutions (multi-turn), needs to be immediately available.

The absolute information within a revolution is required e.g. as information for the direct start-up of synchronous servo motors, so that from the start the currents can be correctly assigned to the three phases of the winding.

Spindle feed axes are an example of an application with absolute angle information of up to 4096 revolutions. Due to the spindle and the gearbox between the motor and the spindle, many revolutions of the motor are required for the total traverse path of the spindle. For example, with a lead-screw pitch of 5 mm and a gearbox ratio of i = 10, a total of 2,000 motor revolutions are required for a traverse path of 1,000 mm. Here, multi-turn absolute encoders offer the option of driving immediately to the target position without homing and starting production immediately. This reduces the turn-on time of a machine quite significantly.

For realisation, as well as the incremental Sin-Cos track, single-turn absolute encoders also have additional tracks on the code disk which permit the detection of the absolute angle. The absolute angle is transmitted via a serial interface to the servo inverter. The following interfaces have become established for data transmission:

- SSI – synchronous serial data transmission,
- HIPERFACE – asynchronous data transmission,
- EnDat – synchronous data transmission,
- BiSS – synchronous data transmission.

Multi-turn absolute encoders have an additional gearbox and additional sensors which allow them to directly distinguish between up to 4096 revolutions. The position is transmitted serially in the same way as for single turn encoders.

In recent years, a digitalisation of the interface has taken place in angle encoder systems. The classic ABZ interface has been expanded with digital signals. HIPERFACE communicates via the Z-track with the encoder, allowing absolute information to be exchanged. With EnDat 2.1, there is a synchronous serial interface in addition to the ABZ interface which allows parallel reading of the encoder.

A purely digital interface is suitable for the control of a servo motor, if the synchronicity of the data is assured and the interface can supply the data fast enough in the controller cycle (e.g. 16 kHz). The EnDat 2.2 development enables transmission rates of 4 Mbit/s, which means that the analogue ABZ interface is no longer required, resulting in cables with fewer cores which are therefore more cost-effective and – due to the digital transmission – more immune to interference [Ku06].

The open standard BiSS has been developed in competition with EnDat and is now used by numerous encoder manufacturers in their products.

3.3.7   Motor brakes

Dr. Sven Hilfert

A three-phase AC motor can only build up torque if it is supplied with electric energy. Brakes are used in applications in which drives need to be mechanically braked or they need to hold loads even when the power supply has failed. These brakes are mounted in the motors as required.

Tasks for motor brakes. The drive tasks to be performed by electric motors are very diverse, and as a result wide-ranging demands are also placed on the brakes of the motors. The main areas of application are detailed below. In practice, a brake will have to perform more than one of the indicated tasks in most applications. This needs to be taken into account particularly when selecting the brake type and for dimensioning of the brake. The brake thus usually needs to meet several dimensioning criteria.

The primary fields of application for motor brakes are:

- The *holding brake* for static holding e.g. of a position during the down-time of a robot, traversing drive, synchronising drive or a hoist drive. The basis for dimensioning in this case is the safety factor in relation to the load torque which needs to be held. The pure holding function is characterised in that no friction work is performed. On holding brakes, depending on the application the backlash of the brake can be an important feature, as it directly influences the positioning accuracy of the drive during down-time.

- The *emergency stop brake* is used to shut down rotary or linear moving masses, e.g. during dynamic braking of traversing drives or hoist drives in emergency situations. These are exceptional situations which only occur sporadically. During an emergency stop, kinetic energy is converted into friction energy and thus heat. The brake should be primarily dimensioned with regard to braking time, braking distance and deceleration. In addition, it also needs to be checked whether the permissible friction work of the brake is not exceeded during an emergency stop, and whether the required service life of the brake will be attained, as only a small amount of wear due to friction energy is usually permissible for these brakes. Another criteria for the selection of an emergency stop brake is the permissible maximum braking torque, which can lead to damage on downstream drive elements like gearboxes, mechanical plant systems etc.

- An *operating or service brake* is used for controlled shutdown of rotary or linear moving masses and is designed to perform friction work. The main difference to an emergency stop brake in terms of application is that a braking process in which friction energy is converted is not a sporadically occurring event, but instead a normal and cyclically repeated event. The heat generated during every braking process and the associated wear to the friction surface are thus among the key variables which need to be considered for dimensioning, alongside the braking time, braking distance and deceleration.

Brake types. Two different types of brake are used in electric motors depending on the structure of the motor and the target application:

- permanent magnetic brakes (PM brakes),
- spring-applied brakes.

These two brakes differ from each other in terms of how they work and in terms of their specific behaviour, and they are described in more detail below.

Permanent magnetic brakes. In a PM brake the force required to generate a braking torque is generated through permanent magnets, e.g. neodymium-iron-boron or ferrites. A schematic overview of the layout can be seen in Fig. 3-37, while Fig. 3-38 shows the installation of a brake in a servo motor.

On the designs used most commonly today, the armature plate of the brake is connected to the brake rotor via leaf springs. As a result, the backlash – which is defined by the stiffness of the springs – is very low for this type of brake (no torque play). Once installed, the brake rotor in the embodiment shown is supported on the inner ring of the ball bearing, so that the springs are pre-loaded in axial direction as a result of the attractive forces of the permanent magnets, and an air gap is set between the armature plate and the brake rotor when the brake is released.

The brake rotor is connected rigidly to the motor shaft, the stationary pole to the bearing shield of the motor. As both the brake rotor and the stationary pole are connected rigidly to components of the motor, it may be necessary to take into account mechanical tolerances of the motor or thermal expansion of components which could affect the air gap of the brake in the design of the motor.

Engagement of the PM brake: The brake operates according to the closed-circuit principle, i.e. the brake is closed in its de-energised state. The magnetic field of the permanent magnets is directed via the inner and outer poles to the armature. The armature, which is made of magnetic steel, is pulled by the magnetic field, whereby the force F_M of the permanent magnets is higher than the restoring force of the springs F_F ($F_M > F_F$), and the braking torque is generated through friction between the fixed poles and the rotating armature plate.

The braking torque which is established depends upon the coefficients of friction of the friction surfaces, the spring pressure and the magnetic forces. The remanence induction of the employed PM magnets has a negative temperature coefficient, as a result of which the magnetic force – and therefore the torque of the brake – decreases with increasing temperature (typical values of around -10 to -12% if the operating temperature changes from 20 to 120°C).

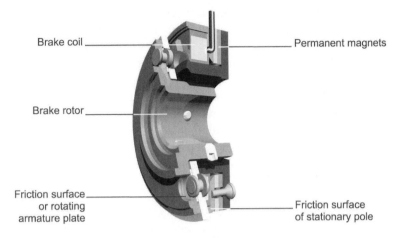

Brake coil _____ Permanent magnets

Brake rotor _____

Friction surface _____ Friction surface
or rotating   of stationary pole
armature plate

Fig. 3-37. Layout of a permanent magnetic brake

Fig. 3-38. Installation of a permanent magnetic brake into a servo motor

Releasing the PM brake: When a current is applied to the brake coil, the field of the permanent magnets is offset in the region of the armature plate ($F_M \sim 0$). Due to the restoring force of the springs F_F, the armature is released free of residual torque from the poles.

If the voltage applied to the brake is increased above the permissible supply voltage, then the magnetic flux in the region of the armature plate becomes non-zero, and a force is exerted against the armature plate again – the brake closes again. PM brakes therefore have a so-called release window, in which the brake is open. Depending on the design of the magnetic circuit, for PM brakes at room temperature a release window between approximately 0.7 and 1.3 V_O is typical, while on more recent designs this can be as much as around 2 V_O. If the operating temperature of the brake is changed, then the limits are shifted due to the temperature sensitivity of the winding resistance and the remanence induction of the magnets, so that across the entire operating temperature range a typical window from 0.85 to 1.15 V_O can be used.

This brake type has a friction surface and can be supplied with or without an additional organic friction lining. Through the use of an organic, non-magnetic friction lining, an additional air gap is created between the armature plate and the pole faces, which reduces the magnetic force on the armature plate and therefore the attainable braking torque. PM brakes are therefore designed predominantly without friction lining in order to reach the highest possible braking torques. The friction system is based on steel-steel friction.

Due to the wear on the friction surfaces the air gap of the brake increases. When the wear limit is reached the air gap becomes so large that the magnetic force is not sufficient to attract the armature plate. The brake can then no longer close ($\tau_B = 0$).

PM brakes are predominantly designed as holding brakes with an emergency stopping function. Firstly, this is because of their friction system, and secondly this is because of their restricted serviceability. These brakes are typically integrated in the inner part of the motor, which means that the entire motor needs to be removed and disassembled when the brake is worn.

Spring-applied brakes. With this type of brake, the force required to generate the braking torque is created by springs. The basic structure is shown in Fig. 3-39, while Fig. 3-40 shows how one of these brakes is attached to a three-phase AC motor.

In the example shown, the magnet housing is screwed to the B end shield of the motor, therefore has a fixed connection. There are a number of possible options for the friction surfaces: on high-end designs, the brake is either equipped with its own end shield or a friction disk (friction plate) is attached to the B end shield, or otherwise the B end shield serves directly as the counter-running surface. The armature plate is supported via sleeve bolts or guide bolts to prevent rotation on the magnet housing. The brake rotor can be axially displaced on a driver hub arranged on the motor shaft.

Due to the clearance on the support of the armature plate and the gear teeth backlash of the brake rotor, by design these brake designs have a backlash in their engaged state (not backlash-free). Typical values are in the range from 0.5 to 1.0 degrees. The air gap between the friction plate, brake rotor and armature plates is set by the sleeve bolts or with the aid of corresponding spacer bolts. As the rotor can be displaced on the motor shaft, the air gap and its tolerances are only determined by parts of the brake, so that mechanical tolerances of the motor or thermal expansion of the motor shaft do not have any influence on the air gap of the brake.

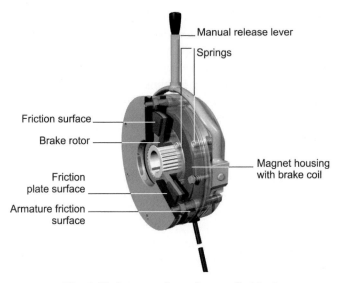

Manual release lever

Springs

Friction surface

Brake rotor

Magnet housing
with brake coil

Friction
plate surface

Armature friction
surface

Fig. 3-39. Layout of a spring-applied brake

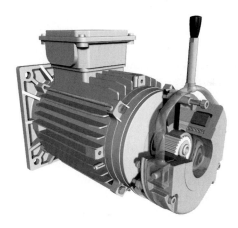

Fig. 3-40. Attachment of a spring-applied brake to a standard three-phase AC motor

Brakes, engagement of the spring-applied brake: The brake operates according to the closed-circuit principle, i.e. in its deenergised state the brake is closed. The magnetic force of the brake coil is zero ($F_{Sp} = 0$). The spring pressure F_F forces the armature plate against the friction lining and the friction plate. The braking torque is generated due to the friction between the fixed friction plate, the fixed armature plate and the rotating brake rotor.

So that the rotor can be clamped in between the two counter friction faces (friction plate and armature plate) without opposing force, axial displacement of the rotor needs to be possible. When the brake is closed the rotor is axially displaced.

The established braking torque depends upon the coefficients of friction of the friction surfaces (friction lining) and the spring pressure. By changing the spring pressure values, it is very straightforward to adapt the braking torque of this brake type.

Releasing the spring-applied brake: When the brake coil is energised, the armature plate is pulled against the spring pressures through the magnetic field of the coil towards the magnet housing. To release the brake, the magnetic force of the coil needs to be significantly higher than the spring pressure ($F_{Sp} \gg F_F$). As the rotor is not actively displaced during release and remains applied against the friction plate or end shield, the brake is not torque-free. When the brake rotor rotates, it floats depending on the brake size and design typically above speeds from around 100 to 300 rpm on an air cushion, so that the operation of the brake is then completely torque-free. The floating of the rotor is influenced among other things by the mounting position of the motor (horizontal or vertical). Every start-up process therefore also involves the conversion of a small quantity of friction energy (until the rotor starts to float). On drives which run very slowly it is also possible for a small torque to occur continuously, which causes conversion of friction energy and thus wear to the friction surfaces and the friction lining.

This brake type is generally designed with a friction lining, usually an organic friction lining. Depending on their composition, organic friction linings often only develop their maximum braking torque when a small amount of friction energy is converted, or at operating speeds of approx. 100 rpm (depending on the friction lining and the size of the brake). When this type of brake is used as a pure holding brake, it may be necessary to take into account the potential reduction of torque (typically around 10 to 20%) for drive dimensioning, or suitable friction linings may need to be selected.

Due to the wear to the friction surfaces or the friction lining, the air gap of the brake is increased. When the wear limit is reached, the air gap becomes so large that the magnetic force of the coil is no longer sufficient to attract the armature plate. The brake then no longer opens (initially there will be a small residual torque, which will increase progressively as the wear increases; $\tau = 0$ to 100% τ_O).

Spring-applied brakes in the form described here are used both as holding brakes with an emergency stopping function and as service brakes. This is firstly down to their friction system with organic friction lining and secondly down to their good serviceability. Typically, these brakes are mounted on the B end of the motor, so that if the friction linings or counter friction faces wear it is possible to replace the brake or the brake rotor or adjust the air gap without removing and replacing the complete motor.

There are also special embodiments of spring-applied brakes for installation in servo motors which operate according to the same basic principles, but are comparable to PM brakes in terms of serviceability. These brake are preferably used as holding brakes with emergency stop function.

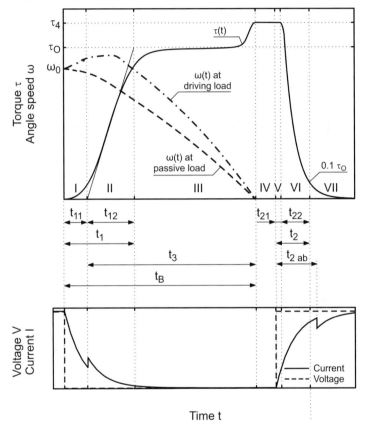

Fig. 3-41. Definition of operating times following DIN 0580 and VDI 224 Sheet 1

Torque behaviour, definition of operating times. The operating times of the brake are defined in DIN 0580 and VDI 224 Sheet 1. The individual operating times and phases during the build-up of the braking torque (braking, engagement) and during opening (release, disengagement) of the brake and the angular velocity during the braking process are shown in Fig. 3-41.

When the voltage is switched off, initially the energy stored in the inductance of the brake is dissipated. Once the current injected by inductance has dropped below a minimum value, the armature plate start to move (spike in the current curve). This delay time is referred to as the response time during engagement t_{11}. It needs to be taken into account particularly when calculating braking distances or braking times for hoists, as if the hoist is moving downward during this time it will experience an acceleration due to the effects of gravity (angular velocity curve for a driving load). A hoist moving upwards will experience a negative acceleration due to gravity, i.e. the load is already decelerated during t_{11} (braking load).

Depending on the magnitude of this acceleration, this will result in an angular velocity curve similar to a passive load (small negative acceleration), or the load may already come to a standstill within the time t_{11} if there is a large negative acceleration. With a passive load (e.g. horizontal moving masses) there will be a deceleration due to the friction in the system. If we ignore the effects of friction, the speed or angular velocity does not change during the time t_{11} (refer to the angular velocity curve for a passive load). During the time t_{12} the torque of the brake increases to the static rated torque τ_O. For a pure holding brake, it is typical to indicate only the engagement time t_1 ($t_1 = t_{11} + t_{12}$), after which the static rated torque of the brake is available. The engagement time of the brake is strongly influenced by the control of the brake, and with AC switching or wiring of the brake with a freewheeling diode the engagement time is increased by a factor of around 5 to 10 in comparison to DC switching. The time between t_{11} and the standstill of the load ($\omega = 0$) is referred to as the slipping time. At the time t_B the braking torque increases to the transmissible torque τ_4.

When the voltage is applied the magnetic field starts to build up immediately, so that the time t_{21} becomes nearly zero and can hence be neglected. During the time t_{22} the braking torque decreases to $0.1\,\tau_O$, and the sum of t_{21} and t_{22} is the disengagement time t_2. The lift-off of the armature plate at the end of the time $t_{2\,l\text{-}o}$ can be seen as a spike in the current curve.

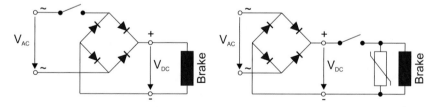

Fig. 3-42. AC and DC switching of the brake

Control of brakes. PM brakes and spring-applied brakes are usually designed for DC control. Accordingly, control gear is required in order to operate a PM brake or spring-applied brake from an AC supply network. With spring-applied brakes, versions are also available for AC control. These are less widespread though, and are therefore not covered here in any more detail.

Table 3-9. Assignment of coil voltages of the brakes to mains voltages

Mains voltage V_{Mains}	Coil voltage V_O of the brake	
	Bridge rectifier	Half-wave rectifier
110 V	103 V	
230 V	205 V	103 V
277 V	250 V	127 V
400 V		180 V
460 V		205 V
480 V	Coil voltages > 250 V normally not used	215 V
500 V		225 V
550 V		250 V

Half-wave and bridge rectifiers are mainly used to generate the DC voltage. The following relationships apply between the mean value of the DC voltage and the r.m.s. value of the supplying AC voltage:

$$\text{Half-wave rectifier: } \overline{V}_{DC} = 0,45 \cdot V_{AC} \qquad (3.66)$$

$$\text{Bridge rectifier: } \overline{V}_{DC} = 0,90 \cdot V_{AC} \qquad (3.67)$$

Together with these rectifier circuits, the standard brake coil voltages stated below are obtained for the most common mains voltages (Table 3-9).

The supply voltage of the brake can be switched on and off on the AC or DC side (Fig. 3-42).

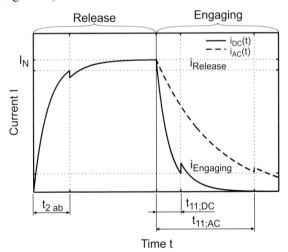

Fig. 3-43. Current curve for AC and DC switching of the brake

In the case of AC switching, the engagement time t_1 is around 5 to 10 times longer than for DC switching. With DC switching, the current flow is interrupted immediately. A very large induction voltage occurs, requiring spark suppression e.g. by means of a varistor. Without spark suppression, the relay contacts will erode very quickly. With AC switching, the diodes of the rectifier form a freewheeling path, so that the current gradually decays and the brake only closes (engages) when it drops below the incident current (Fig. 3-43).

Table 3-10. Variants for operation with a bridge/half-wave rectifier

	Variant 1	Variant 2	Variant 3	Variant 4
Ratio of coil voltage to mains voltage V_{Coil} / V_{AC}	0.45	0.9	0.6–0.8	0.45–0.9
Excitation voltage V_E	$2 \cdot V_{Coil}$	V_{Coil}	$1.2–1.5 \cdot V_{Coil}$	$1.0–2.0 \cdot V_{Coil}$
Operating voltage V_{Op}	V_{Coil}	$0,5 \cdot V_{Coil}$	$0.6–0.75 \cdot V_{Coil}$	$0.5–1.0 \cdot V_{Coil}$
Notes	Shortened disengagement time, increase of the maximum wear air gap	Shortened engagement time, reduced power input (to 25%)	Shortened disengagement time and engagement time, reduced power input (to 36-60%)	Insensitive to voltage fluctuations, nevertheless reliable function
Example				
Coil voltage	103 V	205 V	180 V	127 V
Mains voltage	230 V	230 V	230 V	140–280 V
Excitation voltage	207 V	207 V	207 V	127–252 V
Operating voltage	104 V	104 V	104 V	63–126 V

A spring-applied brake has a power loss of 20 to 200 W. On self-ventilated standard motors which are operated at low speeds with a frequency inverter, this leads to a significant build-up of heat, which causes the performance of the brake to deteriorate. With an intelligent rectifier, which switches electronically between a half-wave rectifier and a bridge rectifier (a so-called bridge/half-wave rectifier), depending on the assignment of the coil voltage of the brake and the mains voltage it is possible to reduce the disengagement time, the engagement time or the power input with open brake (Table 3-10). An example current characteristic for variant 3 is shown in Fig. 3-44. In order for the advantages shown to be effected, the magnetic circuit of the brake must be designed for over-excitation.

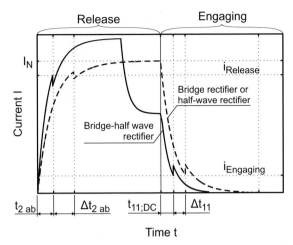

Fig. 3-44. Current curve with a bridge/half-wave rectifier

Another option for achieving a similar behaviour is to equip the brake with two separate coils or to use a coil with a centre tap. During disengagement, only the first coil or the first part of the coil is actuated (excitation current), and it switches to the second coil or the entire coil when the brake is released (operating current). Depending on the design of the coils, it is possible to achieve both an over-excitation and a reduction of the winding resistance and the inductance for disengagement of the brake, thus offering a faster current pick-up. The engagement of the brake can be influenced by corresponding design of the winding which is effective at the operating current.

With closed-loop controlled drives, in many cases there is a 24 V supply in the control cabinet for the supply of the PLC, inverter and other components. In this case the brakes are frequently also equipped with a 24 V coil and fed directly from this voltage. In the case of PM brakes, this is the most commonly used coil voltage. Due to the relatively narrow release window on PM brakes, it is also necessary to distinguish between control with smoothed and unsmoothed DC voltage, which requires a different adjustment of the brake or design of the brake coil. Operation of PM brakes with a half-wave rectifier is usually not permitted for the same reason.

Dynamic and static braking torque of brakes with and without friction lining. In relation to the static and dynamic braking torque, there are differences between brakes with (organic) friction lining and brakes without friction lining (steel-steel), and depending on the particular application and the requirements for the brake these may need to be taken into account for drive dimensioning. A typical characteristic for the dynamic braking torque for brakes with and without friction lining is shown in Fig. 3-45.

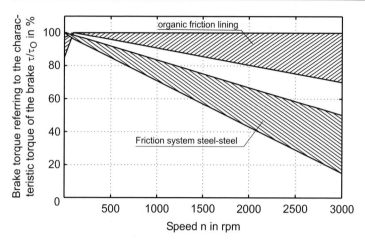

Fig. 3-45. Typical characteristic of dynamic braking torque

The speed-dependent variation of braking torque is dependent upon the engagement speed and, when using friction linings, upon their specific behaviour. Therefore the characteristics of the brake being used should be taken into account in each specific case. For larger motors – and therefore for higher speeds than those indicated in the diagram – the described reduction of braking torque can already occur at significantly lower speeds.

At lower speeds the dynamic braking torque approaches the holding torque of the brake.

In particular when calculating stopping distances and stopping times, e.g. for emergency stops or when dimensioning hoists, knowledge of the dynamic behaviour of brakes is very important. It may e.g. be possible that a brake can safely hold a hoist statically, but that – during an emergency stop from high speeds – dynamic braking of the load is no longer possible, as the dynamic braking torque is lower than the torque resulting from the load due to the weight.

In cases with a load torque of $\tau_L = 0$, the mean dynamic braking torque τ_{1m} with a given mass inertia J is calculated as follows:

$$\tau_{1m} = J \cdot \frac{d\omega}{dt} = J \cdot \frac{\Delta\omega_0}{t_3 - t_{12}} \tag{3.68}$$

The mean dynamic braking torque depends on the magnitude of the mass inertia and the angular velocity ω_0 from which the braking takes place, and is therefore not a generally valid parameter of the brake. Instead, it can only be indicated for defined operating conditions.

Dimensioning of brakes. The dimensioning of a brake is largely defined by the required braking torque. Other important factors which need to be taken into consideration include:

- the mass inertias which are to be decelerated,
- the relative speeds,
- dynamic braking torques,
- braking times and stopping distances,
- the operating frequency, friction energy and service life of the brakes.

Now that the motors and their built-on accessories (encoders and brakes) have been presented in detail, the next component of a drive system to be covered is the inverter, which controls the exact behaviour of the motor in open and closed-loop control.

3.4   Inverter

Dr. Edwin Kiel

Having covered motors which convert electrical energy into mechanical energy in the previous chapter, we shall now take a look at inverters. These components are used to supply electric energy to the motor of a drive system. Here, the inverter is a control gear for the motor which controls the energy conversion in order to influence the behaviour of the motor shaft and the mechanical process being driven by the motor as required by the application. The information with which the inverter controls this energy conversion is received from the master control. This makes the inverter the actuator of the automation system. In order to ensure the optimal division of functions between the control and the inverter, in many cases the inverter already performs the motion control that generates the variations in time of the speed and the angular position of the motor [Sc98].

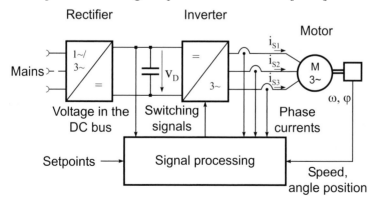

Fig. 3-46. Structure of an inverter

The electric hardware of an inverter comprises the following two parts (Fig. 3-46):

- the power conversion from the mains to the motor,
- the signal processing which controls the power conversion.

As this book focuses on the power range up to 500 kW, we will only look at the pulse width modulation inverter with DC voltage link and IGBT power switches. Other designs only play a marginal role in this power range and are therefore only outlined in brief.

Inverters are divided into the following two groups:

- *Frequency inverters* control the motor without any measurement of angular position or speed (this is also referred to as open-loop control). Asynchronous motors are generally used, and accuracy is limited. In the simplest case, a voltage-frequency control (V/f control) is used, which requires only little computing power and is very robust. In order to achieve an improved drive behaviour, a sensorless vector control is also used.
- *Servo inverters* evaluate an encoder for the angular position and speed, which enables highly dynamic and precise control of the motor (closed-loop). Field-oriented control with a lower-level current control is generally used here. This can be used together with a synchronous or an asynchronous motor, usually as a servo motor.

A comparison of the main features of frequency inverters and servo inverters is shown in Table 3-11.

Table 3-11. Comparison of frequency inverters and servo inverters

	Frequency inverter	**Servo inverter**
Speed measurement	No	Yes
Operation	Open-loop control	Closed-loop control
Speed setting range	Up to 1 : 100	1 : 10,000
Speed accuracy	Typically 1 – 5%	Better than 0.01%
Acceleration time	Typically 0.1 – 10 s	Typically 10 ms – 1 s

Today, signal processing in inverters is performed almost exclusively with the aid of microcomputers. Accordingly, this is the only technology covered in this book.

The software which controls the power conversion and therefore influences the motor behaviour can be divided into two parts:

- The drive control, which measures the instantaneous values of the motor (torque, speed, angular position) and adjusts the values towards the set-points.
- The motion control, which defines the time characteristics of these status variables (in particular the speed and angular position), via which it executes the application functions. Here, positioning functions are usually implemented in the drive.

Due to the steep switching edge of the switched power conversion of an inverter with DC voltage link, they represent a significant interference source. Chapter 3.4.8 shows how safe and reliable operation of the plant and compliance with limit values for electromagnetic emissions can be ensured not only through measures in the inverter, but also through the installation of the drive system.

These individual aspects of the inverter are described in detail in the subchapters below.

3.4.1   Power conversion

Power conversion is the heart of every inverter. There are different basic topologies, and the first one we shall look at is the pulse width modulation inverter with voltage link. This is the most commonly used topology in the power range up to 1 MW.

On a voltage link converter, the power conversion from the mains to the motor takes place in two stages:

- the rectifier converts the AC voltage into a DC voltage,
- the inverter then converts the DC voltage back into an AC voltage.

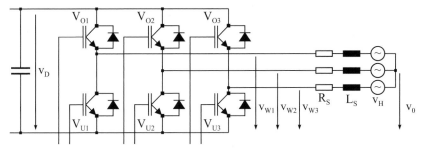

Fig. 3-47. Structure of the power inverter

Structure of the inverter. The task of the inverter is to create a three-phase output voltage for the connected motor. This output voltage must be variable in terms of its amplitude and frequency (or instantaneous phase angle) so that it can be used to control the behaviour of the motor. As high power values are converted and therefore a high efficiency is necessary, only switched power components can be used. The inverter comprises six power switches which can connect the two poles of the DC voltage v_D in the DC link to the three output phases (Fig. 3-47). This means that each of the three output phases $v_{W1...3}$ can assume the positive or negative potential of the DC link. Losses in this configuration are created by the voltage drop when the motor current flows through the power switch or the anti-parallel diode and when a power switch is switched on or off.

In the power range we are looking at here, IGBTs (Insulated Gate Bipolar Transistors) have become established as the main type of power switches. They offer a high switching frequency, easy controllability, a small semiconductor area (and therefore lower component prices) as well as lower transmission losses and switching losses. Power MOSFETs have a much larger semiconductor area for mains-operated inverters and are therefore more expensive than comparable IGBTs. In the low-voltage range MOSFET transistors offer clear advantages over IGBTs, as in their case the low bulk resistance provides a very low voltage drop [Gi92, Zw92].

Each of the six switches must enable a bi-directional current flow between the DC link and the motor. As IGBTs only permit uni-directional current flow, a freewheeling diode needs to be connected in anti-parallel.

Design of the inverter. An inverter is initially designed according to its rated power. All components must be dimensioned for this power to ensure that the maximum limits of components are not exceeded. In addition to the permissible voltages and currents, this particularly also includes the maximum temperatures. For the inverter, this means that its entire thermal behaviour needs to be analysed and optimised in great detail. The power semiconductors of the inverter are usually integrated in power modules which contain an electrically isolating but thermally conductive layer. Several connections are created which are exposed to high thermal loads. These include the bonding wires on the upper side of the power semiconductors, the connection between the power semiconductors and the substrate surface and the connection between the power module and the heatsink.

The power loss in the power semiconductors arises at the junction in the semiconductors. From here to the heatsink, a thermal chain is created containing heat-storing elements and thermal contact resistances. In order to achieve a long service life, the following dimensioning rules must be satisfied:

- The maximum junction temperature of the semiconductors must not be exceeded. It is important here to take into account the maximum ambient temperature at which the inverter will be operated.

- The temperature cycles at the connections must not exceed limit values, as otherwise the different thermal expansion coefficients of the materials could cause continuous loads which could result in premature wear.

In terms of dimensioning, torque which needs to be applied by the drive at standstill or low inverter output frequencies is critical. Inverter manufacturers perform extensive dimensioning calculations to ensure that reliability can be ensured even in these operating states.

As well as the rated power, which is the power that can be converted continuously by the inverter, the maximum output current is important for dynamic processes like acceleration and braking. The maximum output current is determined by the size of the power semiconductors. Frequency inverters and servo inverters are usually designed slightly differently here, in that the maximum peak current on servo inverters corresponds to 2 to 4 times the rated current, whereas the maximum output current of frequency inverters is less than twice the rated current. This reflects the fact that servo drives are often used for dynamic drive tasks which require high acceleration capability but do not need to deliver high continuous power. In any specific application it is necessary to check the extent to which the maximum output current and the rated current of an inverter match the requirements of the motion sequence of the application.

The influence of switching frequency. The magnitude of the switching frequency has a significant influence on the behaviour of the drive. For example, the switching frequency determines the dynamics of the motor control. The smooth running performance also improves with increasing switching frequency.

Noise considerations also need to be taken into account. Audible noise is generated in both the inverter and the motor according to the switching frequency, and as a result a minimum switching frequency of 16 kHz should be used in any application in which this noise would cause a disturbance (e.g. fans in living areas or offices). On the other hand, in many industrial applications the noise level is already so high due to the production process that a low switching frequency does not have a significant impact on the overall noise levels. For this reason, industrial inverters are usually operated at switching frequencies of 4 to 8 kHz, while inverters for noise-sensitive applications tend to run at 16 to 20 kHz.

A higher switching frequency also causes stronger EMC emissions, which then need to be suppressed by corresponding EMC measures.

The switching losses of the inverter also increase with the switching frequency of the power switches. In many cases, when inverters are operated with a higher switching frequency, derating (i.e. a reduction of the output power) is necessary due to the higher switching losses. As the switching losses also depend on the magnitude of the mains voltage (and therefore the voltage in the DC link), derating is also required in some cases at higher mains voltages (e.g. on devices which are designed for a voltage range of 400 to 500 V and are operated at 500 V).

Principle of operation of the inverter. The two power switches of one of the three half bridges of the inverter must only be switched on in anti-parallel. It also makes no sense to switch off both power switches simultaneously, because then the potential of the output phase would only be determined by the sign of the instantaneous output current. This means that each half bridge can adopt two states (upper or lower switch closed), and the entire inverter can consequently adopt eight (2^3) states [SchSt64].

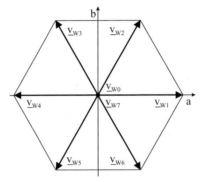

Fig. 3-48. Output voltages of the inverter

If these eight states of the inverter are represented as vectors of the three-phase system, they form a voltage hexagon in the complex plane (Fig. 3-48) [Ge88]. Whenever one or two upper switches are closed, a voltage phasor with the length of the DC link voltage and a phase angle of $n \cdot 60°$ arises at the output of the inverter. If all three upper or lower switches are closed, then the three motor phases are short-circuited and the instantaneous voltage is zero. These two states are referred to as zero vectors.

Table 3-12. Output vectors of the inverter

Vector	Phase 1	Phase 2	Phase 3
v_{W0}	Low	Low	Low
v_{W1}	High	Low	Low
v_{W2}	High	High	Low
v_{W3}	Low	High	Low
v_{W4}	Low	High	High
v_{W5}	Low	Low	High
v_{W6}	High	Low	High

v_{W7}	High	High	High

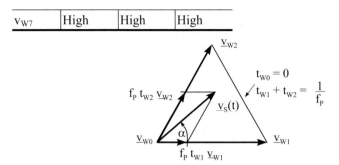

Fig. 3-49. Calculation of the turn-on times

In order to be able to adjust the output amplitude and phase of the inverter, we exploit the fact that the connected motor has an input inductance and that, as a result, in a time interval (e.g. the switching period $1 / f_P$ of the power switches) the voltage which acts is made up of the sum of the individual output vectors, weighted by their relative turn-on time. Fig. 3-49 demonstrates this geometrically.

For the example shown, the turn-on times of the vectors are therefore calculated as follows [JeWu95]]:

$$\underline{v}_S = f_P \sum_{n=0}^{7} t_{Wn} \, \underline{v}_{Wn} \tag{3.69}$$

$$\frac{1}{f_P} = \sum_{n=0}^{7} t_{Wn} \tag{3.70}$$

$$t_{W1} = \sqrt{3} \, \frac{\sqrt{2} \, V_S}{f_P \, v_D} \sin\left(\frac{\pi}{3} - \alpha\right) \tag{3.71}$$

$$t_{W2} = \sqrt{3} \, \frac{\sqrt{2} \, V_S}{f_P \, v_D} \sin(\alpha) \tag{3.72}$$

$$t_{W0} = \frac{1}{f_P} - t_{W1} - t_{W2} \tag{3.73}$$

As well as the instantaneous phase angle α of the output voltage vector $\underline{v}_S$, the turn-on times also incorporate its amplitude V_S and the value of the DC link voltage v_D. The maximum output voltage which can be set for each phase angle, is obtained from the circle which fits in the voltage hexagon. The maximum value of the output voltage is therefore:

$$V_S \leq \frac{v_D}{\sqrt{2}\sqrt{3}} \tag{3.74}$$

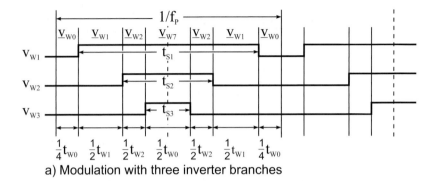

a) Modulation with three inverter branches

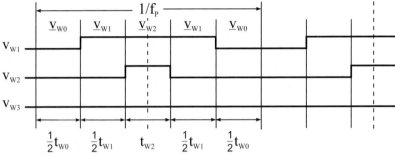

b) Modulation with two inverter branches, V_{U3} closed

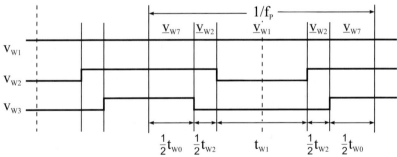

c) Modulation with two inverter branches, V_{O1} closed

Fig. 3-50. Modulation methods

A modulation diagram converts the times for the individual voltage vectors into a switching sequence for the power switches [Ho92, JeWu95]. Two methods are commonly used:

- All three half bridges are switched in each switching interval of the power switches. This method results in the lowest motor current harmonics, but it does also incur the highest switching losses (Fig. 3-50 a).
- In each switching interval only two of the three half bridges are switched, while the third remains at the potential of the higher or lower

DC link for a period of 60° of the output frequency. In comparison to the modulation with all three phases, this method only incurs 2/3 of the switching losses, but the downside is higher motor current harmonics and hence less smooth running performance of the motor (Fig. 3-50 b and c).

In some cases it is possible to switch between these methods. Together with the setting of the switching frequency, particularly in the higher power range it is then possible to find a good compromise between the operating characteristics of the inverter and the power loss.

The power transistors generally have a switch-off delay which is longer than the switch-on delay. During the switchover of a half bridge, a dead time therefore needs to be observed in which both of the power switches should be switched off. The exact switchover time for the output phase depends on the sign of the output current. This results in a fault in the output voltage, which particularly for low output frequencies (< 5 Hz) leads to a torque ripple six times the output frequency of the inverter. There are several methods which can be used to compensate for this effect. As a result, the smooth running performance particularly in the open-loop controlled operation of frequency inverters, which do not have closed-loop current control, is significantly improved at lower output frequencies.

Mains supply. The voltage link of the inverter is supplied with power from the mains via the mains supply. The following methods are used here (Fig. 3-51, Table 3-13) [BaDi06]:

- Uncontrolled diode rectifiers, which only enable power flow from the mains to the motor and generate high harmonics in the mains (Fig. 3-51 a).
- Inverters which are pulsed at mains frequency and therefore enable a bi-directional flow of energy, although the currents to the mains are in block form (Fig. 3-51 b) [Go06].
- Pulse width modulation inverters, which like the inverter are pulsed at high frequency to the motor and therefore generate both sinusoidal currents as well as a bi-directional flow of power (Fig. 3-51 c).

The uncontrolled diode rectifier is used most commonly, as it is the least expensive and the most efficient option. As it is not capable of feeding power back into the mains, suitable measures must be put in place for the braking energy of the motor (see below).

Furthermore, current harmonics can be significantly reduced through the use of a mains choke. This is explained in more detail in chapter 3.4.6. Mains chokes are generally used in higher power ranges. The use of a choke in the DC link is also standard in some cases.

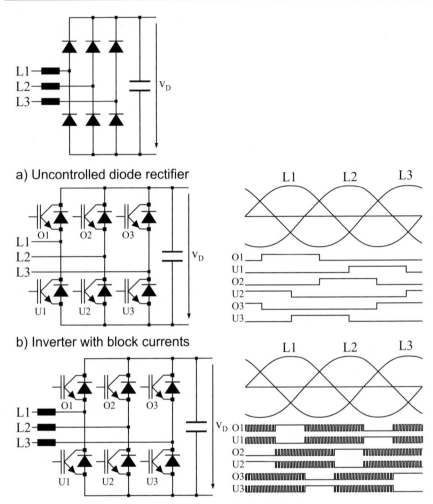

a) Uncontrolled diode rectifier

b) Inverter with block currents

c) Pulse width modulation inverter with sine currents

Fig. 3-51. Mains supply

With the aid of a supply via an inverter, which is pulsed at the basic frequency of the mains network, it is possible to establish an inverter which is capable of power recovery and is only slightly more expensive than an inverter with diode rectifier, but does not require any mains chokes or brake transistors and brake resistors.

In terms of technical properties, the best – but also the most expensive – solution is to provide the supply via a pulse width modulation inverter which is pulsed at high frequency. In all cases this will require a filter to the mains network with chokes in order to reduce the pulse frequencies and to have a defined inductance. As this inverter works in the same way as a three-phase step-up converter, it is possible to use it to control the DC link voltage to a value which is above the rectified voltage. In practice, this is exploited in order to make the voltage for inverters on the motor side and therefore the maximum motor voltage independent of the mains voltage. In this way the acceleration capacity would then no longer be dependent upon fluctuations in the mains voltage. Particularly machines with a high production power (e.g. machine tools) can be operated at a defined operating point like this.

Table 3-13. Mains supply method

	Uncontrolled diode rectifier	**Inverter with block currents**	**Pulse width modulation inverter with sinusoidal currents**
Capable of power recovery	No	Yes	Yes
Mains current	Non-sinusoidal	Non-sinusoidal	Sinusoidal
Power loss	Low	Moderate	High
Costs	Low	Moderate	High

DC link. The entire power flow to the motor is implemented in the voltage link. In principle, the power flow of a three-phase system is constant to both the mains and to the motor.

With the aid of the inverter, at low output voltages the motor current is only supplied for short periods from the DC link, while it is short-circuited by the zero vectors for the remaining time. The mean value of the current which the inverter takes from the DC link thus corresponds to the active power which flows to the motor. On the other hand, the peak value of this current is equal to the motor currents, and its frequency spectrum is made up of multiples of the switching frequency. Capacitors in the DC link must make available the energy for this pulsating current flow.

If a rectifier or a block-switching inverter is used for the mains supply, then the resulting rectified voltage will have a ripple on the DC link side. With a single-phase mains connection, this ripple has twice the frequency of the mains supply and voltage dips reaching down to 0 V. With a three-phase supply, the basic frequency corresponds to six times the mains frequency, and the dips reach down to 87% of the peak voltage.

By using a capacitor in the DC link, it is possible to smooth the rectified voltage. As a result, the minimum voltage in the DC link is reduced during the course of the mains phase, which also determines the maximum output voltage for the motor. The capacitor must store and retrieve energy during the course of the ripple of the rectified mains voltage.

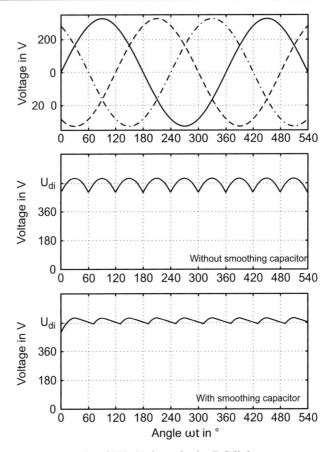

Fig. 3-52. Voltage in the DC link

There are two basic principles for designing the course of the DC link voltage across the mains period:

- In the case of a DC link with electrolytic capacitor, enough energy is stored so that the voltage in the DC link only collapses slightly during times when the diodes are non-conducting. This increases the maximum output voltage of the inverter. The capacitances are significantly higher than with a slim DC link. Electrolytic capacitors are dimensioned according to their service life, as the internal resistance and the AC current load lead to a power loss in the capacitor, which allows the electrolyte to evaporate. The DC link with electrolytic capacitors is the most commonly used type because it provides the highest output voltage (Fig. 3-52).
- In a slim DC link, only the AC current loads are compensated for by the inverter. The capacitances are small, and film capacitors are normally used. The voltage in the DC link has a ripple with six times the mains

frequency and voltage dips reaching down to 87% of the peak voltage (Fig. 3-52). This value corresponds to the maximum output voltage which can be output by the inverter. If an inverter which is capable of power recovery is used with block-shaped mains currents for the mains supply, then this form of DC link must be used.

Particularly in the case of a single-phase supply, the capacitor in the DC link has a high AC current load. Through the use of a mains choke this can be significantly reduced, allowing the service life thus to be significantly extended.

The capacitors used in the DC link need to be charged after switching on. In order to prevent this from causing excessive currents in the mains cables, corresponding charging circuits need to be used here. The following concepts are used:

- In lower power ranges up to 1 kW, charging can be performed via an NTC resistor, which initially has a high resistance (cold state) and limits the charging current. When it heats up after the charging process is complete, it reduces its nominal resistance and its power loss. The disadvantage here is the power loss in operation, but the low costs are an advantage. The frequency of the mains switching cycles is limited, and in addition a voltage drop remains which reduces the maximum output voltage.
- In the medium power range up to approx. 30 kW, charging resistors are usually used. Once charging is complete, they are short-circuited via a relay. This method represents a good compromise between function and cost. In order to prevent thermally overloading the charging resistors, the maximum number of mains switching cycles per time unit have to be limited.
- In the high power range above 30 kW, either thyristors can be used in the rectifier and operated via a generalised phase control during charging, or special charging circuits with their own power electronics can be used.

Braking energy. During operation of a motor in generator mode, energy flows from the motor to the inverter. There, this energy is converted (Fig. 3-53 a). Two fundamental operating modes can be differentiated by the quantity and duration of the regenerative energy:
- Regenerative energy is only generated for a short period during braking of the motor, and it corresponds to the kinetic energy stored in the drive train ($E_{kin} = \frac{1}{2} \cdot J \cdot \omega^2$).
- Regenerative energy is generated during a longer period. Typical applications include the lowering of loads (dissipation of potential energy), unwinding and load drives in test benches.

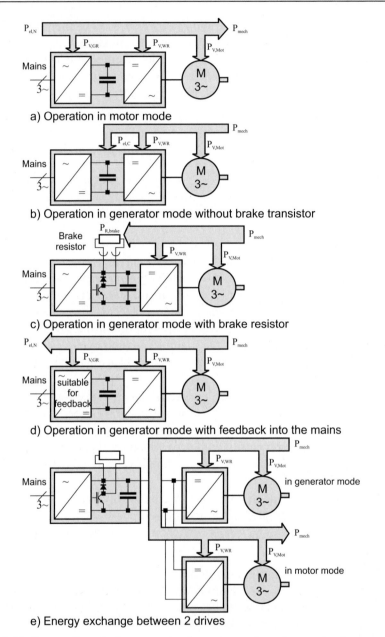

a) Operation in motor mode

b) Operation in generator mode without brake transistor

c) Operation in generator mode with brake resistor

d) Operation in generator mode with feedback into the mains

e) Energy exchange between 2 drives

Fig. 3-53. Energy flow during operation in motor mode and operation in generator mode

There are different methods for using this braking energy (Fig. 3-53):

- In many cases, no additional provisions are required, because the braking energy is converted into work energy in the process and power losses in the motor, or it is absorbed by the DC link capacity (Fig. 3-53 b). By adjusting the braking time of the drive, it is possible to control how much braking energy is returned to the DC link in order to prevent the maximum voltage from being exceeded as a result. Operation without further measures in terms of the braking energy is generally suitable for drives for pumps and ventilating fans, as well as for some low-powered servo drives.

- If, during braking or in normal operation so much energy is returned to the DC link that an extension of the braking time is not sufficient and the voltage in the DC bus will become too large without further provisions, then a brake transistor with brake resistor can be used (Fig. 3-53 c). The brake transistor is often already integrated in the inverter, while the brake resistor is connected externally. In other cases a braking unit is connected externally to the terminals for the DC voltage link. For the brake resistors, there are designs for a short-term load, which are used for dynamic braking, as well as designs capable of converting higher power into thermal energy for longer periods of time. These are provided e.g. for hoist applications. The brake resistor can become too hot if its maximum thermal capacity is exceeded. Accordingly, it needs to be monitored.
- If the rectifier is capable of power recovery (inverter with block or pulse operation to the mains), then the braking energy can also be fed back into the mains. This is particularly beneficial for applications in which a lot of feedback energy is generated (Fig. 3-53 d). The overall energy balance is better, and there is no power dissipation at the brake resistor.
- If several drives are working in a machine and the overall energy balance is such that individual drives require energy while others are being braked, then this energy can also be exchanged by connecting the DC links between the drives (Fig. 3-53 e). This is often used in process lines where the unwinder feeds back energy which is then used by the other drives. Test benches e.g. for gearboxes with drives which supply energy to the object under testing and others which place a load on gearbox outputs can also be equipped in this way.

However, it must always be taken into account how the kinetic energy stored in the drive trains is handled when a mains failure occurs. In this case, either uncontrolled braking (coasting) of the drives must be permitted, or a sufficiently large braking unit must be present which absorbs the braking energy, or the entire plant must be shut down in a controlled manner (mains failure control). In this case the coordinated shutdown of the drives is controlled in such a way that the energy which is fed back keeps the entire system running for long enough time until all of the drives have reached a standstill. This method makes sense particularly for textile machinery and process lines.

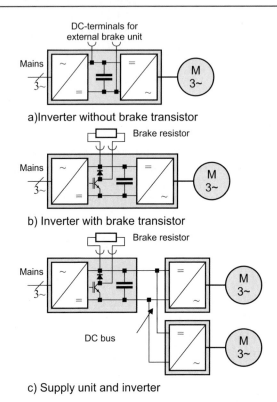

DC-terminals for
external brake unit

a)Inverter without brake transistor

b) Inverter with brake transistor

Brake resistor

DC bus

c) Supply unit and inverter

Fig. 3-54. Device concepts for single-axis and multi-axis applications

Device concepts for single-axis and multi-axis applications. The full range of power units comprises the following components:

- inverters for the motor,
- rectifiers and various types of inverters capable of power recovery,
- the DC link,
- braking units with brake transistor and brake resistor,
- the connection between several inverters via the DC link.

These components are combined as follows to form device concepts (Fig. 3-54):

- *Inverters without brake transistors* comprise a diode rectifier and an inverter. A braking unit may need to be connected as required via the DC link terminals (Fig. 3-54 a).
- *Single-axis inverters with a brake transistor* also comprise the brake transistor in addition to the diode rectifier and the inverter, meaning that only the brake resistor needs to be provided externally (Fig. 3-54 b).
- For *multi-axis applications* there are inverters on the motor side without supply, as well as a central power supply unit which supplies several in-

verters (Fig. 3-54 c). The power supply unit comprises either a rectifier with brake transistor or a mains inverter capable of power recovery, although the latter also requires a braking unit in case of mains failure.

All of these device concepts are quite common and offer advantages for different applications which are covered in chapter 4 [HeKa04].

Further functions in the power unit. As well as the functions listed above, the power unit must also include the following circuits:

- The different internal circuit components need to be equipped with a power supply. Switched-mode power supplies are used for this, and they are either connected to the DC voltage link or supplied from an external control voltage supply (often 24 V_{DC}). The transformers in the switched-mode power supply also provide the necessary insulation between the different potential levels.
- Different quantities need to be measured. The most important ones are the motor currents and the voltage in the DC link. As well as these, it is also important to measure temperatures for monitoring reasons. Different measuring points are used for the current measurement (two or three output phases, the base points of the three half bridges, the DC link) and various measuring principles (current shunts with direct evaluation or isolation amplifiers, magnetic current measuring elements with direct or compensating measuring techniques) are used. The measurement of the voltage in the DC link is performed via voltage dividers or an additional winding of the transformer of the switched-mode power supply.
- The IGBT power switches need to be controlled via gate drivers. These are implemented either with insulating optocouplers which already contain the gate driver, via non-insulating high-voltage switching circuits or via discrete circuits.

Additional inverter topologies. The inverter topology described up to now – i.e. a hard-switching six-pulse voltage inverter with IGBT power switches – is by far the dominant structure in the power range up to 1 MW. However, for reasons of completeness the alternative concepts should also be mentioned at this point.

With a *matrix inverter*, all three mains phases are connected to the three motor phases via bi-directional power switches (Fig. 3-55) [Zi99]. Advantages of this topology include full power feedback capability and the omission of the voltage link, but disadvantages include the higher number of power switches (18 IGBTs, as two IGBTs need to be used in anti-parallel for each bi-directional switch) as well as the filter circuitry on the mains side. In addition, the control of the nine bi-directional power switches is more complex compared to an inverter with DC voltage link.

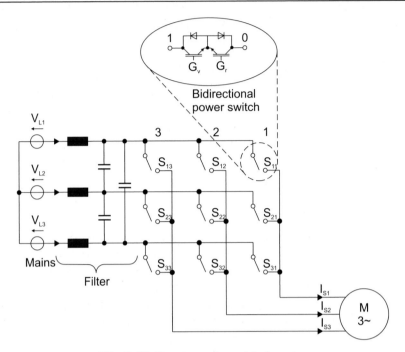

Fig. 3-55. Structure of a matrix inverter

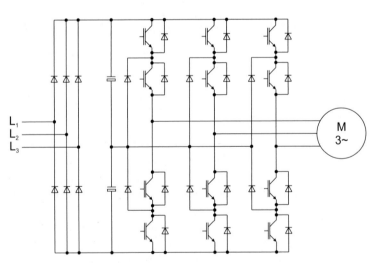

Fig. 3-56. Structure of a three-level inverter

With a *three-level inverter*, four power switches are used in series for each half bridge (Fig. 3-56). This allows the output voltage to the motor to be applied at half the amplitude of the DC link voltage. Accordingly, the voltage load of the motor is significantly lower. Three-level inverters are frequently used in applications with higher power or higher voltages. In low-voltage applications (400 or 480 V_{AC}), the increased number of IGBTs (12 instead of 6) is disadvantageous [He+04].

In addition there are also topologies in which no hard switching takes place. In these topologies, resonant circuits are achieved via additional inductances which allow the power switches to switch almost loss-free across zero voltage. The lower switching losses and EMC emissions are advantageous, but the higher number of components and the restrictions in terms of control are disadvantageous.

3.4.2   Mechanical layout

Inverters are electronic devices which are either installed in a control cabinet or mounted on or in the immediate vicinity of a motor. All of the components of the inverter must be placed in the housing, which must perform the protective functions required for the location. In addition, straightforward mechanical and electrical installation must be assured. The devices must also incorporate diagnostic elements which allow easy diagnosis of the devices' state.

Unlike other electronics devices, in which a printed circuit board for the electronic components and a plastic housing are sufficient, the mechanical design of an inverter is more complex:

- Power losses are generated during the power conversion, which need to be dissipated by a cooling unit.
- The power modules, the capacitors in the DC link and the magnetic components (chokes, transformers) are bulky and need to be housed efficiently.
- The high currents in the power circuit either require high cross-sections on the printed circuit board or even dedicated busbars.
- It must be ensured that the power unit, the control connections and the signal processing are properly insulated from each other.
- In many cases additional modules are used for the extension of the inverter.

Overall, the mechanical design of an inverter is a complex challenge in which many seemingly conflicting requirements need to be met [Ki00]:

- ease of installation,
- inexpensive components,
- compact volume, i.e. high power density,
- safe thermal behaviour,
- ease of use.

Fig. 3-57. "Bookcase" and "box-size" inverters

This has led to the development of a large number of configuration concepts, which can only be described in brief here.

Inverters for installation in a control cabinet. The majority of inverters are installed in a control cabinet (> 80% of all industrial inverters). The most popular enclosure is IP 20, which permits electrical connections from the top and from the bottom side. In general, two configuration concepts have proved most successful (Fig. 3-57):

- With a "box-shaped inverter" there is a heatsink on the rear panel, to which a power board with the power module is initially attached. On the front panel there is a control card with the control connections. The device is designed for compactness and ease of manufacturing, but not for minimisation of a particular dimension.
- The aim of a "Bookcase inverter" is to minimise the width of the device. This is advantageous for all applications in which a large number of inverters are housed in a control cabinet. With this concept, internally positioned heatsinks and a power card which is positioned perpendicular to the control cabinet mounting surface are used in some cases.

The key factor for the design of the device is its cooling, as this can take up as much as 30% of the volume of the device. Air cooling is generally used, as it does not require any additional infrastructure (e.g. a water cooling system with a heat exchanger). In order to keep the volume of the device to a minimum, pure convection cooling (i.e. no fan) is used up to powers of approx. 1 kW. At higher ratings, fans are used which significantly increase the air speed in the heatsink and therefore the amount of heat which is dissipated. These fans must be suitable for the intended service life and the ambient conditions. Aluminium is used nearly always as the material for the heatsink. Most heatsinks are either extrusion-moulded or die-cast.

Apart from the heatsink, the enclosure of the device is the most important element of the design. Plastic housings are used for most applications, except in more high-power applications which use a sheet-metal casing.

Mechanical mounting is performed with the aid of fastening elements which are used to screw the inverter to the rear panel of the control cabinet. If very fast replacement is required in the event of a fault, then it would also make sense to use a mounting unit into which the inverter is inserted (Fig. 3-58).

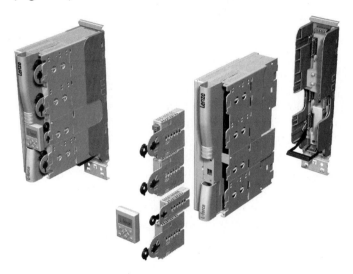

Fig. 3-58. Inverter with mounting unit for fast replacement

The electrical installation is performed via the electrical connections of the inverter. The control connections are laid out as terminals with cross-sections of 1 or 1.5 mm². Screw terminals and screwless cage-clamp technology are used. The terminals can be fixed or pluggable, whereby pluggable terminals are advantageous for quick component replacement.

With the power connections, the cross-section depends on the current. Up to a power of around 55 kW screw terminals are available. These can also be supplied as pluggable versions (up to around 11 kW). Threaded bolts are required at higher ratings.

Efficient connection of the shielding of the motor cable is also very important for an easy installation. There are now a number of methods which ensure quick and reliable shield connection with good EMC properties.

Fig. 3-59. Inverters with a higher degree of protection

Inverters with a higher degree of protection. For locations outside the control cabinet an IP 20 enclosure is not sufficient. For example, IP 21 or NEMA 1 is required for use in a central control room, which also includes protection against water dripping from above. Here, all connections must be made from the bottom, and the upper side of the device must be completely closed. This is largely possible for inverters with a box design, although some require an additional cover for the upper side.

If the degree of protection IP 54 is required then there must not be any slots in the housing. The fan must also comply with this degree of protection. All entry points for electric cables must be sealed.

Inverters with the degree of protection IP 65 do not permit the use of external fans. Dissipation of the power losses must be ensured through the provision of sufficiently large heatsinks.

As soon as inverters are used outside a control cabinet, their mechanical robustness also needs to be increased (Fig. 3-59). Inverters for pumps and ventilating fans are usually still mounted on a wall, while inverters for conveyor drives tend to be installed on the mechanical systems of the plant and are therefore in the access area of personnel and fork lifts. For this reason it makes sense to place this type of inverter in a robust and often metal housing which offers better mechanical strength than a plastic enclosure. All electrical connections must have a high degree of protection.

Inverters for mounting on a motor. If the inverter is mounted on the motor or structurally combined with it by other means, then the resulting unit is referred to as a mechatronic unit (Fig. 3-60). There is no motor cable and, in some cases, no connection for the encoder from the motor to the inverter. EMC filtering is made easier by the short connection to the motor. In addition, particularly on large plants with many drives, a number of control cabinets are no longer required. One disadvantage is the higher expense for the inverter, which only balances out in a general assessment of overall costs. With motor inverters, a large number of controlled drives can be operated with just two cables: firstly the power cable and secondly the bus cable. Together with suitable installation technique, this approach saves costs in larger plants (chapter 5.2.3).

Fig. 3-60. Combination of a geared motor with a motor inverter

3.4.3   Control electronics and software

The signal processing in an inverter takes place almost exclusively digitally [SchHe85, Ki94a], with one or more microprocessors being used. This means that an inverter is also an embedded system (Fig. 3-61). This system has the following interfaces:

- to the power unit, comprising the control signals for the power switches and the measuring signals from the power unit,
- to the control terminals of the inverter, generally comprising analog and digital input and output signals (I/Os),
- to encoders (if the inverter supports their evaluation),
- to communication interfaces,
- to integrated operating and control functions.

In terms of the specific selection of a microprocessor, a large variety of computer families are available to choose from, offering a wide range of computing power. 16 and 32-bit architectures are used most commonly, while 8-bit computers do not usually offer enough computing power. Special signal processors support the algorithms of the drive control which require high computing power.

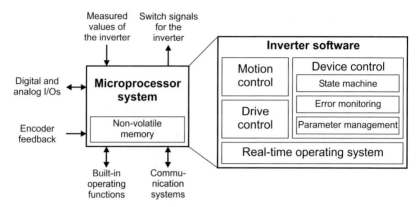

Fig. 3-61. Control electronics and software

In addition to the actual microcomputer system, which comprises a CPU and a memory, special peripheral functions are required. For example, the pulse-width modulated signals for the power switches are implemented with special counter and timer circuits. Analog-digital converters are required for the measurement of current, voltage and temperature values as well as analog input values [Ki94b].

Evaluation of the encoders. The evaluation of signals from angle sensors (resolvers, incremental encoders) requires special evaluation circuits. Various methods which are currently used in practice are described elsewhere [Ki94a] and do not need to be covered here in more detail. For the application it is important that the instantaneous angular position is determined with high precision, so that the drive control has access to accurate actual values. During the measurement it is also important that the cycle time is accurately met, so that the rotational speed value obtained through differentiation has minimal digitisation noise.

As well as measuring the actual values for the drive control, it is also necessary to determine exact positions in the mechanical system and relate them to the instantaneous position of the drive. This is necessary both for homing of the mechanical system (chapter 4.4.3) and for the acquisition of markers (chapters 4.6.3 and 4.8.3). For example, if it is necessary to determine positions with an accuracy of 0.01 mm at a speed of 500 m/s, then a time resolution of 1.2 µs is required. So-called *touch-probe inputs* are used for this purpose, which capture the time instant of a sensor signal with high precision and high resolution and, at the same time, determine the instantaneous position of the encoder at this time.

Highly-integrated microelectronics components. Highly-integrated microcomputers are available for implementation of the control electronics in motor control applications. As they have all of these functions integrated, the control electronics can be set up with one highly-integrated circuit and just a few external components. In other cases, a programmable logic circuit (FPGA) or an application-specific integrated circuit (ASIC) is used to implement the periphery functions. Integrated system-on-chip solutions, which integrate the microcomputer with specific periphery functions, are also common. All of these integration approaches serve to implement the required functions of the control electronics as inexpensively as possible [Ki94a, KiSch94a, KiSch94b, KiSch95a, KiSch95b, KiKrSch01].

Parameter memories. A large number of parameters are used for the setup of the application function, and they need to be stored in a non-volatile memory. Corresponding memories are used for this purpose (EEPROMs, flash memories). In order to be able to easily transfer these parameters to a substitute device for replacement purposes, in some cases these parameter memories are located on an interchangeable memory module.

Basic software functions. In order for an embedded system to function correctly, it requires – in addition to the control electronics – a number of basic software functions together with the microcomputer. These functions are only covered in brief here:

- The *operating system* ensures that the individual program parts are executed. Due to the stringent real-time requirements, a real-time operating system is required here which also supports program tasks which are executed with a repetition rate of 32 μs (servo control) or 250 μs (frequency inverter).
- The overall status of the drive is managed by a *state machine*. For example, after startup there are initially a number of initialisation routines which need to be performed, before the drive can be started. During operation, a distinction is generally made between standstill of the drive, normal operation and error states. The transition between the states is performed via commands from the operator or from the master control, as well as through internal recognition of error states.
- Many internal variables of the inverter are monitored for *error states*. This includes errors relating to temperatures, output currents, the voltage in the DC link (undervoltage and overvoltage), the presence of the mains supply and signals from the encoder, as well as internal states of the drive control and motion control (e.g. following error monitoring). In some cases, a warning is generated first. If impermissible values are then exceeded and no direct counter-measures can be initiated, the system switches to the error status, in which the inverter is switched off and the drive slows down. It must always be ensured that such error messages are not issued too early, as the reliable operation of the drive has highest priority. On the other hand, an inverter needs to be capable of

protecting itself against being destroyed, e.g. as the result of a short circuit on the motor cable.

- The adjustment of the inverter settings requires a large number of *parameters*. For this purpose, the software must also provide parameter management functions. Every parameter has – in addition to its instantaneous value – a number of other attributes (e.g. limits, display units, a name). Parameter setting is performed either by means of an integrated or pluggable diagnostic unit, via a diagnostic interface by a PC or via a communication interface by the master control.
- If a drive performs dynamic processes (e.g. positioning sequences), then it also makes sense for it to be able to record internal signals as a function of time (*oscilloscope function*), which are then displayed on a PC.

Today, an inverter will have a quite large amount of software. It must be ensured with the aid of corresponding development methodology, including comprehensive software testing, that the software is very reliable [Be05, LiRo05]. Inverters are one of the most important elements for the safe operation of a machine, and therefore they play a key role in defining its reliability and productivity.

In the following chapters we will be taking a look at the most important parts of the inverter software. We will cover the drive control first, followed by the motion control of the drive.

3.4.4 Drive control

Dr. Edwin Kiel, Dr. Carsten Fräger

The drive control is the heart of the inverter software. It has the task of controlling the power conversion to the motor so that it follows the setpoint. In most cases, the setpoint is a speed setpoint. For other applications, setpoints for the torque or angular position are also often applied. Interrelated values for all three mechanical variables can also be used for precise motion control of a drive, which the drive control is then required to follow as accurately as possible. Overall, the drive control is designed to fulfil the following tasks:

- Direct following of the setpoints (command action).
- Compensation of disturbances as quickly as possible, e.g. for load torque changes in speed control mode.
- Safe guidance of the drive. The control over the drive must not be lost at any working point (the drive must not run away). If the drive operates outside its limits (e.g. there is a speed setpoint jump which is too fast and cannot be matched by the acceleration capability), then the drive must work at its operating limits (e.g. maximum drive torque).

- The drive must be protected, and it must not be damaged due to over-loading. For this purpose, there are limits for the maximum currents in the inverter and in the motor, maximum voltages in the DC link and maximum temperatures of the components in the power unit, all of which need to be adhered to.

Operating modes for motor control. Frequency inverters and servo inverters use different operating modes for motor control:

- In a frequency inverter, the motor operates in an open-loop control without measurement of the actual speed. A voltage-frequency control *(V/f control)* is common. In order to avoid a number of restrictions of this operating mode, many frequency inverters also offer a *sensorless vector control*.
- In servo inverters, the speed and angular position are measured so that a very accurate and dynamic control of the motor can be performed. This *servo control* is performed in accordance with the principles of field-oriented closed-loop control, which can be employed for both synchronous and asynchronous motors.

In the following we will first take a look at V/f control. This is followed by an introduction to field-oriented closed-loop control with feedback. The explanation of sensorless vector control is saved until the end, as it is based on the theories of field-oriented closed-loop control.

Voltage-frequency control. With V/f control, the output frequency of the inverter is determined from the setpoint for the motor speed, and the output voltage is derived from this value via a fixed ratio calculation. Here, the output voltage follows the output frequency proportionally until it reaches the rated voltage of the motor at its rated frequency or until the maximum output voltage is reached, which is defined by the mains voltage (Fig. 3-62). It is not possible to increase the voltage beyond this point, and the motor is then operated in the so-called field weakening mode, in which the attainable torque decreases and the power remains constant.

As the motor has a voltage drop due to the resistance in the stator, the characteristic needs to be boosted slightly at small frequencies in order to compensate for these voltage drops. This boost takes place according to the rated data of the motor (and is therefore matched to the rated power) and does not take into account the actual current. As a result of this, the voltage applied to the motor does not correspond to the optimum at every operating point. The magnetisation of the motor changes with the load. Consequently, a drive with a strongly varying load torque cannot work at the optimum operating point in all cases.

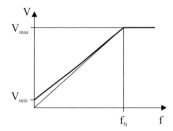

Fig. 3-62. V/f characteristic

From an energy standpoint, it makes sense to work with a square-law characteristic in applications with an operating characteristic which results in only a low torque being required at low speeds (e.g. pumps, ventilating fans), as then the magnetisation and therefore the current demand are reduced even at lower speeds.

The V/f control initially offers no protection if an excessively high torque – and therefore an excessively high output current – is demanded during an acceleration. Therefore a current limitation is added, which in this case delays the acceleration or even decelerates it once the final speed has been attained if a static overload is present. This enables operation at the current limit of the motor and acceleration with the maximum torque.

The V/f control also does not compensate for the speed drop of an asynchronous motor under load (the so-called slip). It is however possible to conclude that a load has been applied by measuring the active current, and to then increase the field frequency so that the motor speed remains approximately constant even under load (slip compensation). As a result of this, the motor has good speed stability in a steady-state load scenario.

As the V/f control is a voltage control, several asynchronous motors can be operated in parallel from a single inverter.

Overall, the V/f control is a very robust technique which only requires minimal data for the motor (the nameplate data). Its advantages always come into play in applications where the motor load is not dynamic and largely free of disturbances. As it does not take into account the dynamic behaviour of the motor, the weaknesses of this technique are exposed in all applications which require high dynamic performance or higher operating ranges (> 1:10). In summary, the V/f control has the following restrictions:

- The torque yield in the low speed range (< 10% · n_N) is limited, because the voltage does not take into account the load condition of the motor and therefore the current-dependent voltage drops at the stator reactances and resistances.
- The torque rise time is slow, because the dynamic model of the motor is not taken into account.

These restrictions can be countered by using a sensorless vector control, which is covered after the servo control.

Closed-loop control of servo drives. The objective of a servo control is to ensure that the motor follows the setpoints of the control as accurately as possible. The servo inverter effectively supplies the motor with exactly the currents required to provide the torque and the speed for the drive task.

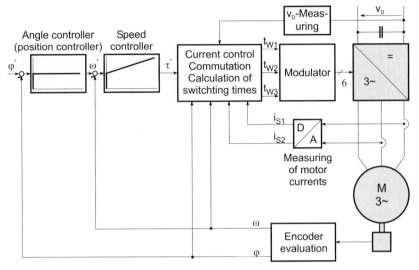

Fig. 3-63. Cascade control

Cascade control. Cascade control is the typical closed-loop control structure of a servo drive (Fig. 3-63). This means that the motor current required to generate the torque is generated in the innermost closed-loop control circuit. The inverter, the motor and the current detection are involved in this [Le97].

The speed control is superimposed above this inner circuit. Its input variables are the speed setpoint and actual value (ω^*, ω). Its output variable is the torque setpoint m^* for the inner control loop. The outer circuit is the position or angle controller, which compares the setpoint angle and the actual angle (φ^*, φ) and sets the setpoint value for the speed controller.

The advantage of cascade control is its ease of operation and commissioning: the control loops can be adjusted one-by-one from the inside outwards. In this way, stable and dynamic operation of the drive can be attained within a short time. By choosing smaller or larger amplification factors, a more dynamic setting for fast motion or a more moderate setting for a smoother operation is both possible.

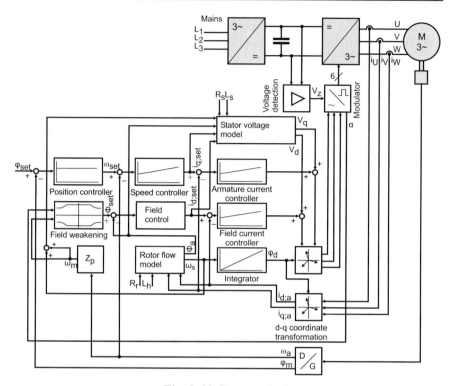

Fig. 3-64. Servo control

Torque control and current control. The innermost control loop for torque generation must take into account the properties of the electric motor, so that the torque setpoint is transferred to the actual torque in the best possible way [Qu93]. It is expedient for the closed-loop control to take place in d-q components (the field coordinates) (Fig. 3-64). This makes the transformed currents and voltages of the stator zero-frequency quantities within the closed-loop control for the steady-state operation.

The inverter outputs a voltage to the motor, resulting in the motor current and the generation of a torque. With field-oriented control, the torque and the magnetisation in the motor are also controlled via current setpoints (which corresponds to the rotor position on a synchronous motor). The current control therefore has the task of calculating the voltage according to the setpoints for i_q and i_d so that these currents will be applied without delay.

For i_q and i_d, two PI controllers are used which are pre-controlled with an inverse feedforward model of the motor (i.e. of the stator resistance, the stator impedance and the induced voltage). As these variables change in operation (the resistance changes due to the temperature and the inductance changes due to saturation effects of the iron), in some cases parameter corrections need to be used so that the current controllers always have the optimum dynamic performance [SchHo05b].

Adjustment of the current control settings requires very good knowledge of the motor data. These data can either be identified during commissioning, taken from corresponding information supplied by the motor manufacturer or even saved in electronic nameplates in the motor. This last method is the best one, as its data are then determined during the manufacturing of the motor, saved in the motor and electronically transmitted to the controller.

The switching frequency of the inverter plays the key role in determining the dynamic performance of the current control, as it defines the cycle rate of the current control. With the switching frequencies used today of 4 to 20 kHz, torque rise times between 0.5 and 2 ms are attained. For most applications this is totally sufficient, as the mechanical dynamic performance of both the drive and the application are significantly lower.

Flux model and field weakening. On an asynchronous motor, the slip frequency and the current value for the magnetisation need to be determined from the currents. Ultimately, this is an observer algorithm for the internal behaviour in the motor (which cannot be directly measured). Motor data are required here again. In order to achieve a precise determination of the flux, it is necessary to also take into account the saturation characteristic of the iron in the motor.

The setpoint for the current i_d is determined by the field controller. Its target value takes into account the field weakening starting from the rated speed of the drive. At higher speeds it is necessary to reduce the magnetisation, so that sufficient voltage is kept in reserve for the closed-loop control of the currents. As a result, the attainable torque is reduced by around the square of the speed.

In order to achieve a higher torque utilisation on synchronous servo motors by using the effects of field weakening (chapter 3.3.5), a non-zero setpoint is set for i_d depending on the speed and the torque. As a result of this, compensation of the inductive voltage drop due to the stator inductance and an improved voltage yield are achieved.

Speed control. The next highest quantity of the drive which needs to be controlled is the speed. In the case of a servo control with encoder, it is measured, so that the determination of the actual value does not pose any problems. The controlled process here is the torque generation (which acts like a delay or a low pass) and the mechanical integration of the torque into the speed. The load torque is acting as a disturbance value. According to chapter 3.1.1, it comprises friction, process forces and other influences. The mass inertia of the drive determines the speed variation which results from the applied torque.

PI controllers which have been adjusted according to the optimum magnitude are normally used for the speed controller. The defining variable for the attainable limit frequency is the deceleration (or equivalent time constant) in the torque generation. Bandwidths of 500 to 1,000 Hz are possible with the switching frequencies in use today. This limit frequency only applies to the low-level signal behaviour, i.e. for small changes in the setpoint. In the event of larger changes, the overall dynamics (i.e. maximum torque and mass inertia) of the drive train are effective.

The adjustment of the speed controller settings depends to a large extent on the mass inertia of the drive. As this parameter can change during operation (changing loads, kinematics), it either first needs to be set to a value which is stable in all operating modes, or the parameters will need to be adjusted.

Angle control and position control. In applications with position control, in addition to the speed control an angle control is also needed. This makes the drive stiffer when holding a position, in that the control already reacts to small measured angular differences. Due to the stability criteria, only a P-controller can be used for the position control. For dynamic positioning without the use of feedforward controls, this controller always causes a following error (difference between setpoint position and actual position).

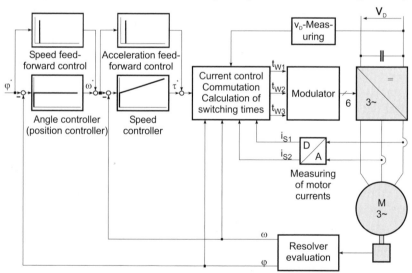

Fig. 3-65. Feedforward control of the position and speed controller

Torque feedforward control for increased dynamic performance. The primary task of the drive control is to ensure stable operation with simultaneously high dynamic performance of the drive. The drive control follows the instantaneous value of the setpoint and initially has little knowledge about the remaining time curve.

In a positioning application the operation changes between accelerating, running at constant speed and braking. The acceleration and braking phases in particular create a deviation to the speed control, which it has to compensate for.

Alternatively, it is possible to apply the necessary acceleration or braking torque determined from the periodic change of the speed to the output of the speed controller. With the aid of this feedforward control the speed controller is relieved, and the build-up of the required torques takes place more quickly. Thus the deviation between the setpoint position and the actual position (the so-called following error) is noticeably reduced. A feedforward control of the position or angle controller can be implemented in similar fashion. Fig. 3-65 shows the associated block diagram for a cascade control with the feedforward controls.

The variables for the feedforward control of the position and speed controller can either be determined directly from the changes in the setpoints, or alternatively they are already calculated and output by the motion control, which then passes on associated values for the position, speed and torque to the drive control.

If the dynamic change of these values is within the limits that can be followed dynamically by the drive, then the load on the controllers is significantly reduced. They then only have to compensate for disturbances resulting from process influences (load torque) and as a result of deviations between actual variables and assumed values (e.g. resistances and inductances in the motor).

Filters for the actual values to improve synchronism. In many applications, drives need to run with angular synchronism to a master drive at a fixed speed. A typical example is printing machines. Here, all of the drives rotate synchronously to the paper track for a perfect print image. Angular differences lead to offsets between the individual print images, which become visible in the overall printing result. In order to manage this task, filters are used in the actual value channel. They remove fluctuations from the actual values which could otherwise cause oscillations in the plant [Sch06]. This filtering can significantly reduce the angle error. However, the drive loses dynamic performance.

Sensorless control of an asynchronous motor. The theory of field-oriented control of an asynchronous motor can also be used to improve the behaviour of a motor operated without a speed sensor in comparison to V/f operation [Jo89]. The primary goal here is to significantly reduce the previously stated restrictions of the V/f mode by taking into account the dynamic behaviour of the motor.

Thanks to the motor model, which is also used for control with sensors and which calculates the instantaneous value of the magnetisation and the slip, all of the variables for control are known even without measurement of the speed. However, the accuracy depends on the exact knowledge of the motor variables. If these go into the control with incorrect values, then the control will be imprecise and – in the worst case – unstable. Accordingly, a sensorless control must always be supplemented with methods for identification of parameters and parameter correction.

As described above, the current control of a servo control comprises current controllers, which are pre-controlled with feedforward voltage values from an inverse motor model. As there is no speed sensor in a sensorless control which would supply the actual value to the speed control, a comparison of the control types in a practical application shows the following:

- The use of current control in a sensorless vector control results in a high dynamic performance for the torque build-up, but also speed variations and lower stability in terms of the control. The computing power requirements are also higher.
- The approach of only setting the voltage via an inverse motor model leads to a less dynamic performance, but also delivers improved smooth running and better stability, as there are fewer controllers in total which define the dynamic behaviour. Through a correction of motor variables (in particular the stator resistance R_s), which can be based on a comparison between the measured and assumed values for the flux-generating current i_d, this method also offers high accuracy with changing motor parameters.

Compensation of inverter errors is also decisive for the safe operation of a sensorless control, see chapter 3.4.1 for more details. Overall, a very robust sensorless vector control for an asynchronous motor can be set up with the following elements (Fig. 3-66):

- a speed and torque controller,
- a flux model for the asynchronous motor, which as well as the motor flux also defines the slip frequency of the motor,
- a field controller,
- an inverse motor model for calculation of the voltages from the setpoint currents,
- good inverter error compensation,
- correction of the motor variables through a comparison of setpoint values and actual values for i_d,
- a parameter identification during commissioning which determines the motor parameters [Sch+01b].

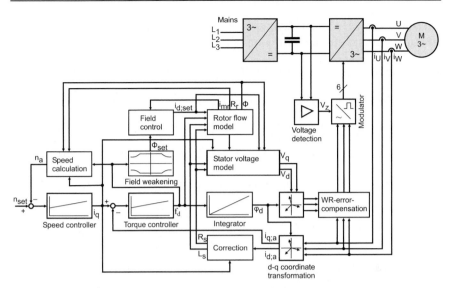

Fig. 3-66. Sensorless control of an asynchronous motor

This type of control has the following advantages in comparison to V/f control:

- In the low speed range a good torque yield is achieved, and the motor voltage is always dynamically controlled to the value required according to the load.
- A torque step change is quickly compensated for. This now depends to a very large extent on the cylce time of the control, and therefore on the computing power used. With a cycle time of 250 μs, torque rise times of 10 to 20 ms are attained, whereas these are around 5 ms with a cycle time of 62.5 μs.
- The speed is held approximately constant when the motor is subjected to a load.

Sensorless vector control can be used favourably in comparison to V/f control wherever the work process has high torque fluctuations or where a high starting torque (breakaway torque) is required. Due to the high dependency on the motor parameters, it is however more complex in terms of application and optimisation compared to V/f control.

3.4.5   Motion control

Now that the drive control has been covered in depth in the previous chapter, we will now take a look at motion control, which provides the drive control with setpoints. This means that the motion control represents the interface between the automation and the controlled drive. The motion control must satisfy the following requirements overall:

- The drive must be controlled in the way required for the mechanical process. Accordingly, the motion control has a high dependency on the specific application and is therefore configured specifically for each application.
- The change in setpoints must not take the drive outside its permissible operating range. This is characterised by the limits for speed, torque (and therefore the acceleration capability of the drive), thermal limits and the strength of the drive components. The rules for drive dimensioning are explained in chapter 3.7.2.

In the following we would like to present a number of general forms of motion control which have a more general character.

Ramp function generator for acceleration and braking. If a drive is accelerated or braked, a corresponding torque needs to be applied to do this. In some applications this process needs to take place as quickly as possible (high dynamic performance), whereby the maximum torque of a drive is always limited. There are two ways to comply with these torque limits in the drive:

- in the drive control – through restriction to the maximum output current,
- in the motion control – through restriction of the rate of change of the speed setpoint.

The restriction in the drive control has several disadvantages here:

- In the case of a servo control with PI speed controller, the speed overshoots if a sudden change in setpoint is applied to the controller. By contrast, a ramp-shaped setpoint change does not display this overshoot, and the final speed is reached cleanly.
- With a frequency inverter, a very effective current limitation must be implemented (in the end, this also limits the rate-of-change of the speed setpoint).

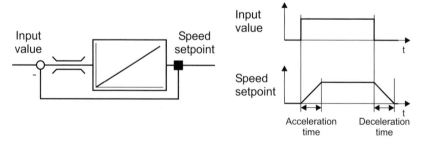

Fig. 3-67. Ramp function generator

- The specific acceleration behaviour depends on many application parameters, such as the load moment of inertia and the load torque. Therefore it cannot be calculated in advance.

As a result of these points, in practice the acceleration is usually limited via a ramp function generator (Fig. 3-67). This operates in such a way that an integrator is placed between the input value and the setpoint of the speed controller, the maximum rate of change of which and therefore the maximum angular acceleration are limited. Different limits for acceleration and braking are normal. Parameter setting is usually performed via the selection of times within which an acceleration from zero to the maximum speed or the rated speed has to take place.

Jerk-limited motion control. The transmission path from the motor to the mechanical system is frequently elastic and therefore capable of oscillation, as was described in chapter 3.1.8. For this reason it makes sense to limit the maximum rate of change of the torque and therefore the jerk as well.

Either s-shaped speed ramps or trapezoidal acceleration ramps are used for rate-of-change limited motion control.

Positioning sequence. In many applications the drive is to perform a positioning process [KiFr06]. The task of a positioning process can be described as follows:

- an object for positioning with the mass m is to be moved in the time t to a position s, whereby the speed at the start and end is zero,
- the target position is to be met with the accuracy Δs.

The positioning process is divided into three phases (Fig. 3-68):

- acceleration from zero to maximum speed,
- traversing at maximum speed,
- braking to zero to reach the target position.

Due to the acceleration, traversing and braking, positioning is a dynamic process. The positioning process therefore requires precise drive control in all four quadrants. For this reason, servo drives are used in most cases.

Whereas for the implementation of a positioning drive the acceleration to final speed and traversing at this speed still correspond to the process which is followed by every drive when it is accelerated, the target position needs to be taken into account for the deceleration ramp. Starting from the braking distance calculated in advance, the setpoint speed is reduced according to the deceleration ramp during the traversing motion until the target position is reached.

In practice there are a large number of other variants for positioning sequences – these are described in detail in chapter 4.4.

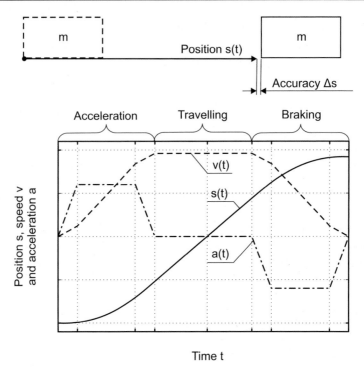

Fig. 3-68. Positioning sequence

Special operating states. The functions described so far for motion control relate to automatic operation, in which a control sets out the functions which need to be performed by the drive. As well as this, there is also a number of further operating states for which an inverter must offer suitable functions:

- If the drive needs to be shut down as quickly as possible because of a fault, then it must be controlled with the aid of a corresponding ramp to a speed of zero. Drives are usually equipped with a *quick stop* function for this purpose, which is tripped via a corresponding command like a digital input signal. Unlike a standard deceleration ramp, its deceleration ramp can be adjusted separately.
- Many angle sensors do not measure the absolute position of the drive. Accordingly, after they have been powered up, a *homing* run needs to be performed first in order to determine the absolute position of the mechanical system it is driving in the machine. The function and the sequence for this homing run are described in detail in chapter 4.4.3.
- When setting up or eliminating faults, it may be necessary for an operator to set the drive in motion on site. Functions for *manual jogging* are available for this purpose. In many cases two buttons are used which enable a motion with a limited speed in the positive and negative direc-

tion. A drive should contain the necessary software functions required for this. It may also be useful to define a motion along a defined path after tripping by pressing a button (inching of the drive).

- After a mains failure, if an inverter is switched to a motor which is possibly still turning then it will first need to determine the speed of rotation of the motor. If a frequency inverter is used without a speed sensor then this will need to be implemented in the software. This function is referred to as a *flying restart circuit*. It is used particularly on large ventilating fans, which coast to a standstill for a long period of time after a mains failure. In order to determine the speed, the inverter runs through the speed range of the motor with its output frequency and a small output voltage and monitors the current in the process. The frequency at which the current has its minimum corresponds to the instantaneous motor speed. The frequency inverter can then boost its voltage at this frequency to the value appropriate for the actual speed in order to take control of the drive.

A number of different operating modes are often used on a drive, between which the system switches back and forth during operation. The motion control software must also ensure continuous and – if necessary – rate-of-change limited motion sequences for these changes.

Interaction between Automation and drives. As was explained in chapter 2.3.1, there are two fundamental architectures for implementation of the motion control:

- The motion control is performed in the drive in a decentralised structure; the control transmits commands (e.g. start/stop) to the drive.
- The motion control is calculated centrally in a control system for several drives; the drives receive setpoints and then only execute the drive control.

The definition of this concept is very closely linked to the automation architecture on the one hand (selection of the control, distribution of functions) and to how closely the functions of the individual drives are coupled to each other on the other hand.

Very close coupling is always given in applications in which several drives work together to move the material being transported, the work piece or a tool, or if they bring it to each other into a required position. This is the case for all continuous production processes which contain multiple drives, as well as all fast-cycled production processes which are based on distributed single drives. Coordinated multi-dimensional motion also requires close coupling of the drives.

If drives are coupled closely to each other, then there are two situations we need to distinguish between:

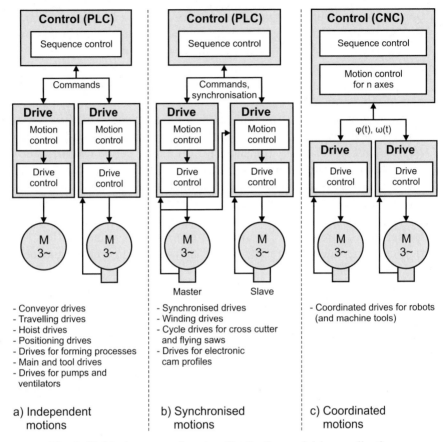

Fig. 3-69. Motion types, function distribution and drive applications

- A master motion in the machine is followed by the drives. This is usually the motion of the work piece, to which the individual processing steps are synchronised. This principle is often encountered in machines which would have traditionally been equipped with a line shaft and mechanically coupled drives in the past.
- Multi-dimensional motions in three-dimensional space are performed by multiple axes. This is for example the case with a robot.

In the first configuration it is not necessary for the motion profiles to be calculated in a central location for all of the axes. A master motion is defined and then distributed to the axes, where motions derived from it are generated decentrally. The master motion does not depend on the motion of the axes, and there is a master-slave relationship. In this type of configuration, synchronised drives (chapter 4.6), winding drives (chapter 4.7) and electronic cam drives (chapter 4.9) are generally used. The drives for cross cutters and flying saws also follow this structure (chapter 4.8). The motion control takes place in the drives, while at higher level a control performs sequence control functions.

In the second configuration, it does not make sense for the motion sequence to be calculated decentrally for each axis. In order to determine the motion sequences, multi-dimensional motion data records are interpreted, coordinate transformations are calculated and – from this – the motion sequences of the individual axes are derived and interpolated (chapter 2.3.1). This task is best performed by a central control. Coordinated drives are used for the drives (chapter 4.5).

If drives are not closely coupled then they can be controlled by a sequence control, usually a PLC control. This is the case on conveyor drives (chapter 4.1), travelling drives (chapter 4.2), hoist drives (chapter 4.3) and positioning drives (chapter 4.4). Often, positioning functions like approaching a target position are executed in the process as well. Drives for forming processes (chapter 4.10), main and tool drives (chapter 4.11) and drives for pumps and ventilating fans (chapter 4.12) are controlled in this way. When pumps and ventilating fans are driven in process engineering, they often receive their setpoints from process control systems or process controllers.

These relationships now lead to three basic configurations for the interaction between the automation system and the drives [Ki04]:

- Drives which are not closely coupled are controlled by a sequence control. Positioning functions are performed decentrally as and when required (Fig. 3-69 a).
- Drives are synchronised with each other through a master motion, which is distributed via a digital frequency bus or a communication system. The machine is controlled at higher level by a sequence control (Fig. 3-69 b).
- A central control calculates the setpoints of the drives which appear coordinated and perform a multi-dimensional motion in three-dimensional space (Fig. 3-69 c).

The connection between the drives and the control is often implemented via communication systems, which are covered below.

3.4.6   Communication systems in inverters

Around half of all industrial communication circuits used today is used by inverters. At the same time, more than half of all inverters are connected via a communication system to higher-level automation. On the one hand, the reasons for this lie in the technical and financial advantages of this structure, but on the other hand an inverter already offers many of the requirements for digital communications thanks to its structure (micro-computer, high share of software).

As there are several communication systems and the decision in favour of a specific system generally depends more on the control being used than the inverters, drive manufacturers must support a number of different systems. Communication modules are usually used for this purpose in order to keep inverter diversity to a minimum. It only makes sense to integrate the communication system as a fixed part of the inverter for high-volume applications with tailored solutions or for systems which are used very frequently.

Communication systems for inverters must perform the following three tasks:

- Process data need to be transmitted. As well as the setpoint values and actual values of the motor control and control information for the motion control, this usually also includes a control and status word for the state machine of the drive. The scope of the transmitted process data is adopted to the specific drive application.
- The time response of multiple drives can be synchronised with the aid of the communication system. This is always necessary when multiple drives have to work together to execute a high-precision task. This is the case for synchronised motion (line shaft substitution, process lines, cam plate applications) and for coordinated three-dimensional movements (robots, machine tools).
- The drive-internal parameters can be read and modified via a service data channel. As drives can perform different tasks and therefore have a large number of parameters, this task is very important. The service channel is used during commissioning of the drive, during adaptation of the function to the specific production task of the plant and for diagnostics. Drives can even be incorporated in remote maintenance schemes via the service data channel, with the aid of which the remote diagnosis of plants and automation components can be performed by the machine manufacturer via worldwide communication networks.

A number of widely used systems are listed below to show how these systems can be used for drive tasks.

CAN and CANopen. The CAN system originates from the automotive sector and can be implemented very cost-effectively as it is already integrated in many microcomputers. It offers a transmission rate of up to 1 Mbit/s and a publisher-subscriber principle, in which each node can receive data sent from the other nodes.

The CANopen standard uses this system for automation systems. For this purpose, the services for process data, service data and the network management are defined via communication profiles. The specifications are fairly comprehensive, so when they are used a high level of interoperability is assured.

CANopen also supports the synchronisation of nodes on the communication system. To do this, a high-priority synchronisation object is defined and regularly transmitted, and all of the nodes can synchronise themselves to this object. The CAN system itself causes a jitter in the process, which can however be largely compensated for through software PLL measures at the nodes.

With the aid of the publisher-subscriber-principle it is also possible to set up communication relationships between drives, e.g. for the transmission of the angle of a master axis to several slaves as replacement for a line shaft. The service data services of CANopen in particular are comprehensive and act as a role model for other communication services. Overall, CANopen is highly suitable for integration in automation and drive technology.

CAN and CANopen have a limited bandwidth of a maximum of 1 Mbit/s. This means that, in a 1 ms cycle, a maximum of 7 telegrams can be transmitted. CAN operates according to the CSMA-CA process (Carrier Sense Multiple Access Collision Avoid), in which arbitration of the telegram is performed within a single bit period. This method restricts the cable length and the transmission rate. It is not possible to increase the transmission rate above 1 Mbit/s. CANopen therefore reaches its limits if many nodes, short cycle times or long cable lengths between nodes are required. In these cases an alternative system will need to be used instead.

AS-i. AS-i is a communication system which was developed for the connection of simple sensors (proximity switches, photo electric sensors) and actuators to a PLC control. Here, only a few digital bit values are exchanged for each node. In general, the communication services of AS-i are not sufficient for use with inverters. AS-i is only used with inverters for very simple applications requiring only a few control bits (e.g. start/stop, direction of rotation, two speeds).

PROFIBUS. PROFIBUS is the most commonly used industrial communication system. It is used predominantly for the connection of a PLC to its periphery, including inverters. PROFIBUS is standardised in DIN 19245. The specification of the communication system is not as far-reaching as CANopen (particularly in terms of the service data services).

In addition, PROFIBUS is also available in different versions, whereby PROFIBUS-DP is the most widely used version. It offers a process data exchange. With the variant DP-V1 there is also a standardised service data channel, whereas previously manufacturer-specific solutions were used in some cases for this purpose. The communication relationships run from the master (the control) to the slaves. Initially, PROFIBUS-DP does not offer synchronisation functions or lateral communications.

PROFIBUS-MC addresses additional drive-specific functions above and beyond PROFIBUS-DP. For example, it offers synchronisation and lateral communications. However, PROFIBUS-MC has so far only taken a small share of the overall PROFIBUS market.

INTERBUS. INTERBUS is a communication system which is also designed for the connection of a PLC to its periphery. It is based on a ring system, whereby the send and return lines of the ring are routed in the same cable, thus implementing the wiring in a line topology. Not only does the ring principle make highly efficient use of the communication channel, but all of the nodes also address each other via their position in the ring, as a result of which no address switch is required on the devices. Communication is generally cyclic and does not take place at fixed time intervals. It is also possible to achieve a transmission at equidistant time intervals with INTERBUS. INTERBUS specifies a communication channel via which parameters can be transmitted.

INTERBUS communication modules are available for many inverters, and they are used for communications between the controller and the inverters.

DeviceNet. DeviceNet is another system which is used for the connection of a PLC to its periphery. It is used most commonly in the USA. The transmission physics are based on the CAN system. Services are defined for the configuration of the communication system and for the transmission of parameters.

SERCOS. SERCOS was developed for drive technology from the start. It offers strict synchronisation of nodes, a cyclic transmission of process data and a well-specified service data channel, and it uses optical fibre as its transmission medium in order to attain a high level of interference immunity [KiSch92]. The communication relationships go from the control to the drives, and there are no provisions for lateral communication. SERCOS is used a lot for the connection of numerical controls (motion controllers) to servo inverters. There are comprehensive specifications of the application functions for this purpose. SERCOS is standardised in IEC 61491.

Ethernet-based systems. Since the end of the 1990s, many companies have also started to develop and offer Ethernet-based systems. The reason for this is that Ethernet is by far the most widespread communication system in commercial information technology, so there are plenty of inexpensive and widely available components for it. In addition, it is also possible to establish communications on the basis of TCP/IP services with Ethernet.

The Ethernet standard 100 BaseT uses galvanic insulation with the aid of a transformer between the nodes. In the EMC-laden environment of inverters this is an absolutely essential requirement, and it achieves a high level of interference immunity.

Ethernet was originally not developed for real-time applications in automation engineering. Ethernet works according to the CSMA-CD principle (Carrier Sense Multiple Access Collision Detect), in which the risk of collisions is accepted. However, collisions are an obstacle in relation to real-time capability. This is also the biggest stumbling block for use of this technology in automation engineering. Nonetheless, a number of different systems have now found ways which enable them to attain this real-time capability. The different real-time upgrades for Ethernet are based on an arbiter which uses different methods to assign transmission rights to the nodes in the bus system.

ETHERNET Powerlink. Powerlink uses its own network segment to ensure real-time capability. A managing node manages the data traffic in this segment. In each communication cycle it sends a telegram to synchronise all of the nodes. Afterwards, the individual nodes are addressed via a PollRequest telegram, to which they then respond with a PollResponse telegram. At the end of the cycle there is still an asynchronous time slot in which – after assignment by the managing node – a TCP/IP telegram can be used [Sch03, He03].

Powerlink performs many of the mechanisms and specifications of CANopen. For example, the network is set up with hubs, which means that each node receives the telegrams from all the other nodes. This also makes it possible to set up publisher-subscriber communication relationships similar to CANopen [Mi06].

Powerlink also has comprehensive specifications for network management, which is used to ensure interoperability.

EtherCAT. EtherCAT uses a sum telegram which passes through all nodes starting with the control. Here, each node takes its input data from the telegram and replaces them with its output data. Once the telegram has arrived at the last node, it is sent back to the control on the return channel. This creates a logical ring with a line topology (the return line is included in the same line) [Ro03].

By using the sum telegram and the ring principle, EtherCAT makes best possible use of transmission capacity. Lateral communications need to be implemented via the control (or via the topology), and in terms of its basic structure the system is a single-master system. Synchronisation is included. Service data services from CANopen or SERCOS are used.

PROFINET. Similarly to PROFIBUS, different versions of PROFINET are also available. For automation engineering, PROFINET-IO RT is initially the most important one. Here, communications are implemented via a standard Ethernet network. Cycle times in the region of around 10 ms can be achieved with low jitter (a few ms). This is sufficient for the I/O communications in a PLC, but not for synchronised drives. The communication services are very similar to PROFIBUS.

For applications with synchronised motions and hence demanding real-time requirements there is the variant PROFINET-IO IRT. Here, special circuits are used to set up a network topology in which the switch modules manage an isochronous time slot, in which the real-time communications then take place. This enables the synchronisation of all nodes with a jitter of less than 1 µs and the synchronous exchange of process data with cycle times of less than 1 ms.

The other communication systems described above like SERCOS and DeviceNet have also defined Ethernet-based systems which perform the basic communication services (SERCOS-III, EthernetIP). These are not covered here in any more detail.

Diagnostic interfaces. Whereas the communication systems described so far have the task of providing the connection to the control, the purpose of a diagnostic interface is to provide a connection to a PC. The PC can then write and read parameters. This type of interface is usually only used during commissioning and for servicing purposes.

A number of different interfaces are commonly used for this. They are usually governed by the interfaces installed in the PC:

- In the past, a serial RS-232 connection was often used. Now that this type of connection is no longer provided in many PCs, it is becoming less widespread.
- The USB interface is used. For safe isolation between the PC and the drive it makes sense to use an electrically isolated converter to a serial protocol.
- The Ethernet interface can also be used for diagnostic purposes. Here, no real-time data are transmitted. By using inexpensive switch components, several inverters with integrated Ethernet interfaces can be connected to a PC at the same time for commissioning and service purposes.
- The use of wireless interfaces also makes sense for these purposes, as it allows data to be read from the inverter inside the plant with a mobile operator device.

As well as the communication systems, safety functions are important elements of an automation system. In the following we take a look at which safety functions are implemented directly in inverters today.

3.4.7   Safety functions in inverters

Stefan Witte

The movements generated by a drive in a machine can represent a danger to personnel. Accordingly, shutting down drives is a key part of the safety concept for the machine.

Chapter 2.3.6 explains how the safety concept of a machine needs to be designed, which standards need to be taken into account and how a modern automation concept can be implemented with full integration of the safety engineering. The focus of this chapter lies on the integration of safety functions into the inverter.

Conventionally, the safety engineering for a drive is solved with the aid of additional components, using for example mains contactors or motor contactors as an electromechanical means to interrupt the electric flow of power.

Alternatively, a safe shutdown path can also be implemented in the inverter as the basis for safety functions which are executed by the inverter and which form part of a machine safety concept. In comparison to conventional solutions, there is a large cost-cutting potential in terms of previously required additional components and a significantly improved cooperation between the safety functions and the automation functions of the drive.

Safe shutdown path in the inverter. If the motor is actuated by an inverter, then the supply to the optocouplers, which transmit the trigger signals of the power switches, can also be used for safe interruption of the power flow to the motor. This means that the shutdown is performed on the low-voltage side. This type of shutdown has the advantage that it is power-independent and can be easily implemented using electronic components (Fig. 3-70) [Wi05b].

Redundancy in the shutdown paths is used to ensure that the safety function is still assured in the event of a fault. The safe shutdown paths must already be taken into account during the development of the power unit, as design measures for the avoidance of errors need to be implemented.

The safe shutdown path in the inverter is a type of release cord which is used to bring the drive into a safe state. This safe shutdown path can be used in the following way:

- It is controlled by an external safety logic. All logic operations with the safety sensors are implemented externally. The safe inputs of the inverter only ensure that – even in the event of a fault in the inverter – no motion is generated if the safe inputs are not released.

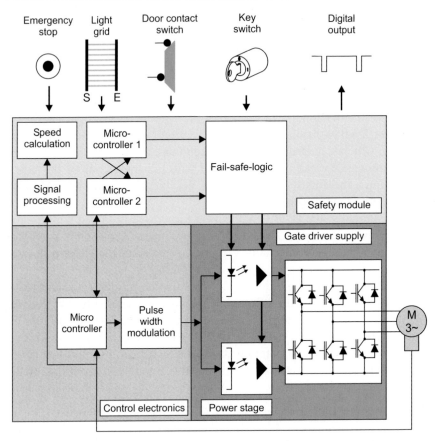

Fig. 3-70. Safety functions in an inverter

- The safe shutdown path is connected to a safe microcomputer system comprising two microcomputers which work redundantly and monitor each other. These then evaluate further inputs – including the encoder – on the motor, perform logic operations and monitoring, and activate the safe shutdown path to the inverter in the event that they detect a deviation from the required behaviour.

Functions. The main purpose of all safety functions is to restrict motion safely on demand or in the event of a fault. In this respect, the most important basic functions are standstill and safely limited speed.

Standstill functions. Figs. 3-71 a and b show two functions which are designed to ensure that the drive is safely at standstill. These functions do not require any measurements or safe evaluation of the speed of the drive. The safe shutdown path is always activated when the motor is no longer allowed to move.

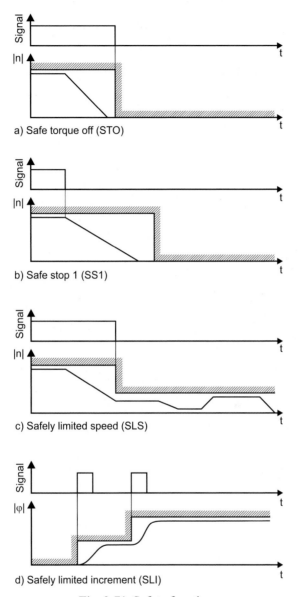

a) Safe torque off (STO)

b) Safe stop 1 (SS1)

c) Safely limited speed (SLS)

d) Safely limited increment (SLI)

Fig. 3-71. Safety functions

- *Safe torque off (STO) (also known as "safe standstill") (Fig. 3-71 a).*
  Here we have the immediate shutdown of the motor according to stop
  category 0 of EN 60204. This function can be implemented on a purely
  hardware basis. The motor stops generating torque after this function
  has been activated. It coasts to a standstill if it is not shut down first. In
  the plant, a mechanically operated brake should for example be used to

ensure that driving loads (e.g. a hoist) cannot perform any impermissible movements.

- *Safe stop 1 (SS1) (also known as "safe shutdown") (Fig. 3-71 b)*. This is the immediate shutdown of the motor along a deceleration ramp and subsequent shutdown of the motor torque according to stop category 1 of EN 60204 (STO).
- *Safe stop 2 / safe operational stop (SS2 / SOS)*. This is the immediate shutdown of the motor according to a deceleration ramp (SS2) with subsequent active holding of zero speed in accordance with stop category 2 of EN 60204 (SOS).

Motion monitoring systems. Safe motion monitoring systems (Figs. 3-71 c and d) require safe evaluation of a encoder. If a defined value is exceeded then the safe shutdown path is activated, which then forces the drive to coast to standstill.

- *Safely limited speed (SLS) (Fig. 3-71 c)*. A configured speed limit must not be exceeded. Several mutually independent speed limits can be implemented for different operating modes or machine functions.
- *Safe maximum speed (SMS)*. A configured maximum speed must not be exceeded.
- *Safe tip mode (STM)*. A movement of the motor in manual operating mode must only be enabled through an enable switch and a further travel command.
- *Safely limited increments (SLI) (Fig. 3-71 d)*. After an enable, an incrementally defined increment must not be exceeded. When the increment is reached the motor must stop automatically.
- *Safe direction (SDI)*. The motor must only be moved in the approved direction of movement.
- *Safely limited position (SLP)*. The motor can only be moved within safely configured absolute positions.
- *Safely limited torque (SLT)*. The maximum torque at the motor shaft is safely limited.

Safe transmission of information. For interaction with further safety components the safety logic can also read and output further signals.

- *Safe input and output signals* . Safe input and output signals should be used in order to safely activate and monitor functions. They operate on two channels or use cyclic pulse patterns to detect fault-free operation.
- *Safe speed monitor (SSM)*. Compliance with a safely limited speed is safely reported.
- *Safe cam (SCA)*. Safe feedback signal that the motor is within a defined range.

- *Safe cascading.* The request for a safety function is forwarded via a cascading ring. In this way, one sensor acts directly on a number of coupled drives.
- *Safe brake control (SBC).* As well as safely controlling the motor brake, this must also detect any faults in the brake. This can be done through cyclic testing of the braking torque. Alternatively, a redundant second brake can be used.
- *Connection to a safety bus.* The transmission of safety information via a communication program places special requirements on the protocol and the evaluation of the received data. The transmission itself on the single-channel communication system is initially non-safe. With the aid of various mechanisms (check sums, cyclic transmission of data and time monitoring, numbering of data packets and checking for compliance with the correct sequence) it is however possible to ensure that faults on the communication channel are detected and that, subsequently, a safe status is initiated via the shutdown path. Many of the communication systems used in automation engineering also have modes and specifications for the transmission of safe data (CANopen Safety, PROFIsafe via PROFIBUS or PROFINET, INTERBUS Safety, AS-i Safety, ETHERNET Powerlink Safety). The use of communication systems which can also transmit safe information is the key element to integrating automation and safety engineering – refer to chapter 2.3.6 for a more detailed description.

Cost-cutting potential. Due to the safety technology integrated in the inverter, a number of circuitry components can be omitted [Gr06]:

- *Mains and motor contactor.* Unless specified otherwise in machine-specific standards, significant cost savings can be achieved particularly in high-power electrical installations by the omission of contactors.
- *Safety switching device.* A safety switching device usually monitors connected passive sensors (e.g. emergency stop buttons, door contacts) and the (re)starting inhibiting circuit. With the aid of delayed and undelayed output contacts, it is possible to bring a connected inverter in controlled fashion to a standstill before the safety switching device switches off the supply to the motor. These functions can be performed by the safety engineering integrated in the inverter. As well as saving on the safety switching device and the associated cables, the interaction between this safety function and the drive control is better coordinated.
- *Second encoder.* Frequently, a monitoring device with connected proximity switches (which are controlled via a gear) is used for speed monitoring. This measure can be omitted, as two-channel status is achieved with just one encoder thanks to the redundant evaluation of the encoder

on the motor and a number of fault detection measures, which is sufficient for category 3 according to EN 954-1.

The omission of the above components also means of course that costs can be saved in the control cabinet and in terms of wiring outlay. It is also easier to make subsequent changes to the safety functionality or to upgrade it later on.

Connection of the safety engineering to the inverter control. In order to be able to implement safety functions which extend beyond safe torque off, there must be an interface between the safety functionality and the software for the motion control. The inverter control, which initially is non-safe, then performs the functions required by the safety function.

For example, the demand for a safety function such as the safely limited speed from the safety engineering is implemented via activation of a sensor, e.g. a door contact. The safety circuit forwards the demand for the function to the motion control, where the setpoint generation and speed limitation take place. The safety engineering only intervenes if the limit values (braking time, actual speed) are exceeded, at which point it switches off the flow of energy to the motor via the safe shutdown path. The safe final state is thus always the standstill of the motor.

Implementation of a safe communication system. Safe and non-safe data communications from the inverter only take place via a bus connection. The safe part of the data telegram is transmitted to the safety logic of the inverter, where it is redundantly evaluated.

Redundant parameter sets. In order to implement consistent redundancy right the way through from the sensor via the processing to the actuator, the parameters required for the safety engineering (e.g. speed limits, deceleration ramps) must also be available on two channels. This is done by saving two identical parameter sets in different memory locations. These parameter sets are checked for conformity.

Safe parameter setting. Special software is needed in order to generate a safety-oriented parameter set. The software must contain components in order to safely generate, transmit and save the parameters and subsequently verify that they are free of faults. Ideally, the software should be integrated in the operating software of the inverter.

Diagnostics. Data from the drive-based safety can be displayed for diagnostic purposes via existing communication or diagnostic interfaces. This is a clear step forward in comparison to conventional safety engineering, which generally does not offer this function.

3.4.8   Electromagnetic compatibility

Rolf Sievers

The operation of electronic devices can be impaired as a result of electro-magnetic interference caused by other devices. Due to the switching in-volved in their operation, inverters generate this type of interference. In order to design the interplay between the electronic components in a ma-chine or plant in a way which ensures the highest possible functional safety and reliability, aspects relating to electromagnetic compatibility (EMC) need to be taken into account. These are discussed below.

The term "electromagnetic compatibility" is used to describe the follow-ing: an electric device is only permitted to emit interference up to a maxi-mum limit which is defined in standards and is still acceptable for other devices in the vicinity. At the same time, the devices must also have ade-quate immunity to interference in this environment. The requirements in terms of immunity to interference are also specified in standards.

As far back as 1892 a first law was passed in Germany relating to tele-phone networks. Due to the continuous proliferation of inverter drives, the EMC standards and laws are becoming increasingly more important. To-day, reliable drive engineering can only be developed with full provision for an EMC-optimised concept.

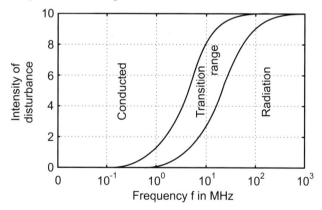

Fig. 3-72. Coupling paths for interference

Interference ranges. Due to the way in which they work, inverters are strong sources of interference. The types of interference effect caused by them can be classified as follows:

- Due to the non-sinusoidal mains currents, the mains rectifier generates mains current harmonics in the frequency range of up to around 2.5 kHz.

Table 3-14. Overview of interference ranges of frequency inverters

	Mains current harmonics	Interference emission	
	Conducted	Conducted	Radiated
Frequency range	0–2.5 kHz	150 kHz–30 MHz	30 MHz–1,000 MHz
Cause	Non-sinusoidal mains current	High-speed switching of output stages and switched-mode power supplies. Their electrical or capacitive connection results in the injection of interference at the mains input	The switching edges in the power output stages with a high du/dt contain high-frequency harmonics which emit interference via the motor cables which act as aerials
Effect	Increased effective mains current, additional heating of mains supply transformers	Interference injection on the supply side to other consumers connected to the same mains network (galvanic connection)	Radiated interference from inverter and motor cables to other nearby high-resistance control signal cables
Counter-measures	Mains choke, active or passive harmonic filter	RFI filter on the supply side (internal/external), filter on the motor side, isolating transformer on the supply side	Shielding of inverter and motor cable, no shielding interruptions

- The switching edges of the inverter with a typical slope of 5 kV/µs are a high-energy source of interference. Together with the coupling capacitances of the motor cable they generate high, needle-shaped charge/discharge currents, which – if the cables are not correctly routed – can cause significant interference voltages.
- Other high-frequency interference voltages are generated by the switched-mode power supply and the microcontrollers of the digital signal processing.

If the frequency of the interference signal is low, then spread of the electromagnetic interference will only be conducted. With increasing frequencies and thus shorter wavelengths, coupling via electromagnetic waves increases. As well as the frequency, it also depends on the intensity of the interference whether the coupling is conducted or radiated (Fig. 3-72).

This means that, depending on the frequency, three interference ranges can be defined for inverters. We will have a look at these interference ranges in more detail below (Table 3-14).

Mains current harmonics. If the mains supply is provided via an uncontrolled diode rectifier (chapter 3.4.1) then mains current harmonics will be generated. These increase the r.m.s. value of the mains current and cause losses in the transformer of the mains supply. Since the start of the 1990s, greater importance has been attributed to these negative effects by the electricity supply companies.

Uncontrolled mains rectifiers generate harmonics due to their non-sinusoidal currents. The frequencies of these harmonics correspond to the integer multiples of the mains frequency which are not divisible by two or three. In the case of three-phase rectifiers, these are the 5th, 7th, 11th, 13th... harmonics, corresponding to 250 Hz, 350 Hz, 550 Hz, 650 Hz ... at a mains frequency of 50 Hz.

The frequencies of the harmonics on single-phase inverters are all uneven integer multiples of the mains frequency. Here, the 3rd harmonic is particularly important, as on a number of inverters it is not compensated for in the neutral conductor, but instead is added arithmetically. This may in some cases result in overloading of the neutral conductor.

Measures for reducing the harmonic current. The following measures can be used to reduce the harmonic currents:

- Use of a mains choke.
- Harmonic filter (active or passive).
- Use of a controlled pulse width modulation inverter (chapter 3.4.1) for the mains supply. This will then generate virtually sinusoidal currents and therefore only very small quantities of mains harmonics.
- 12-pulse supply. Here, with the aid of a transformer which is connected in a delta layout on the primary side and a star layout with centre tap on the secondary side, a 150° phase-shifted voltage system is generated which supplies the inverter via a second rectifier.

Table 3-15. The magnitude of the 5th harmonic and the r.m.s. mains current of a 4 kW inverter after different measures have been applied

	5th harmonic	R.m.s. mains current
Without any measures	54%	13.5 A_{eff}
Mains choke	33%	9.1 A_{eff}
Harmonic filter	4%	8.7 A_{eff}
12-pulse supply	12%	9.2 A_{eff}

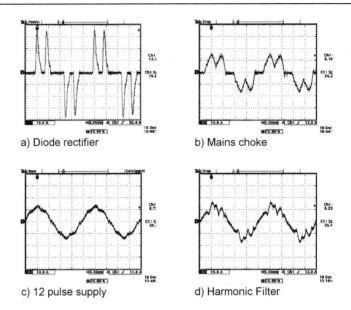

a) Diode rectifier

b) Mains choke

c) 12 pulse supply

d) Harmonic Filter

Fig. 3-73. Mains current forms after various measures have been applied to reduce the harmonic current

Table 3-16. Amplitudes of the harmonic current with different mains choke

Mains current harmonic		% amplitude		
Harmonic number	Frequency in Hz	Without mains choke	With mains choke $uk = 4\%$	With mains choke $uk = 6\%$
1 (fundamental wave)	50	100	100	100
5	250	64	30	26
7	350	39	9.0	7.6
11	550	1,6	5.7	4.8

For an inverter with a three-phase supply, the 5th harmonic is dominant. The magnitude of this harmonic is compared for the different measures as a way of evaluating their effectiveness (Table 3-15).

The current forms of the different measures are shown in Fig. 3-73. The most commonly used measure here is the use of a mains choke. The disadvantage of this choice is the reduction of the maximum output voltage of the inverter as a result of the voltage drop at the choke. Table 3-16 shows the reduction of the harmonics for different chokes.

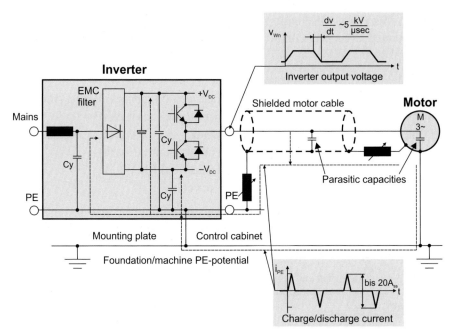

Fig. 3-74. Paths on which the charge/discharge currents spread

In comparison to chokes, higher effort is required for harmonic filters. Because of the design changes to the inverter (two rectifiers) and the required transformer, a 12-pulse supply is only used in the high power range above 1 MW.

Interference emission. Due to the inverter output voltage, which is modulated with the switching frequency (4 to 16 kHz), an interference current flows across the parasitic coupling capacitances of the shielded motor cable and the motor windings to the PE potential during every switching edge [HoKi01]?

The interference currents are charge/discharge currents from the parasitic capacitances which are caused by the rate of rise of voltage of up to 5 kV/µs of the individual switching edges of the hard-switching inverter. In order to prevent these charge/discharge currents, which can adopt values of up to 20 A_{pp} and flow across all of the PE connections between the motor and the inverter, from having a disturbing influence, it is important for there to be a current path which is defined and carries the largest part of this current due to its impedance. This is the shield of the motor cable, which needs to be connected both to the inverter housing as well as to the motor housing with a connection which has a large surface area and therefore has a low impedance. Within the inverter the interference current is fed back to the DC link via Y-capacitors (Fig. 3-74).

Due to this charge/discharge current a high-frequency interference voltage is generated, which is conducted on the mains supply line. Its magnitude depends on the following influencing variables:

- the length of the motor cable,
- the capacitance of the motor cable (between the phases and the shield),
- the number of parallel motor cables (multiple motor drives),
- the amplitude of the switching frequency,
- the edge slope of the inverter power output stages (dv/dt),
- additional filters on the motor side (sinusoidal filters),
- the quality of the shield contacts (motor cable).

Most of these influencing variables are not defined by the design of the inverter, but instead by the specific operating conditions (e.g. the length of the motor cable and the switching frequency).

Table 3-17. Categories and environments for the interference levels

Category	c1	c2	c3	c4
	$V_N < 1$ kV	$V_N < 1$ kV	$V_N < 1$ kV	$V_N > 1$ kV
Environment	Environment 1 (living areas, business areas and commercial areas)		Environment 2 (industrial)	

Standards. The EMC product standard for inverters IEC/EN 61800-3 defines the permissible levels for the limit values of conducted and radiated interference emissions in the frequency range from 150 kHz to 1 GHz. These levels depend on the category and environment in which the inverter is used (Table 3-17).

For conducted emissions, corresponding limit values apply for the radio interference voltage according to the assigned category (Figs. 3-75 and 3-76). Here, there are always two limiting characteristics for the average and maximum values (quasi peak) of the measurement.

Measures for reducing conducted emission. On an inverter, a substantial part of the cost is accounted for by measures required for compliance with conducted emissions at the mains connection. RFI filters are used for this purpose. These can be either integrated in the inverter or provided as a separate component, which is combined e.g. in the form of a footprint filter with the inverter. Combinations of internal and external filters are also common.

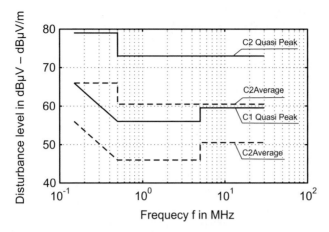

Fig. 3-75. Category C1 and C2 interference emissions

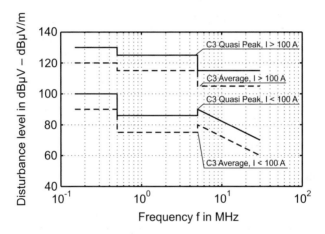

Fig. 3-76. Category C3 interference emissions with I < 100 A and I > 100 A

Basic circuits for RFI filters. RFI filters for reducing conducted emissions are designed for the frequency range from 150 kHz to 30 MHz. They always use an LC low-pass circuit, whereby the parallel capacitances are connected both between the voltage-carrying conductors (type X) and between the external conductors and PE (type Y) (Fig. 3-77 and 3-78).
As well as the filter circuits shown here, the use of a current-compensated choke in the DC link of the inverter is also common.

Due to the large number of possible influences which can have an effect on the magnitude of the radio interference level, the conditions under which inverters satisfy the limit values assured by the manufacturer must be clearly defined. This includes e.g. the type and maximum length of the motor cable.

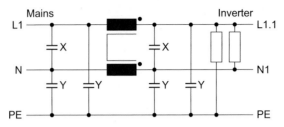

Fig. 3-77. Basic circuit for a single-phase filter

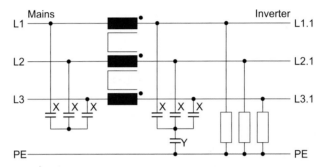

Fig. 3-78. Basic circuit for a three-phase filter

Radiated interference. As, in contrast to conducted emissions, radiated emission occurs in a much higher frequency range (10 to 1,000 MHz), suitable shielding measures are the only way to ensure that defined limit values are met.

The measures required for compliance with the required interference level (Fig. 3-79) according to the category can be divided in to measures which are implemented on the device side (i.e. the responsibility lies with the manufacturer) and measures which are implemented on the installation and wiring side (i.e. the responsibility lies with the user and operator).

Measures on the device side. The effort for device manufacturers in order to sufficiently limit the radiated interference of the inverter is significantly less than the filter measures for the conducted emissions.

The radiated interference on the device side is almost entirely down to the fast-switching transistors in the inverter (including the brake transistor) and in the switched-mode power supply.

Through corresponding design of the layout of the printed circuit boards (shielding surfaces) and the housing design (shielding sheets or metal housings), a major contribution can already be made to reducing the radiated emissions on the inverter side.

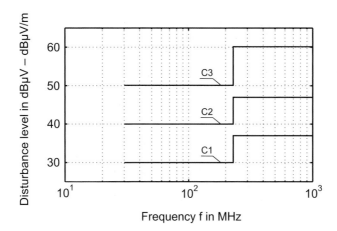

Fig. 3-79. Limits for the radiated emissions in categories C1, C2 and C3

Measures on the user side. The bulk of the shielding work required in order to limit radiated emissions is associated with EMC-compliant wiring. As these measures fall under the responsibility of the user, comprehensive and detailed documentation is needed from the inverter manufacturer for EMC-compliant wiring.

Shielding through installation in a control cabinet. In the case of inverters which are installed in control cabinets, it may be necessary – unlike motor inverters – to incorporate a metal, earthed control cabinet in the measures to limit the interference radiated from the inverter, as this provides an additional level of shielding.

It should be noted here that the control cabinet base plate should not be painted, should offer good conductivity and should have an EMC-compliant connection with a large surface area to the cabinet.

Shielded, EMC-compliant wiring and connection systems. In order to achieve optimum shielding results in the control cabinet and machine wiring, there is a long list of points which needs to be taken into account. The design of the motor cable has the highest importance here. The following individual aspects should be noted:

- Motor cables with a shield made of tinned or nickel-plated copper braid should be used, whereby the overlap rate should be at least 70%. This material offers good long-term stability for the shielding effect.
- Shield connections on the inverter and motor side must only be made using the recommended connection systems, such as shield clamps and shield screw fixtures.
- Avoid interruptions in the motor cable through terminals or switches.

- Additional supply lines for other systems – e.g. brake control – should only be carried inside the motor cable with an additional, separate shielding.
- Unshielded cable ends of the motor cable should be kept as short as possible (approx. 40 mm).
- Only use conductive control cabinet base plates (e.g. zinc-coated sheet steel).

Inverter manufacturers also offer complete concepts and solutions, not only for the shield connection on motor cables, but also for the shield connection on control cables, so that these connections can be made very efficiently and reliably. Users should implement these concepts in order to ensure EMC-compliant wiring.

Bearing currents. The effect of bearing currents is also associated with the hard-switching method of operation of inverters, which itself is responsible for the EMC interference. The bearings of the motor are lubricated with grease and therefore initially non-conductive. However, the high-frequency charge/discharge currents, which are created due to the steep switching edges of the inverter, can also create currents due to capacitive coupling within the motor. These currents then flow through the bearings of the motor. This can result in a arcing which gradually destroys the bearing grease. As a consequence, the bearing will ultimately fail.

The potential risk due to bearing currents depends on the motor frame size, the wiring and the inverter. No destructive bearing currents are observed on motors below 5 kW. A suitable earthing concept usually reduces bearing currents to a potential which does not have a prematurely destructive effect. If necessary, it may be necessary to insulate the bearings in the motor so that the flow of current through the bearings is prevented or at least reduced to a safe level.

In this chapter we have taken a detailed look at inverters. Together with a motor, it builds a controllable drive with large setting ranges for speed and torque. However, the most favourable operating range of a motor from a commercial point of view is often found at higher speeds than the speeds required by the work process. Gearboxes are used here in order to find a meaningful compromise. The use of direct drives, which is also technically possible, is significantly more expensive in many applications than the use of a gearbox, and is therefore not cost-effective. The way in which the gearboxes work is described in the following.

3.5 Gearboxes

Ralf-Torsten Guhl

A gearbox transmits or transforms mechanical energy to match the requirements of the application in the machine. This matching is necessary, as the technically and economically optimum operating point of motors is generally at higher speeds than the application requires.

3.5.1 Application areas and designs

The majority of applications employ toothed gearing systems as torque/speed transformers, as this represents the optimum solution for most mounting conditions and performance requirements. Alongside excellent reliability and safety, toothed gearing systems occupy little space and offer good efficiency.

Gears are today manufactured across a vast size range. These include tiny gears of around 1 mm for applications in fields such as medical engineering, and stretch to gears with an outer diameter in excess of 30 meters for applications such as mining.

The automotive industry is the largest producer of gearboxes in Europe. The other gearboxes are predominantly units for stationary applications with constant gear ratio, which we will examine in more detail in the following. Multi-speed gearboxes, which are not used in engineering applications very often, and mechanical variable speed drives, which have now generally been replaced by electronically controlled drives, will not be covered here. Applications in production technology and logistics require torque in the range of 100 to 10,000 Nm. Any torque higher than this is generally the domain of process engineering. The large-scale gearboxes required for this are also not covered in this book.

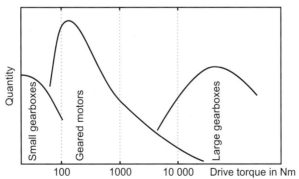

Fig. 3-80. Gearbox market based on quantities used

Breaking down the gearboxes based on the quantities used shows that the greatest requirements are in the range of 100 to 300 Nm (Fig. 3-80). Geared motors, with a direct combination of gearbox and motor but no coupling, are generally used here. With large drives of over 10,000 Nm, the gearboxes are produced separately and then connected to the motor using a coupling. Drives with output torque lower than 100 Nm are often designed for special applications.

In the classic geared motor sector, on the other hand, the appropriate drive is generally configured from the available combinations of the manufacturer to match a specific application. The correct ratio for each application can be selected from an extremely large range here, since various motor frame sizes can be combined with the respective gearbox in each case. This leads to a large amount of potential variants, which is why geared motors are typically assembled to order. However, to ensure fast delivery for customers, geared motor ranges are kept modular, in some cases with pre-assembled assemblies being used.

In addition, gearboxes can be classified according to the characteristics of the type of toothing used and the internal design. You can see a breakdown of the sales figures of stationary toothed gearing systems in Table 3-18.

Table 3-18. Market shares of the various gearbox types (source: VDMA)

Gearbox type	Proportion of sales
Planetary gearboxes	44%
Helical gearboxes	30%
Bevel gearboxes	17%
Worm gearboxes	9%

If we take into account the significantly lower revenue per unit of helical gearboxes compared to planetary gearboxes, helical gearboxes have the largest share of units sold.

The proportion of bevel gearboxes among right-angle gearboxes has grown significantly in the last few years compared to worm gearboxes. And this trend is likely to continue, in particular due to the better efficiency offered by these units.

For a special application case, the key question is not which gearbox type to use, but rather how well the drive (motor and gearbox) can be integrated in the system – i.e. what design is best suited to the application. Fig. 3-81 shows the basic options.

Type of gearbox	Helical gearbox (Solid shaft and shaft mounted)	Bevel gearbox and worm gearbox	Planetary gearbox
Type of shaft			
Solid shaft	 coaxial	 angled	 coaxial
Hollow shaft	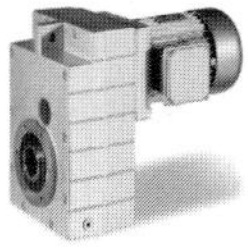 axial	 angled	end-to-end hollow shaft not possible

Fig. 3-81. Gearbox designs

A gearbox design with hollow shaft allows the drive to be integrated into the machine in a very compact way. Hollow shaft gearboxes can be fitted directly onto the machine's drive shaft and the torque can, for example, be transmitted using a friction-type connection. Shrink disc connections are particularly well suited connections. With these connections, the torque is transmitted via frictional resistance through elastic deformation of the hollow shaft. The principle of shrink disc connections is detailed in chapter 3.6.6. The integration of the drive into the machine must be set up in such a way that the reaction torque of the drive can be taken up by the chassis of the machine. Torque plates, which can be combined with the gearbox housing, are available for precisely this application.

If the gearbox type only permits a solid output shaft, a coupling is needed to connect the drive shaft to the application. Other applications, such as roller conveyor drives, are often designed with a gear ratio downstream of the gearbox, for example in the form of a belt or chain drive. A belt pulley or sprocket is fitted directly onto the driven shaft of the gearbox in this case.

It is clear that we must examine the total costs of all individual components throughout the entire drive train if we wish to optimise the costs of a drive. Hollow shaft gearboxes enjoy a share of approximately 30 to 35%, despite the fact that these gearboxes are significantly more expensive than gearboxes with a solid shaft.

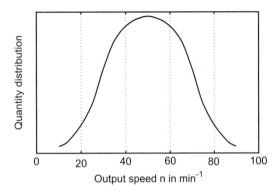

Fig. 3-82. Output speed distribution for torque of 100 to 300 Nm

Overall, a gearbox performs the following functions, which must be taken into account when making a selection:

- converting the speed and torque of the motor for the application,
- carrying radial and axial forces on the driven shaft which are generated either by the work process or other drive elements,
- supporting the reaction torque,
- matching with the motor to create the geared motor unit.

Gearbox ratio. Since motors tend to operate more economically at high speeds, yet applications generally do not need such high speeds, a gearbox ratio is vital in most cases. Where an application requires machines to be driven directly, i.e. without a gearbox, direct drives are normally used. However, the total costs for these solutions tend to be higher than for a geared motor and, as such, they are only used in special cases.

Generally the ratio of a gearbox describes the ratio between the input speed and the output speed.

An analysis of output speeds used in the torque range from 100 to 300 Nm shows a peak at approximately 50 rpm (Fig. 3-82). We can observe a general trend toward higher speeds and dynamics in the industry here. Since low masses are generally moved faster and large masses more slowly, in future we are likely to see applications with low output torque move toward higher output speeds and applications with high output torque move toward lower output speeds.

With four-pole three-phase asynchronous motors (the most common in the industry) at a rated speed of approximately 1,400 rpm, the most commonly used output speed of 50 rpm requires a gearbox ratio of 28. Assuming use of a conveying belt with driven rollers with a diameter of 200 mm, this produces a speed of around 0.5 m/s for the transported products.

Fig. 3-83. Two-stage helical gearbox

3.5.2   Helical gearboxes

Helical gearboxes are characterised by the drive shaft and driven shaft being aligned in parallel (Fig. 3-83). Alongside the housing, the shafts and bearings, the lubricant and the sealing elements, the spur gears are the main components in the gearbox.

The evolvent tooth profile is almost always used as the tooth shape of the spur gears. The geometry of the evolvents comes about from rolling a straight line on a base circle. The evolvent tooth profile has the following advantages:

- not sensitive to deviations in the distance between axes,
- smooth movement transmission, i.e. low-vibration running,
- extremely precisely calculable due to well established design tools based on comprehensive practical experience and testing,
- ease of manufacture, which leads to low costs.

However, these advantages are opposed by the following disadvantages of the evolvent tooth profile:

- In the case of external gearing, convex edge parts run against convex edge parts, which limits the load capacity.
- When using a small number of teeth, the teeth are undercut due to the manufacturing process, i.e. weakened.

The positive-fit torque transmission from a driven wheel (pinion) to the output gear generally takes place via several pairs of teeth that are meshed simultaneously. In Fig. 3-84 you can see the meshing of one pair of teeth.

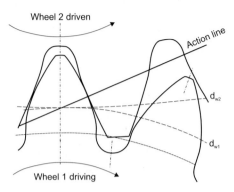

Fig. 3-84. Meshing of two teeth with evolvent tooth profile

One tooth each of the driving wheel and the driven wheel contact each other at a point on the mesh line. This then resembles straight line contact across the width of the tooth when viewed from above. Pure rolling of the teeth only takes place at the point when the pitch circles (d_{w1} and d_{w2}) form an intersecting point with the mesh line. In the areas before and after this, a relative movement takes place between the tooth flanks.

The teeth of helical gears are generally designed for infinite service life, whereby manufacturers must take the following criteria into account.

Tooth root strength. If the permissible loads are exceeded, the teeth tend to break off at their base. The tension that occurs at the root of the teeth primarily depends on the length of the tooth, i.e. the lever arm and shape of the tooth in the root area.

Tooth flank load capacity. If the maximum tolerable pressure of tooth flanks in mesh is exceeded, parts of the tooth flank break off, leaving behind recesses that resemble pitting – hence the term tooth flank pitting. Flank damage is not permissible if the number and size of the pitting occurrences increases with runtime and then leads to failure of the gearbox.

Scuffing load capacity and wear load capacity. Scuffing of the tooth system describes situations when tooth flanks are briefly and repeatedly welded together and then separated again as a result of the lubricant film failing. This damage can be avoided by selecting a corresponding lubricant with high scuffing load capacity and designing the tooth system with low slip.

Wear occurs in the form of abrasion on the tooth flanks when slip takes place with mixed or dry friction.

Lubricant. The lubricant therefore plays an important part in the gearbox. Up to a circumferential speed of the gears of approximately 15 m/s, oil bath lubrication is used. With this kind of lubrication, at least one gear is dipped in oil and the teeth are then wetted with oil. When the arched tooth flanks glide on top of one another, this forms a wedge-shaped gap, into which the oil is pressed. This enables hydrodynamic lubrication with fluid friction. At the same time, the oil dissipates the high temperatures in the mesh area and also allows heat exchange with the gearbox housing.

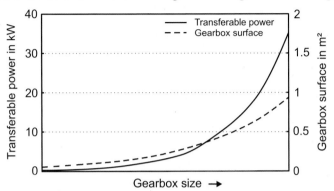

Fig. 3-85. Transferable power and the surface of the housing on helical gearboxes

Power loss. When calculating the entire power loss, which must be dissipated as heat, alongside toothing losses it is also important to take into account losses in the bearings and the shaft sealing ring as well the churning loss which occurs when teeth are dipped into the oil bath. On small gearboxes, heat dissipation via the gearbox housing is often relatively unproblematic. The steady-state temperature that establishes itself as equilibrium between the generated and dissipated heat is in a non-critical range here. However, with high output torque and high output speeds, the steady-state temperature can rise above the permissible temperature range of the lubricant or the sealing elements. In this case, the transferable power of the gearbox must be limited for thermal reasons. In Fig. 3-85 you can see the size of the housing surface and also the transferable power for a gearbox ratio of i = 25. The figure clearly shows that the transferable power of a gearbox rises more sharply as gearbox size increases, meaning that the ratio of power loss that occurs and heat dissipation through the gearbox housing become more critical factors. Once power output exceeds 50 kW, gearboxes are generally fitted with an active cooling system to dissipate the power loss.

3.5.3   Planetary gearboxes

Alongside their high power density, planetary or epicyclic gearboxes excel through their low moment of inertia and backlash. They are therefore particularly well suited for highly dynamic drives. When combined with servo motors, very compact drives can be created. With their weight advantage, planetary gearboxes are also well established for extremely high output torque and for mobile applications.

The most common form is standard epicyclic gearboxes (Fig. 3-86). The internal geared wheel is permanently attached to the housing here. The driving central or sun wheel drives the revolving planets, which run against the internal geared wheel and thereby drive the pinion cage. The driven shaft is permanently attached to the pinion cage here.

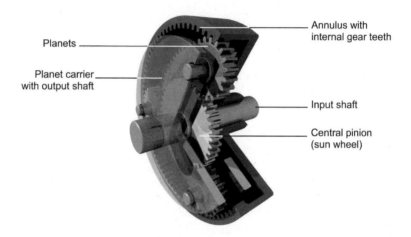

Planets _____

Planet carrier_____
with output shaft

Annulus with
internal gear teeth

Input shaft

Central pinion
(sun wheel)

Fig. 3-86. Standard epicyclic gearboxes

With this design, the maximum ratio that can be achieved in a single stage is approximately i = 12. Larger ratios are then achieved by adding further downstream stages.

A prerequisite for splitting power is ensuring that the load distribution on the planets is as even as possible. When using three planets, an optimum distribution is given, but the use of four or five planets is also common.

On planetary gearboxes, helical gearing is increasingly being used. Alongside the increased load capacity, the idea here is also to further reduce noise emissions. However, the additional axial forces require special design measures for the planet bearings.

3.5.4 Bevel gearboxes

Bevel gearboxes are used to transmit and convert torque and speed between axes that intersect and cross one another. Fitting bevel gears is much more complex than fitting spur gears, since the exact axial position must be secured for the parts of the tooth system. This is the only way to achieve the correct noise behaviour and reliable transmission of the torque.

Fig. 3-87. Hypoid wheel set with curved profile

Bevel gear sets typically have a maximum ratio of i = 6. Many bevel gear-boxes are therefore combinations with spur gear stages. And bevel gear-boxes are also used as pre-stages for planetary gearboxes.

Alongside straight-toothed and helical-toothed bevel gears for applications with low requirements in terms of carrying capacity and smooth running, angular-toothed or spiral-toothed bevel gears are also used. The advantage of higher load capacity is based on the fact that convex flanks mesh with concave flanks.

If the driving and driven shaft has an offset a_V, this is referred to as a hypoid wheel set (Fig. 3-87). The advantage of higher load capacity and better slip must be offset against the disadvantage slightly poorer efficiency with this type of tooth system.

The history of bevel gear toothing has been shaped by the development of the manufacturing possibilities and machine tools used. A curved tooth shape whose tooth thickness and tooth height taper to form an apex is the most commonly produced form today. Another tooth system is characterised by an evolvent tooth profile across the width of the teeth, whereby the height of the teeth remains constant.

A large proportion of all bevel gear toothing systems are today produced in the classic manufacturing sequence of milling, hardening and then lapping. During lapping, the teeth run against each other on dedicated machines, whereby the use of special pastes and specific alignment of the machine axes ensure that the tooth flanks suffer minimal material wear. Driven by the automotive industry, the trend here is moving toward finishing by polishing.

Fig. 3-88. Cylindrical worm gearbox (globoid wheel and cylindrical worm)

3.5.5   Worm gearboxes

On worm gearboxes, the axes intersect at an axis spacing a, in contrast to hypoid bevel gearboxes, where the axes intersect with an offset a_V. In Fig. 3-88 you can see the most commonly used form of gearbox, the cylindrical worm gearbox. Here, a globoid wheel is used with a cylindrical worm, since the opposite configuration with a globoid worm is expensive and is only used for high performance gearboxes.

Alongside rolling slip of the tooth flanks, worm gearboxes also have longitudinal slipping. At average circumferential speeds, a polished steel worm is combined with a high-load bronze worm wheel. The worm gear is subject to heavy and continuous wear, which must be taken into account when using worm gearboxes.

The efficiency of a worm gearbox is based on the ratio and drops sharply as the ratio increases (Fig. 3-89). At the same time, wear to the worm also increases. Due to the combination of materials, slightly higher torque is required at start-up for a short time. This is why we differentiate between the start-up efficiency and the efficiency in operation on worm gearboxes. Depending on the ratio, the start-up efficiency can be some 20 to 30% below the efficiency in operation.

Worm gearboxes can implement ratios up to 50, in some cases even higher, in a single gearbox stage. If the efficiency is below 50%, which is possible with high ratios, self-locking kicks in when reversing the direction of force. This can be used positively for several application cases (for example swarf conveyor), as it allows the system to work without a holding brake.

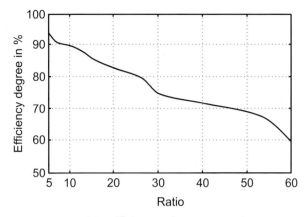

Fig. 3-89. Efficiency of a worm gearbox

Worm gearboxes, and in particular single-stage designs, can often be identified by ribs on the gearbox housing. These ribs offer the housing a bigger surface area to improve heat dissipation.

3.5.6   Combining gearboxes with motors

Combining gearboxes with motors creates geared motors. These geared motors can be created by connecting the motor to the gearbox using a coupling. However, it is much more advantageous to integrate the motor and gearbox directly (see Fig. 3-90).

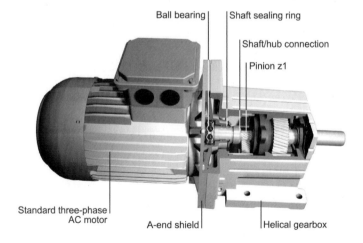

Fig. 3-90. Direct mounting of the motor to the gearbox

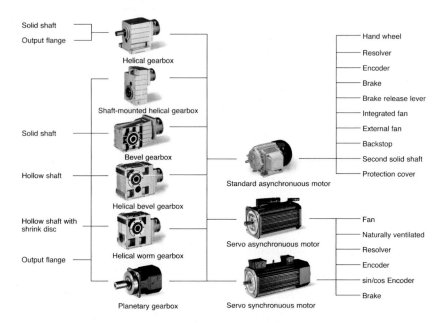

Solid shaft
Output flange

Helical gearbox

Shaft-mounted helical gearbox

Solid shaft

Bevel gearbox

Hollow shaft

Helical bevel gearbox

Hollow shaft with shrink disc

Output flange

Helical worm gearbox

Planetary gearbox

Standard asynchronuous motor

Servo asynchronuous motor

Servo synchronuous motor

Hand wheel
Resolver
Encoder
Brake
Brake release lever
Integrated fan
External fan
Backstop
Second solid shaft
Protection cover

Fan
Naturally ventilated
Resolver
Encoder
sin/cos Encoder
Brake

Fig. 3-91. Construction kits made up of gearboxes and motors with the options

Here, the pinion z1 of the first gearbox stage is connected to the shaft of the motor via a suitable shaft/hub connection. This eliminates the need for bearing mounting the input shaft of the gearbox. If a friction-type connection is selected for this connection, this ensures a high degree of reliability.

In the integrated system, the A-end bearing shield of the motor forms the interface to the gearbox.

Geared motors cannot only be formed together with standard three-phase AC motors, but also with servo motors. This then creates geared servo motors, which are particularly well suited to dynamic applications.

The various gearbox types can be combined with different motor types, each of which has their own options (Fig. 3-91):

- different types of shafts, some of which use connection elements,
- flanges at the output end which can also serve as torque plates,
- dedicated fan or other blower for the motors,
- handwheel,
- backstop,
- motor brakes, expandable to include manual release lever,
- encoders.

The drive train does not end with the gearbox. It also contains further drive elements which continue the movement up to the operating process.

3.6   Drive elements

Olaf Götz

Drive elements are the group of machine elements in the drive train that connect the main components to one another and to the work machine. They perform important tasks, such as transmission of torque and movements, interrupting the drive train if necessary, holding rotary parts, compensating shaft misalignment or transforming rotary movements to linear movements. Drive elements include:

- clutches,
- couplings,
- shaft/hub connections,
- bearings,
- traction drives such as belt, toothed belt and chain drives,
- linear transmission components,
- non-linear transmission components,
- guide systems,
- electromechanical brakes, which were described in chapter 3.3.5.

This chapter only deals roughly with the design or selection of elements. The aim is to describe how they work, their properties and their influence on the active elements of the drive system (inverters, motors, gearboxes). The drive elements must be correctly dimensioned (service life is particularly important) and fitted to ensure that the machine does not get damaged or even fail. You can find further information on many drive elements in [Dub95].

Torsional stiffness and backlash. With most drive elements, the torque τ being transmitted and the stiffness of the elements themselves creates torsion. The stiffness c_φ, sometimes also referred to as spring constant, is given in Nm per angular minute (arcmin). The resulting torsional angle $\varphi_{Torsion}$ is determined using the following equation.

$$\varphi_{Torsion} = \frac{\tau}{c_\varphi} \tag{3.75}$$

The same applies to linear forces (tensile, pressure, bending):

$$s_{Displacement} = \frac{F}{c} \tag{3.76}$$

The reciprocal value of the stiffness $1/c$ is the elasticity.

The total stiffness of n drive elements arranged in series is calculated from the reciprocal total of the individual elasticities. The more drive elements are in place, the "softer" the machine is.

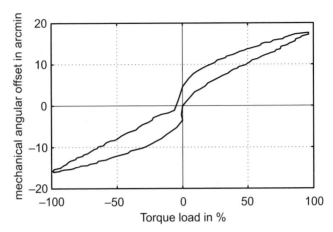

Fig. 3-92. Total mechanical angular misalignment through backlash and torsional stiffness (basic concept)

$$\frac{1}{c_{ges}} = \sum_{i=1}^{n} \frac{1}{c_i} \qquad (3.77)$$

Some drive elements have backlash. This is the angle that is travelled when the direction of rotation is changed. This is easily recognisable if we look at the example of toothed gearing. Within this angle, the tooth flanks are not in contact, meaning that the drive train is interrupted and no torque/rotation is being transmitted.

The entire mechanical angular misalignment results from the sum of the backlash $\Delta\varphi_{Backlash}$ and the angle of rotation, which changes as a function of the load (Fig. 3-92).

$$\Delta\varphi_{mech} = \Delta\varphi_{Backlash} + \frac{\tau}{c} \qquad (3.78)$$

From a control perspective, the driving elements play an important part due to possible elasticities or backlash. The mechanisms should be as rigid as possible so that control processes do not become unstable and also to ensure that the accuracy requirements are met.

Alongside stiffness, the physical location of the components, their size and the system damping also have an influence on the natural vibration behaviour of the machine. It is vital to prevent the mechanical resonant frequency from being excited permanently during operation, as this can lead to oscillation amplitudes, which in turn can damage the machine.

[Dr01] details the vibration response of drive systems.

3.6.1 Clutches

The task of clutches is to interrupt or connect elements of the drive train during operation using a trigger signal. They are broken down into groups based on the way they transmit torque and are classified as mechanical, hydraulic or electromagnetic clutches.

Electromagnetic clutches. Due to their ease of control, electromagnetically operated clutches are predominantly used in mechanical engineering applications. The torque is transmitted without backlash through the use of friction or positive fits, similarly to electromagnetic brakes (chapter 3.3.5), with the exception that both the drive and the output can move freely.

The method of functioning generally follows the open circuit principle. When de-energised, the clutch is open. Prestressed flat springs ensure disengagement free of residual torque. When the coil of the stator is supplied with DC voltage, a magnetic field is created. The magnetic pull causes the armature plate to be drawn against the force of the springs across the air gap and toward the friction surface of the rotor. The torque is transmitted. There are also clutches which work to the closed-circuit principle and are therefore closed when no voltage is applied. When disengaging, the force of the electromagnet works against the force of the springs.

Positive-type clutches, such as tooth clutches, can be engaged at standstill or in synchronism. Some designs can also be engaged at low relative speeds, in some cases as high as 200 rpm. With the positive connection, significantly higher torque can be transmitted without slip than with friction clutches. For synchronisation between the drive shaft and the driven shaft, tooth clutches with one or more lock-in positions can be used (fixed point engagement).

Friction-type clutches. On friction-type clutches (Fig. 3-93), the torque is transmitted by pressure between friction surfaces. These clutches can be engaged while the machine is running at high relative speeds. Depending on the level of friction energy to be transmitted, the working air gap should be checked at certain time intervals and adjusted when necessary.

As is the case with a brake, when engaging a clutch, the engagement time (control delay time and current rise time) until the torque has been established and the disengagement time until the torque has dropped must be taken into account.

The procedure for dimensioning a clutch is as follows:

- The size is specified by the required torque. The following must be taken into account here: moment of inertia, relative speeds, acceleration and braking times, safety factors for extreme operating conditions.

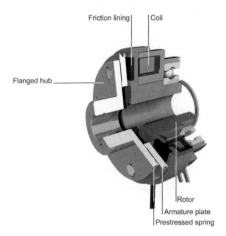

Fig. 3-93. Flange mounted clutch

- The thermal load is checked. The switching energy per switching cycle (Q) and the operating frequency (S_h) determine the thermal load on the clutch. The calculated values must be lower than the maximum permitted values for the size.
- Optionally, the number of engagements until the air gap is readjusted can be calculated.
- The slipping time (resulting acceleration or delay time), during which a relative movement occurs between the drive and output end while the clutch is engaged, can also be determined.

3.6.2 Couplings

Couplings, such as torsionally stiff couplings, connect two shafts permanently with one another.

There can be radial, axial or angular misalignment between the shafts due to temperature influences, bearing clearances or tolerances in assembly and manufacturing (Fig. 3-94). These errors can be compensated using torsionally elastic and torsionally stiff shaft couplings.

Axial offset is longitudinal misalignment of the shafts connecting to one another. It often comes about due to thermal expansion or is a result of the machine structure.

Radial offset is a parallel alignment error of the shafts relative to one another and is virtually always present, although well aligned structures can have values below 0.2 mm.

Angular offset is an inaccuracy of alignment and is generally unavoidable to a certain extent in all applications. Typical values are 1° to 2°.

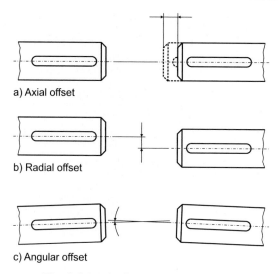

a) Axial offset

b) Radial offset

c) Angular offset

Fig. 3-94. Misalignment of two shafts

A combination of all three types of errors is often in place. When using couplings, particular attention must be paid to the radial and angular offset, as this leads to cyclic stressing, restoring forces and inertia in the bearings of the shafts. Wear caused by this and its influence on the service life must be taken into account when designing the bearings to ensure that the machine can achieve the necessary service life and reliability. Axial misalignment leads only to static forces within the coupling.

Safety and overload couplings. Torque can be securely limited and incorrect directions of rotation avoided with safety or overload couplings, also known as torque limiters. These work according to the positive-fit or friction-type principle. Some applications require the coupling connection to be released quickly when the machine encounters an overload.

Rigid couplings. Rigid couplings are low cost solutions, have extremely rigid transmission behaviour (0.1° angle of twist at rated torque) and are backlash-free. They are therefore well suited to positioning tasks and reversing drives. High levels of torque can be transmitted in a small space, although they allow virtually no axial, radial or angular offset (Fig. 3-95).

The coupling can be attached to the shaft using either individual locking screws or clamping hubs. The latter encompass the entire shaft and can transmit higher torque. Locking screws are sensitive to vibrations, and can work loose. Torsionally stiff couplings are easy to remove and, with rupture points, can perform the tasks of an overload protection device.

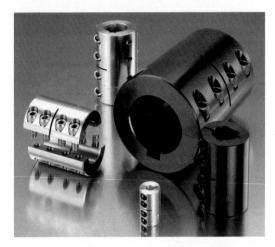

Fig. 3-95. Torsionally stiff coupling

Torsionally flexible shaft couplings. The elasticity is created by springs made of metal or elastomer (rubber, plastics). Their principle goal is to reduce torque peaks, damp rotary vibrations, move resonant frequencies to non-critical operating areas in the drive train and compensate shaft misalignment (Fig. 3-96). Large angles of twist are possible under load (metal-elastic couplings: 2 to 25°; elastomer couplings with average elasticity up to 5°, with high elasticity 5 to 30°).

Torsionally stiff shaft couplings. These are backlash-free and torsionally stiff. Depending on their design and type, they can also compensate axial, radial and/or angular misalignment. They are used where the torsional vibration behaviour should not be changed and precise angular rotary transmission is required (Fig. 3-97).

Fig. 3-96. Jaw coupling with rubber spider

Fig. 3-97. Torsionally stiff shaft coupling

Schmidt couplings are often used for large and changeable radial misalignment. The functional principle with three discs and two guide levels makes for a short and torsionally rigid coupling. The transmission of the angular velocity between drive and output is without errors. These couplings are, for example, used for synchronised and forming drives (Fig. 3-98).

Fig. 3-98. Schmidt coupling

Universal joints are used when standard couplings are no longer adequate or when large amounts of power have to be transmitted. These couplings allow angular offset of up to 40° to be bridged. They are also used for length compensation.

Selection and design. The optimum selection of a coupling is the result of a compromise between many aspects. Since couplings can have a significant influence on the reliability of a machine, the price cannot be the decisive factor when selecting models. The following selection criteria are used:

- Torsionally stiff or flexible coupling? Flexible designs are generally cheaper.
- Is backlash tolerable? Backlash-free couplings are more expensive.
- Determining the torque to be transmitted, taking into account application factors.
- Selection of the coupling based on the rated torque of the coupling
- Checking the maximum permissible speed, the maximum permissible torque and the size of the coupling.
- The topic of explosion protection as well as applications in the foodstuffs and clean room ambient conditions place further requirements on couplings.

3.6.3 Shaft/hub connections

To transmit torque from a drive shaft to a rotary body, such as a pulley, a connection must be established between these two elements. These so-called shaft/hub connections can be created with a positive-fit, friction-type or material connection.

A *positive-fit* transmission of force is based on the shape and design of the shaft and hub. Examples include keyways, dowel pins, polygon profiles or splined shafts. The various versions differ in terms of the forces transmitted and the costs. The elements are not always backlash-free. For safety reasons, positive connections should be used for hoists to prevent slipping.

Friction-fit connections offer better torque transmission properties than positive connections. They allow extremely high power transmission with excellent concentricity. They are better suited to dynamic applications and alternating loads. Friction-type connections include conical interference fits, which are easier to detach compared to cylindrical interference fits.

Fig. 3-99. Hydraulic locking assemblies

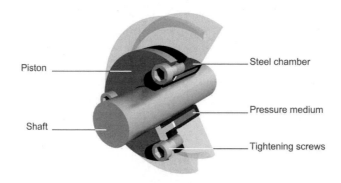

Piston ———

Shaft ———

———— Steel chamber

———— Pressure medium

———— Tightening screws

Fig. 3-100. Functional principle of a hydraulic locking assembly

Friction-fit connections include *locking assemblies*. These are friction drive, detachable shaft/hub connections and are used on smooth and unslotted shafts and boreholes. Tensioning cones creates a friction-type connection. Large torque and axial forces can be transmitted backlash-free. Hydraulic clamping elements (Fig. 3-99 and 3-100) achieve concentricity accuracy of up to 0.6 µm, mechanical clamping elements of 0.02 to 0.05 mm. The elements can be fitted and removed quickly and easily.

Correct fit of the hub on the shaft guarantees smooth running character-istics, so particular attention should given to centering the connection. The connections are divided into self-centering or non-centering. Non-centering clamping assemblies can compensate large differences between shaft and hub.

Shrink discs are friction-fit shaft/hub connections which work with exte-rior tensioning, i.e. the shrink disc is attached to the hub and generates surface pressure between the shaft and hub by reducing its inside diameter. High levels of torque can also be transmitted backlash-free.

Friction-type connections are selected based on the torque or axial force they have to transmit. If axial force and torque are applied simultaneously, tests must be performed to check their interaction.

Material connection transmissions come about by welding, soldering or gluing. However, this kind of connection is rarely used in engineering, since it is extremely inaccurate and difficult to remove. Soldered and glued connections can only transmit low level forces.

3.6.4 Bearings

Bearings support and hold rotary shafts in machines or drive components, such as in motors or gearboxes. We distinguish between sliding bearings and roller bearings. Differentiation is made with regard to the direction of the bearing force, resulting in radial and axial bearings. There are also locating bearings, which take up radial and axial forces in one or both directions. Axial movements can be performed by a floating bearing.

If a shaft is fixed by two bearings a certain distance apart, this reduces the radial force load for the drive components.

Fig. 3-101. Spherical roller bearing as an example of a roller bearing

The bearing consists of two rings (inner ring and outer ring). On sliding bearings, a lubricant forms the sliding surface for the rings. In a roller bearing (Fig. 3-101), balls or rollers run on a track between the inner and outer ring. They are guided by a cage and kept at a distance from one another.

Roller bearings are cheap, virtually maintenance-free, have low friction and low backlash. They are a simple way of carrying bearing loads. Strict standards govern roller bearings, which simplifies their procurement or replacement. During the design process, the service life, maximum speed and forces to be carried are important criteria, since these have a direct influence on the reliability and service life of a machine. They are sensitive to knocks and pollution. Roller bearings can handle higher radial and axial forces than ball bearings.

Sliding bearings can tolerate heavy loads and high speeds. They are also quieter than roller bearings and less sensitive to knocks. Assuming that the relatively intensive lubrication is applied correctly and constantly checked, they allow virtually wear free continuous operation. Their friction is significantly higher than that of roller bearings, particularly at low and high speeds.

The friction torque τ_{fr} which the drive has to apply can be calculated roughly based on the bearing friction μ, the mass m acting on the bearing and the bearing radius r_L as follows:

$$\tau_{fr} = \mu \cdot m \cdot g \cdot r_L \qquad\qquad (3.79)$$

Typical values for bearing friction μ are in the range from 0.002 (ball and roller bearings) to 0.1 (sliding bearings).

3.6.5 Traction drives

The basic task of traction drives is the adjustment of speed and torque between two or more shafts (the shafts must not be arranged coaxially). Compared to toothed geared systems, the shaft spacing can be significantly larger and the construction costs lower (Fig. 3-102).

Fig. 3-102. Traction drive

The power is transmitted in the loaded strand. The returning load-free end is referred to as the return strand.

We differentiate between friction and positive traction mechanisms. *Friction traction mechanisms* include:

- flat belts,
- V-belts,
- round belts (cords).

Initial tension must be applied to maintain the friction. Due to their operating principle, friction traction mechanisms suffer from load-dependent slip.

Mechanical variable speed drives use V-belts or similar. Varying the distance between the drive plates allows steplessly adjustable ratios.

Positive traction mechanisms include:

- toothed belts (synchronised belts),
- chains.

These traction mechanisms also require initial tension to ensure optimum running with a long service life and, if necessary, also to prevent teeth from being skipped. However, the initial tension is lower than friction traction mechanisms. This creates a lower load on the bearings and makes component design simpler and therefore cheaper. Due to the positive connection, the angle of wrap of the belt on the gear does not have to be as large as with V-belts or flat belts. Operation is slip-free.

Use of traction drives. Traction drives form part of the transmission from the motor via the gearbox to the application. They are often used to bridge shaft gaps in installations where space is at a premium and are sometimes used as a rupture point for overload scenarios. They can be replaced quickly and easily at low costs. Higher torque can be transmitted by arranging multiple belts in parallel.

Important properties. Due to their slip, friction traction mechanisms are unsuitable for positioning tasks. The positioning accuracy of a toothed belt drive depends on the stretching of the toothed belt and the backlash between toothed belt and toothed belt pulley. The backlash is caused by four properties:

- the combination of belts and pulleys,
- the torque of the drive,
- the number of teeth on the pulley (the more teeth meshed at once, the better the positioning accuracy),
- the pitch between belt and pulley (More teeth are meshed with small belt pitches).

The stretching of the belt due to dynamic load should only be of interest when the positioning is being measured dynamically. The belts assume their original length again when they are no longer under any dynamic load. Belt connections also help damp shocks.

The initial tension of the traction mechanisms generates radial load on the shafts, which must be carried by the bearings. This means that the motor or the toothed gear, for example, have to be checked for this.

The maximum radial load can be roughly estimated based on the maximum load torque $\tau_{Load,max}$, the pitch-circle diameter d_W and an additional radial force factor f_Z of the transmission component.

$$F_R = \frac{\tau_{Load,max} \cdot f_Z}{d_W} \tag{3.80}$$

The additional radial force factor is dependent on the angle of wrap and, on friction traction mechanisms, also the coefficient of friction. This has been translated to empirical values in Table 3-19.

The torque that results from the initial tension must also be applied by the drive in addition to the stationary and dynamic load of the process.

Table 3-19. Additional radial force factor of various traction drives

Transmission component	f_Z
Gears	1.12
Sprockets	1.25 to 1.4
Toothed belt pulleys	1.5
Narrow V-belts	1.5 to 2

Traction drives may cause vibrations. Among other factors, they create rotary vibration in the adjacent drive and driven shafts. Clamping elements can be fitted to damp the vibrations.

Correct alignment of the belt pulleys and the shafts is decisive for smooth operation of the traction mechanism. Parallel shaft spacing or angular offset should be avoided or, if necessary, kept within tolerances.

Toothed belts have excellent *efficiency* (typical values are 95 to 98%), which means that energy costs can be reduced in comparison with other transmission components.

3.6.6   Linear transmission components

Linear transmission components and also feed drives convert the rotation of the drive shaft to a linear movement. Leadscrews, rack and pinion systems and linear belts (Fig. 3-103) are generally used. Linear motors (Chapter 3.3.4) are also suitable for this task.

Due to their design, virtually all linear transmission components have a limited traverse stroke. One exception to this are rack and pinion systems, with which extremely long distances can be achieved when the rack and pinion elements are arranged accordingly.

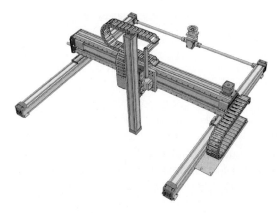

Fig. 3-103. Gantry system with toothed belts as linear transmission components

Within their traverse path, they are particularly well suited to cyclically recurring and highly dynamic tasks. These might be positioning tasks in pick-and-place applications or the synchronous movement of a flying saw to constant material movement. Servo drives are mainly used to meet the requirements. Alongside physical requirements, the cost-effectiveness also plays an important part when selecting the transmission components.

If multiple linear transmission components are combined, gantry systems or X-Y-Z axes systems can be established which allow positioning tasks in a limited space.

Leadscrews. A screw drive consists of a shaft, which has a helical guide (thread) on its surface. This thread is used to move an attachment, the slide (nut), to which the load is coupled (Fig. 3-104). For the screw itself, we differentiate between classic trapezoidal thread screws, circulating ball screws, roller screws and planet roller screws.

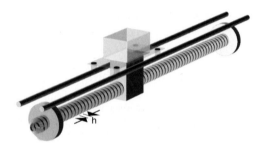

Fig. 3-104. Leadscrews

The leadscrew pitch h determines the transformation ratio from rotation to translation. The pitch h is typically given in mm and is the resulting linear path that the screw travels during one revolution (for example 5, 10, 20 to 50 mm).

$$h = 2 \cdot \pi \cdot \frac{s}{\varphi} \tag{3.81}$$

Leadscrews can be operated horizontally and vertically.

The friction μ_1 must be overcome in the linear guide, which carries the masses to be moved. This is generally rolling friction.

The leadscrew efficiency takes effect between the leadscrew and the slide, and is determined by the coefficient of friction μ_2, which varies according to the type of leadscrew. On leadscrews containing balls or rollers, the efficiency is very high due to the low rolling friction (η up to 0.99). On trapezoidal-thread screws, the screw and nut slide on top of one another, which offers very poor efficiency (typically $\eta = 0.3$ to 0.5). Trapezoidal-thread screws are generally self-locking, since the pitch angle is smaller than the friction angle. A mechanical brake for stopping is not necessarily required. However, the effect is influenced by vibrations and the lubrication, so a check should be performed to assess the need for a brake on lifting screws. Optimised thread profiles are also available to help achieve smoother running on trapezoidal thread screws.

Due to their larger contact area, roller screws have better stiffness and a higher load capacity than circulating ball screws. This also offers a longer service life.

Leadscrews can be made backlash-free and stiffer by applying initial tension. This allows better positioning accuracy to be achieved. The initial tension can, for example, be applied by increasing the diameter of the balls or by tensioning two nuts against one another. However, this initial tension does create higher idling torque.

Due to the way they work, leadscrews rotate very fast. In most cases, no gearbox is needed between motor and leadscrew. In other cases, small gearbox ratios are sufficient to match the speed and torque of the motor and leadscrew.

Axial force is present on the leadscrew shaft. This force has to be taken up by the drive components. In most cases, the moment of inertia of the leadscrew (including load) dominates the inertia of the motor. This requires precise setting of the inverter to achieve accurate control.

Long leadscrews have speeds at which bending becomes critical. These speeds should be avoided. And the resonant frequencies of the leadscrew's linear movement are dependent on the actual position.

Rack and pinion. Rack drives have a gear (pinion) fitted on the drive shaft which drives the rack. Rack and pinions are predominantly used for vertical movements (Fig. 3-105).

Fig. 3-105. Rack and pinion

They are also ideal for long traverse paths, both horizontal and vertical, by mounting multiple rack segments next to one another and using these to generate driving force. The driving pinion can, for example, be located on a carriage, on which the load is then placed. The total stiffness does not depend on the length of the traverse path. It results from the stiffness of the pinion and the pinion-rack pairing.

The diameter of the pinion can be determined by the dimensions of the toothing τ_{fr} and the number of teeth on the pinion.

$$d = z \cdot \tau_{fr} \tag{3.82}$$

The pinion is typically used to transmit high torque and low speeds, which necessitates a gearbox.

Two parallel pinions can be braced against one another to reduce backlash and increase positioning accuracy. However, this increases wear and costs. Helical toothing can also be used to reduce backlash, although this generates additional axial forces which have to be taken up by the drive.

Rack and pinion systems are low cost when there are no strict requirements in terms of accuracy.

Linear belts. Alongside their use as traction drives, toothed belts are also used in linear and transport engineering. This is a positive and slip-free type of power transmission. The key materials used are rubber (neoprene) or plastics (polyurethane). The reinforcement, which transmits the forces, must be flexible and extremely resistant to longitudinal stretching. Steel cord, glass fibre or Kevlar are typically used here.

The standard arrangement is for a circulating toothed belt, on which the load is placed and which is driven by a driving wheel (Fig. 3-106).

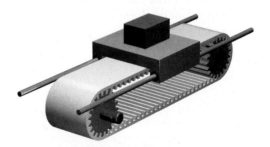

Fig. 3-106. Circulating linear belts

Omega belts represent a different option, which is often used for the vertical axis (z-axis) instead of a rack and pinion due to the lower costs. This type of transmission is also suitable for telescopic applications. The omega arrangement is used for linear units with long distances, where the length of the travel makes circulating belts unsuitable. The reduced belt stretching compared to circulating belts is another advantage. The stretching is only half that of circulating belts, since there is no lower end.

Fig. 3-107. Omega belt with fixed drive

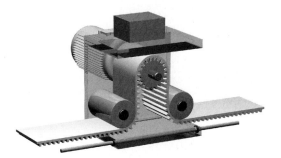

Fig. 3-108. Omega belt with moved drive

Omega belts are available in a version with fixed drive, on which the belt moves together with the guide. This type is standard for omega belts and is generally used for vertical applications. The entire arrangement has lower weight than other linear transmission components and therefore requires less energy (Fig. 3-107).

Omega belts are also available with fixed belt, where the motor is moved with the load. This arrangement is not used as often and is generally only for horizontal applications (Fig. 3-108).

On toothed belts, the diameter of the driving wheel is determined based on the tooth spacing p and the number of teeth z.

$$d = \frac{p \cdot z}{\pi} \tag{3.83}$$

Just like toothed belt gearboxes, linear belts also need to be pretensioned to ensure optimum running and prevent tooth skipping.

If the driving wheel is fitted directly on the drive shaft, a radial force load occurs due to the initial tension of the belt. To relieve the stress on the drive components, an intermediate gear with integral bearing can be used.

Table 3-20. Comparison of various linear transmission components

	Transmission components			
Property	Ball screw	Rack and pinion	Linear belt	Linear motor
Traverse path	Limited	Unlimited	Limited	Limited
Speed	2 m/s	5 m/s	10 m/s	15 m/s
Acceleration	Up to 20 m/s²	40 m/s²	Up to 50 m/s²	Up to 100 m/s²
Feed force	+++	+++	++	++
Positioning accuracy	0.01 mm	0.1 mm	0.1 mm	0.001 mm
Stiffness	+++	+++	+	+++
Costs	+	++	+++	-
Examples of applications	Machine tools, precision machines, printing presses	Gantry machine tools, feed axes	Gantry machines, handling devices	Machine tools, handling tools, high-speed machines

3.6.7 Non-linear mechanisms

On these transmission components, the movement and the torque are transmitted in a non-linear manner. The values at the output end change based on the angle of rotation or the position. Non-linear mechanisms are capable of converting continuous rotation into cyclic motion sequences. The kinematics of non-linear mechanisms determines the form of motion sequences here.

Non-linear transmission components include mechanical linkages, mechanical cam discs, scissor-type lift tables, swivelling tables and pendulums.

These are robust transmission components that can perform low-loss movements and are generally highly resistant to dirty environments. Several elements, such as the slider-crank mechanism, have the advantage that their movement is limited due to their mechanical structure. This prevents movements beyond end positions.

Calculating the transmission behaviour typically requires differential equations to be solved. Simulation tools have proven themselves here, in particular programs that support multi-body simulation.

The following section gives a brief description of the most important mechanical linkages and eccentric gears. Due to the sheer scope, the section will not deal with the design of these driving elements and their drive components.

Mechanical cams are detailed in chapter 4.9. Scissor-type lift tables are detailed in chapter 4.3.

Mechanical linkages. Mechanical linkages can, for example, be used to transform a rotary movement into a rocking or circulating movement. The resulting movement can have a straight-line or curved form. They consist of drive and driven components which are connected to one another by links and joints. The arrangement must contain certain fixed points (e.g. locating bearings) to create the motion. The advantage of mechanical linkages is that complex movements can be performed in the machine. Commonly used mechanical linkages include thrust cranks and four-bar linkages.

Mechanical linkages are characterised by:

- a ratio that changes via the angle of rotation,
- a moment of inertia that changes via the angle of rotation

Due to this interaction, machines using mechanical linkages with uneven gear ratios are significantly more complex when dimensioning drives than machines with linear interaction.

When these kinds of drives need to be controlled dynamically and with precise positioning, the control system needs excellent management of disturbance variables and, if necessary, a torque feedforward control.

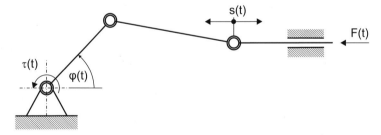

Fig. 3-109. Principle of slider cranks

Slider-crank mechanisms. The rotation of the drive is converted using a slider-crank mechanism into a reciprocating motion. The fixed point in the system is the crankshaft, to which the drive torque is applied. However, the application can be operated the other way round. It is possible to transmit large forces and implement energy effective movements (Fig. 3-109).

The non-linear mechanism results in a changing torque load for the drive despite the continuous speed (Fig. 3-110).

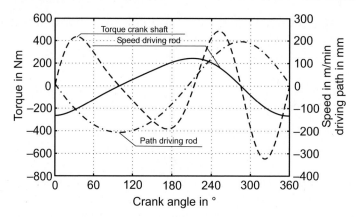

Fig. 3-110. Typical torque curve of a thrust crank for one revolution at constant input speed

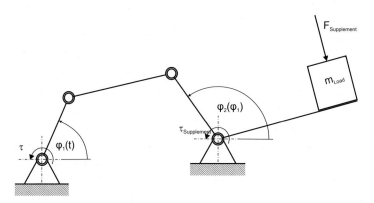

Fig. 3-111. Principle of a four bar linkage

Four bar linkages. The four bar linkage contains two fixed points. The optional arm is actually no longer a component of the four bars system (Fig. 3-111). Swivelling units often use this kind of mechanism. An example of this is body clamps in chassis manufacturing.

Eccentric gears. The eccentric is a special case among mechanical cams. It is a mechanism that is, for example, used to perform fast and small stroke movements. On an eccentric, the rotary point is located half the stroke distance away from the centre of the control cam (Fig. 3-112).

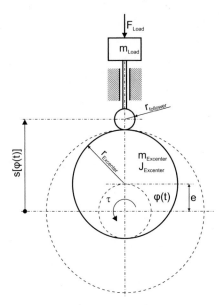

Fig. 3-112. Eccentric gear

3.6.8   Guidance systems

The friction load of the masses to be moved is taken up by the linear transmission components of linear or rotary shaft guides. In addition, the guides are used to specify the movement path. Linear guides can also be designed to allow curves. And wheel-rail systems or rollers can be used for guides and to carry the mass. The drive force to be applied is described in chapter 3.1.4. The following section deals with linear guidance systems in more detail.

Linear guidance systems are generally anti-friction guideways (mounting rails) which use balls or rollers between two guide elements to implement the movement via rolling friction. Slideways are becoming less common due to their relatively high friction forces and tendency to wear.

Guidance systems must have the necessary load capacity, be smooth running and guarantee excellent positioning and running precision throughout a long service life. They must be backlash-free and offer a high degree of stiffness in all load directions. Since the applications often require high speeds, they also need fast running properties. To be economic, the systems must have a simple design and allow easy maintenance.

On anti-friction guideways (Fig. 3-113), the force applied by the drive components is determined by the motion resistance. This is caused by the roller element. Among other things, it depends on the load, the lubricant with its viscosity and the initial tension. The initial tension is required to suppress any backlash that may be present and to increase stiffness.

$$F_{Drive} = \mu \cdot m \cdot g + f \qquad (3.84)$$

μ is the coefficient of friction and f is a specific resistance of the guidance system.

If, for example, two guide rails are fitted in parallel, inaccurate assembly (parallelism inaccuracies and height differences) can increase the motion resistance.

Slideways (and in particular hydrodynamic slideways) can suffer from the so-called stick-slip effect with constant switching between static and sliding friction. This leads to vibrations, increases noise generation (particularly at startup and low speeds) and impairs positioning accuracy. The effect should be avoided at all costs. It can be overcome with proper lubrication, a vibration-damping design or by selecting different material pairs.

Slideways offer better damping than anti-friction guideways.

The driving force to be applied is essentially determined by the coefficient of friction for the sliding friction and the mass to be moved.

Fig. 3-113. Recirculating roller unit as an example of an anti-friction guideway

$$F_{Drive} = \mu \cdot m \cdot g \qquad (3.85)$$

With the representation of the driving elements, all components that make up a drive system have now been presented. The properties of the individual components can, of course, create questions on the best way to fully optimise a drive system. We will go through these in the following.

3.7   Overall set-up of the drive system

Dr. Carsten Fräger, Dr. Edwin Kiel, Johann Peter Vogt

Now that we have looked at the individual components of a drive system, in this chapter we would like to examine global aspects which can help us optimise system set-up. In many applications, simply examining the individual components on their own is insufficient if we wish to find the optimum solution.

The following aspects concerning overall set-up of a drive system are also described in detail:

- selecting the components to match the application,
- overall dimensioning of the components of the drive system,
- optimisation of the motion profile,
- oscillatory loads.

3.7.1   Selecting the components

A controlled drive is a mechatronic system made up of the following components:

- mechanical components, such as the driving elements and the gearbox,
- electromechanical components, such as the motor, the brake and the angle sensor,
- the inverter as the central electronic component,
- software in the inverter for drive control and motion control of the drive,
- connection to the master control via input and output signals or a communication system.

For each of these individual components, a large number of different types and options are available. The overall configuration of the drive system therefore offers great freedom of choice. Within the scope of this choice, a solution should be found that fulfils the following requirements:

- as fast as possible or as necessary,
- as precise as necessary,
- as affordable as possible.

The configuration of a drive system is derived from the requirements of the application. These can also be very diverse and summarised as follows:

- the static and dynamic requirements in terms of torque and speed, which also determine the maximum and effective power output
- the requirements in terms of the dynamics of the drive derived from this

- the required accuracy of the drive,
- the drive functionality,
- the distribution of tasks between the control system and the drive, as well as the connection between the two,
- the necessary safety functions for the machine and the plant.

The overall set-up of the drive, which is based on torque and speed requirements, is described in detail in this chapter.

The drive functionality is the focus of chapter 4, in which the large number of different applications are broken down into 12 groups, for which the requirements and solutions are highlighted. This application often then also determines the distribution of total functionality among the control system and the drive.

You can find information on communication systems for connecting control systems and drives in chapter 3.4.6. Details on safety functions executed from the drive can be found in chapter 3.4.7.

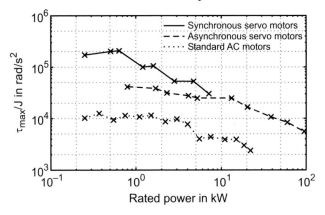

Fig. 3-114. Accelerating power of standard and servo motors

Dynamics of the application. Many applications, and in particular positioning applications, require high dynamic performance to increase the productivity of a plant. The dynamics are essentially determined by the motor type and the ratio of mass inertia to maximum torque that results from this. Synchronous motors have a significant advantage over asynchronous motors here. Fig. 3-114 shows the ratio of maximum torque to moment of inertia, which is one indicator of maximum accelerating power, for various motor types and their output.

For dynamic applications, synchronous motors must be combined with an optimum configuration of gearbox and further driving elements.

Applications which do not require high dynamic performance can also use an asynchronous motor, as this generally offers the same rated power at a lower cost.

Accuracy of the application. For processes that are activated by drives, there are requirements in terms of the accuracy of the speed and the angular position. These vary significantly.

If there are no strict demands on speed accuracy, speed measurements can be omitted. Asynchronous motors, which are controlled by a frequency inverter, achieve speed accuracy of 3 to 5% of the rated speed. By using vector control, this can be improved to between 0.5 and 1% of the rated speed. For many applications (e.g. conveyor drives, driven tools, drives for process engineering, pumps, fans), this speed accuracy is sufficient.

For applications that require higher speed accuracy, a speed sensor (angle sensor) and a speed control must be used. The motor type is not the decisive factor for the speed accuracy that can be achieved by the drive. Particularly in applications that require excellent synchronism (electronic gearboxes, winding drives), asynchronous motors with speed sensors offer advantages over synchronous motors, since they do not have any detent torque, which can hinder speed control.

Many applications require positioning. This in turn always requires the use of position sensors. In the simplest case, these can be proximity switches (photo electric sensors or inductive sensors) which supply on/off type data. This data is received by the drive and the further motion sequence is then controlled in such a way that a defined end position is reached. With this type of positioning, accuracies of several mm can be achieved, which is more than adequate for the stop position of conveyor drives. No angle sensors are needed and the use of an asynchronous motor with a frequency inverter is sufficient.

Many positioning sequences in materials handling require accuracy in the range of 0.1 mm. These can be easily achieved using a resolver as the feedback device. Standard helical gearboxes, bevel gearboxes and toothed belts are also sufficient for the linear movement with these positioning accuracies.

For significantly better positioning accuracy, more accurate angle sensors are required. These are optical incremental encoders with Sin-Cos signals and allow positioning accuracy of up to 1 μm. The drive then has to be equipped with precise drive elements such as planetary gearboxes and leadscrews for the linear movement. Alongside the angle sensor on the motor for motor control, a sensor should also be used for the load (a second rotary transducer or linear encoder) so that inaccuracies of the mechanical path do not limit the positioning accuracy. For applications with the strictest positioning requirements, it can be necessary to use linear motors, since all influences of mechanical transmission elements then disappear.

Table 3-21. Selecting drive components based on accuracy requirements

Speed accuracy	Angular accuracy	Motor	Angle sensor	Control
> 3% n_N	None	Standard three-phase AC motor	None	V/f
1% n_N	None	Standard three-phase AC motor	None	Sensorless vector control
0.1% n_N	± 10 arcmin	Servo motor	Resolver, encoder	Servo control
0.01% n_N	± 2 arcmin	Servo motor	Sin-Cos absolute encoder	Servo control

The rules described in this chapter specify which types of drive components should be used. We then need to determine the correct sizes and further parameters for these components in the following chapter.

3.7.2   Drive component dimensioning

The goal of drive dimensioning is to select the optimum drive train for the application to reliably meet the specified requirements. The components then form a matched mechatronic system which works as a functional unit.

For drive dimensioning, the electrical and mechanical limits of the individual components must be taken into account. The inverter sets the speed and torque limits through its output voltage and the output current. The motor must not overheat, and the gearbox temperature must also be kept within the permissible range to protect the lubricant and seals. Overload can cause mechanical damage to gears, shafts and bearings.

Based on all this information, the dimensioning of a drive deals with four key questions:

- *Drive function*: Can the required speeds, torque and acceleration be achieved with the selected drive?
- *Mechanical strength*: Can the drive transmit the torque and forces in the long term?
- *Thermal dimensioning:* Does the operating temperature remain within the permissible limits to prevent premature ageing?
- *Economic design*: How economical is the drive– in addition to purchase costs, this should also take into account the total costs over the life cycle of the product?

Finding the optimum design for a drive helps avoid drives that are too large or not large enough. For dynamic applications in particular, simply dimensioning the drive based on the rated power is generally not good enough.

The targeted dimensioning of drives is based on defined motion profiles and fixed ambient conditions which are determined by the application. Based on the calculated values required for the process, the combination of inverter, motor and, if necessary, gearbox is selected and checked for suitability.

Since the input speeds of the electric motor generally do not match the desired output speeds, gearboxes are often used for speed reduction. A gearbox also reduces the required torque τ_1 of the drive motor and therefore contributes to a smaller size. The torque of a motor is proportional to the square of the diameter and the overall length (and thereby to the size). The rated speed of a motor n_1, on the other hand, can be adjusted by the winding. Since gearboxes generally have high efficiency η, the power is not affected. The process power determines the motor power requirements, regardless of the ratio i of the gearbox.

$$i = \frac{n_1}{n_2} \tag{3.86}$$

$$\tau_1 = \frac{\tau_2}{\eta \cdot i} \tag{3.87}$$

$$P_1 = \frac{P_2}{\eta} \tag{3.88}$$

The following section focuses on drives with gearboxes. For drives without gearbox, the ratio $i = 1$ should be used and all parameters affecting the gearbox left unchanged.

The selection of inverters, motors and gearboxes is described, which helps users find the right drive for their application in a logical manner. The following procedure is used here:

- determining the dynamic torque and speed curve by determining the maximum and average values,
- selecting the components, typically in the order motor, gearbox and then inverter,
- checking that the components can fulfil their task.

How to determine torque was explained in chapter 3.1. The static components are derived from the proportions not independent of acceleration, such as lifting torque, friction torque and torque resulting from process forces. The dynamic components are those determined by the acceleration of masses and the moment of inertia.

From these values and a known motion sequence, the speed and torque curve can be plotted against time. This can then also be used to calculate the average values for speed and torque using the equations in chapter 3.1.7, which are in turn important when selecting the components based on continuous power and thermal limits.

Selecting the motor. When a drive is used together with a gearbox, it can be assumed that the operating limits of the motor can be adapted to the requirements of the application via the gearbox ratio i. This allows the motor to be initially selected based on the required process power.

On drives without a gearbox, attention must be paid to ensuring that the required speed of the application matches the speed that can be supplied by the motor. The various motor types are offered with various rated speeds, which help users find a solution that matches their application.

Three-phase asynchronous motors are produced with different numbers of pole pairs, which allows their speed range to be adapted. With inverter feed, the nominal speed can be increased by a factor of 1.7 using the delta connection (chapter 3.3.2). Synchronous servo motors are often equipped with different windings, which allow the rated speed to be selected while maintaining the same torque.

However, it is important to take into account that the size of the motor, and thereby its costs, determine the torque. This means that motors with low speeds will always cost more for the same power. It is therefore always a good idea to aim for high motor speeds. Only when a low-speed motor allows a solution to be set up without the need for a gearbox will the complete drive system be likely to be less expensive.

Selecting the gearbox. The following are the key factors when selecting the gearbox:

- necessary ratio for adjusting the selected motor to the application,
- maximum torque on the output end,
- maximum speed,
- if required, ability to carry radial forces on the driven shaft,
- design for easiest integration of the geared motor in the machine.

Selecting the inverter. The inverter must be able to provide the required current. Inverters are generally designed in such a way that they can provide a specific maximum current I_{max} and thermally limited continuous current I_{eff}.

The values required for the motor or geared motor, together with the relationship between motor current and torque, determine the power requirements for the inverter. Synchronous motors have a linear relationship between torque and power requirements, which only changes due to saturation effects at high torque levels.

On asynchronous motors, the relationship between torque and power requirements is non-linear due to the magnetising current required during idling. The current must be determined for the respective torque here.

In dynamic applications, made up of acceleration phases with large torque requirement as well as periods with constant speed and standstill with low torque requirement, the effective power requirements of an asynchronous motor are significantly higher than those of a synchronous motor. This often leads to larger inverters which cost more, although these extra costs can of course be offset against the lower costs for the motor. For defined applications, checks should be performed to determine the best combination in terms of both performance and costs.

The maximum output voltage of the inverter (which is dependent on the mains voltage), the mains voltage tolerances to be taken into account, the design of the DC bus (chapter 3.4.1) and line-side chokes, determine the maximum speed of the drive. In addition, the voltage drop on the stator of the motor reactance and field-weakening effect the torque characteristic when the motor is run faster than nominal speed.

Checking a combination of drive components. Once a combination of motor, gearbox and inverter has been found for a specific application using the methods described, this must then be checked. The following points should be considered here:

- *Drive function*: Are all working points within the speed-torque limit characteristics of the selected drive, even when taking into account the tolerances of the mains voltage of the electrical power system?
- *Thermal design*: Are the temperatures within the thermally permissible limits for the components, allowing for a long service life?
- *Mechanical strength*: Are the gear systems and shafts strong enough to transmit the torque safely?
- *Economic aspects*: Does the drive system that has been selected offer optimum costs throughout the life cycle?

The drives are rated by the specifications of the geared motor, such as the output torque, the ratio, the mass inertia referred to the motor shaft and also by the speed/torque characteristic of the motor-inverter combination.

The drive must offer sufficient torque and sufficient speed for the requirements of the process. Here, the speed-torque characteristic of the motor-inverter combination is transformed to the output shaft with the gearbox ratio i and the efficiency of the gearbox η. The speed/torque characteristic of the application must now be within the maximum torque characteristic.

The thermal limits are observed when the effective power point, which is made up of the effective torque τ_{eff} and the average speed $n_{average}$, is below the permissible characteristic for continuous operation (the so-called S1 characteristic).

To protect the lubricant and the shaft sealing rings, a specified maximum speed may not be exceeded in the gearbox for thermal reasons.

The mechanical strength of the gear system and shafts is determined by the gearbox rated torque. The maximum torque required by the application must not cause any overloads.

When selecting the gearbox ratio, as well as the speed ratio between drive and work machine the control system behaviour must also be checked. For reliable control, the load-matching factor k_J should be in the range from 0.5 to 10 (chapter 3.1.4).

The points to be taken into account when evaluating the life cycle costs and ways in which these can be minimised are described in detail in chapter 5.2.

If the selected drive fulfils these conditions, it can perform the drive task reliably and economically.

3.7.3   Optimisation of motion profiles

Dynamic applications, such as positioning, exploit the ability of motors to deliver a higher level of torque than the continuous permissible torque for a short time. This maximum torque depends on the speed. The result is a maximum torque characteristic.

Selecting the motion profile tailored to this maximum torque characteristic allows the optimum drive for the application to be selected. We will now demonstrate this effect by examining a comparison between a trapezoid profile and a sin² profile.

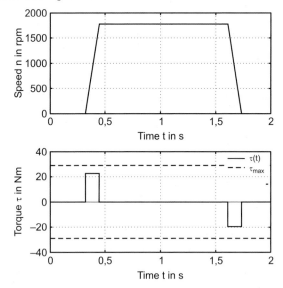

Fig. 3-115. Speed and torque curve of a trapezoid motion profile

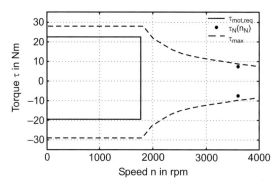

Fig. 3-116. Torque/speed characteristic of a trapezoid motion profile

With a trapezoid profile, the acceleration is built up and reduced with constant torque. During the acceleration and braking times, the speed therefore changes in a linear fashion (Fig. 3-115). At the speed-torque level, this leads to a rectangle (Fig. 3-116). The point with the highest power is the corner point, at which acceleration stops and the curve moves over to constant speed travel. This corner point lies in a range in which the maximum torque of the motor can already drop off.

When a sin² profile is used, the acceleration (and thereby also the torque) is built up continuously (Fig. 3-117). At the torque/speed level, the curve is oval, which fits the maximum torque characteristic of the servo motor much better (Fig. 3-118). A higher traversing speed can be reached for the same maximum acceleration. This motion sequence also automatically limits jerks, resulting that the mechanisms are not jarred as much. This effect is described in more detail in the next chapter.

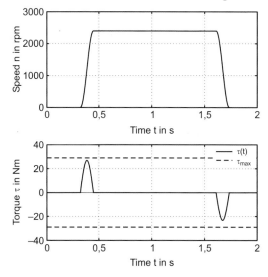

Fig. 3-117. Speed and torque curve of a sin² motion profile

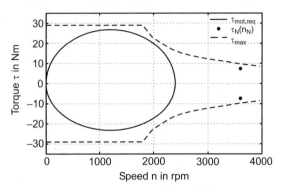

Fig. 3-118. Torque/speed characteristic of a sin² motion profile

3.7.4 Oscillatory loads

The motor is often connected to the work machine via elastic transmission components. The motor, this elasticity and the work machine then together form an oscillating system. The basic forms of interaction were described in chapter 3.1.8. Examples of elastic transmission components include (see also chapter 3.6):

- couplings with elastic elements,
- long shafts,
- belt ratios,
- elasticity in machine frames,
- toothed belts for linear movement.

The following section details the example of an indirect linear drive with a toothed belt. The statements can, of course, also be applied to other cases [Fr06].

Indirect linear drives, which convert the electromagnetic force generated in the motor into a movement of the device to be positioned, cannot be made as stiff as would be desirable. A kind of spring-damper coupling is formed between the motor and the device to be positioned. The damping is generally $d < 0.1$. The natural frequency can be determined from the mechanical design, but in most cases is between 10 and 200 Hz. The natural frequency depends on the position, and is lowest on toothed belt drives when the device to be positioned is located the furthest away from the motor.

The frequency range which the motor control can influence is determined by the torque rise times (typically < 1 ms) and the limit frequency of the speed control loop derived from this (500–1,000 Hz). The motor control is therefore set up for higher frequencies than the mechanisms, and the system is soft-coupled [Sch01a]. The motor control can influence the motor with better dynamic performance than the mechanisms can implement. This leads to a situation where the controlled drive can create changes in acceleration which are above the resonant frequency and which the mechanisms cannot follow. This in turn results in poorly damped vibrations of the load, which have two negative effects:

- the mechanisms are stressed,
- the position overshoot at the target position causes it to take longer to achieve the desired position accurately.

These vibrations should ultimately be avoided. Vibrations always require a stimulus and the most powerful stimulus during positioning procedures is the motion profile itself, i.e. the increase and reduction in acceleration. All other effects (such as gear meshing on the pinion) have significantly less influence.

To prevent vibrations from being excited, it is necessary for the frequency range of the acceleration (i.e. the force or torque) above the natural frequency to have only very small amplitudes. The goal of this is ultimately the same as jerk limitation, which can be achieved either directly or indirectly:

- The motion control can change the acceleration while taking the maximum tolerable jerk into account.
- The dynamics of the speed control can be reduced to a point where the control generates correspondingly slow acceleration curves. This method is selected when the motion control provides jerky curves, although it has the disadvantage that the entire control (i.e. the response to setpoint changes also the disturbance behaviour) becomes very slow. This method is really an attempt to make corrections in an unsuitable way.
- A state-space controller with observer algorithm can control the entire status of the line, i.e. also the spring-mass vibration system. This allows the dynamics of the motor control and the mechanism to be operated in a controlled way. However, determining the actual parameters of the system is extremely costly. In some cases, these parameters change during positioning or are dependent on the mass of the device to be positioned, which itself is not always known. This costly approach is generally only used when extremely dynamic and powerful positioning systems need to be set up [Sch99].

The simplest way to prevent vibrations from being excited in practice is jerk limitation, which was described in chapter 3.4.5.

Jerk limitation is generally described by the time $t_{a,max}$ during which maximum torque is built up. If this is set to a value more than double the period of the first natural frequency f_{res1} of the oscillating mechanism, vibration stimuli are largely avoided:

$$t_{a,\max} > 2 \cdot \frac{1}{f_{Res1}}$$ (3.89)

Second actual speed value. In several applications with oscillatory loads, the system behaviour can also be improved by implementing a second speed sensor (besides the speed sensor integrated in the motor) on the load side as far away from the motor connection as possible. On linear drives with toothed belts, this can be the counter-roller. Supplying the speed control with this measured value allows better system dynamics to be achieved.

Having looked at optimisation aspects for the drive system, we should now turn our attention to the question of how to ensure reliability of this system. This is the focus of the next chapter.

3.8   Reliability of drive systems

Detlef Kohlmeier

The productivity of a machine or plant is measured in units per second or meters per minute. This specification refers to the maximum performance under the assumption that the machine can be operated without any errors. This requires a stable production process and a high degree of technical availability of the production line.

Under ideal conditions, all machine operators expect trouble-free operation throughout the entire service life of their installation. Depending on the industry sector and output (chapter 2), this service life can be between five and ten years for a 1, 2 or 3-shift operation. If we assume an operating time of 24 hours a day for 365 days a year (8,760 hours), the minimum service life for 5 or 10 years of operation comes to approximately 43,000 or 88,000 hours respectively. Components designed to meet these requirements would generally be completely overdimensioned for intermittent operation of around 4 hours a day in a single shift operation (43,000 operating hours in 29 years). This example shows how important it is to understand service life dimensioning, as specifying too long a service life is uneconomical. On the other hand, planned and unplanned maintenance reduces production time and thereby the productivity of a system. Completely trouble-free operation requires 100% reliability of all components used. In practice, however, systems can experience several error mechanisms, which we will now examine further.

This chapter describes how reliability is defined and how we can differentiate error causes. In doing so, we will deal with various aspects of electronics, software and mechanics.

Reliability does not happen by accident. It is the result of a planned design and product development process. In this chapter, we will look at the most important methods and measures that can be employed by manufacturers of mechatronic drives to meet reliability requirements throughout the entire service life of a machine.

Definition of terms. Practice shows that people often have quite different understandings of what the terms reliability, service life, failure rate and wear actually mean. However, for a professional analysis of the topic, we need clearly defined terms.

Reliability: Reliability R is defined in the standard DIN 40041:1990-12 as a measure of the ability of a unit to perform its defined functions under the specified ambient conditions for a fixed time period.

Service life: The service life of a component describes how long a component can perform its task in uninterrupted use without wear-based failures occurring.

Failure / Failure rate: A failure is when a unit can no longer perform its defined functions. Failure rate λ is defined as the probability that a component will fail in the time interval specified.

It is therefore a measure of reliability and is generally given in units per hour. For electronic components, the failure rate is typically given as FIT (Failure in Time). 1 FIT corresponds to 1 error in 10^9 hours. So 1,666 FIT would mean a failure probability of 0.5% in 3,000 hours.

Random failures: Random failures are when a failure is due neither to a systematic design error or overload. The probability of a random failure is not dependent on the age of the unit. The causes of such errors cannot be predicted. The failure rate λ is virtually constant.

Wear: The service life of a machine is determined by the wear behaviour of its components. All mechanical components which are subject to wear, such as bearings, must be designed in such a way that they can meet the dimensioned service life without failing. When wear-based failures occur within the expected service life, these are not random errors, but rather systematic errors, which can be traced back to the design (underdimensioned) or the application (overload) of the drive.

MTBF, MTTF: The reciprocal value of the failure rate λ specifies how much time is likely to pass on average until a unit fails. For maintainable units, the abbreviation *MTBF* (Mean Time Between Failures) is often used to describe the time between two failures. For non-repairable units, the term *MTTF* (Mean Time to Failure) is often used for the time that elapses until an error occurs. *MTBF* and *MTTF* values refer exclusively to random failures and do not take the wear limit into account.

MTTR (Mean Time to Repair) is the average repair time of a unit.

3.8.1   Service life curve

If we evaluate the failures of a large enough number of products, we can see different failure rates during the service life (Fig. 3-119). Premature failures occur at the start, while errors at the end are generally due to wear. During actual utilisation, the failure rate is constant. The development of the failure rate over time $\lambda(t)$ describes the failure behaviour of the component, which is displayed in a graph using the so-called bath-tub life curve.

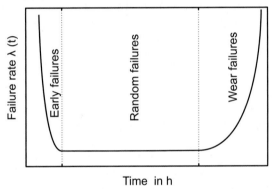

Time in h

Fig. 3-119. Service life curve: Failure rate plotted against time

Phase of premature failures. The failure rate drops rapidly within this phase. Premature failures are typically due to component failures (material failure) or production-based deficiencies, such as defective soldered connections on electronic assemblies. Application errors are also classed as premature failures in the commissioning phase.

The goal of manufacturers is to keep the number of premature failures as low as possible and exit this phase before delivery of the products.

Phase of random failures. This phase is classed as the utilisation phase of the product. No systematic failures occur. The failure rate level is based on the failure rates of all individual components. The duration of this phase is not random. It is based on careful dimensioning and the choice of components by the manufacturer.

As a general rule, every production process is subject to variations, which are due to irregularities in processing by humans and machines, irregularities in materials and deviations in process parameters, such as temperatures. These fluctuations are within tolerances and will allow perfectly good finished products to be created, although these may not be absolutely identical to one another. Unfavourable grouping of these fluctuations, of the individual operating conditions and, if applicable, of ageing effects can then sporadically cause failures during the life cycle of a product which cannot be predicted. These failures are classified as random failures. All manufacturers are anxious to keep the rate of random failures as low as possible by keeping the process parameters within tight tolerances and implementing checks of the production process stages to ensure exact execution. The aim is to use consistent process control and continuous process optimisation to reduce the rate of random failures.

Phase of failures due to wear. In this phase, the failure rate rises above average. The components reach the end of their service life. The failures are caused by wear. The wear time is determined by the design of the devices.

3.8.2 Basis of calculation

In the mid section of the service life curve, the operating range, failures are random and not dependent on the service life (Fig. 3-115). In these cases, the following applies for the ratio between reliability R and failure rate λ:

$$R(t) = e^{-\lambda \cdot t} \tag{3.90}$$

Each machine and each module is made up of a number of individual components. If the failure of an individual component causes the entire system to fail, this is referred to as a serial system. The following applies for the overall reliability $R_{Sys}(t)$ and the overall failure rate $\lambda_{Sys}(t)$:

$$R_{Sys}(t) = R_1(t) \cdot R_2(t) \cdot \ldots \cdot R_n(t) \tag{3.91}$$

$$\lambda_{Sys}(t) = \lambda_1(t) + \lambda_2(t) + \ldots + \lambda_n(t) \tag{3.92}$$

MTBF (or *MTTF*) times are defined as the reciprocal value of the failure rate:

$$\frac{1}{\lambda} = MTBF \tag{3.93}$$

The *MTBF* values for a system are calculated on the basis of the *MTBF* values of the components within the system.

$$MTBF_{Sys} = \frac{1}{\lambda_{Sys}} = \frac{1}{\dfrac{1}{MTBF_1} + \dfrac{1}{MTBF_2} + \ldots \dfrac{1}{MTBF_n}} \tag{3.94}$$

The failure probability of a serial system can only be reduced by increasing the reliability of the individual components.

The availability A of a plant or a system is classed as the ratio between actually available time and total time.

$$A = \frac{MTBF}{MTBF + MTTR} \qquad (3.95)$$

3.8.3   Reliability of mechatronic drive technology

Electronic drives. The service life of an electronic module is limited by two factors – firstly by wearing components and secondly by components subject to continuous ageing.

Electrolytic capacitors, which are used as DC-bus capacitors and also to support internal supply voltages in switched-mode power supplies, have a strong influence on the service life of frequency and servo inverters. Electrolytic capacitors are subject to ageing by drying out, whereby the volume of electrolyte and thereby the capacity of the capacitors is reduced. The service life of a capacitor is essentially dependent on temperature. As a very rough rule, we can say every 10° Kelvin increase in ambient temperature cuts the service life in half. Or to express this positively, every 10° Kelvin reduction in ambient temperature doubles the service life.

Relays and connectors can also have an effect on service life. These are wearing parts with a limited number of switching or plugging cycles as well as electromechanical components, such as blowers with a limited bearing service life.

Geared motors. In contrast to electronic components, random failures can be virtually eliminated in mechanical systems. The motor or gearbox housing, shafts and the toothing of the gearbox are not relevant for service life evaluation, as they are not subject to wear or ageing. However, the roller bearings, the shaft sealing ring, the lubricant and, if applicable, positive shaft/hub connections do contribute to the service life.

In the application, the driven shaft is subjected to load in three ways: output torque, radial force and axial force. These loads are transmitted to the bearings and thereby influence their service life. When dimensioning bearings, these forces are taken into account alongside the type of bearing, the material properties and residual stress of the material. Other factors that influence the service life of roller bearings include the state of lubrication and the ambient conditions.

The forces that occur on the driven shaft are also transmitted to the shaft/hub connection, which establishes the connection between the motor shaft and the gearbox pinion. In contrast to friction-type connections, positive connections are subject to wear. Keyways can get deflected during dynamic positioning applications with alternating direction of rotation and thereby age.

The shaft seal is the most critical component in the gearbox. It ensures that the rotary shaft is sealed. When manufacturing the sealed surfaces, particular attention is therefore paid to ensuring a precise fit. If the surface is too rough, this can lead to excessively high temperatures on the seal lip. During operation, high speed contributes to an increase in temperature, which in turn has an effect on the service life of the shaft seal. The task of the gearbox oil is to dissipate the heat from the seal lip. Lowering the temperature on the seal lip by 10° Kelvin improves the service life by 25%. Seal lips are available in various materials to handle operating temperatures of up to 200°C.

The task of the lubricant in the gearbox is to lubricate the bearings, the tooth system and the shaft sealing ring to ensure that no wear takes place and also to dissipate the heat that is generated. When it comes to the viscosity of the oil, a compromise must be reached. The heat at the shaft sealing ring is dissipated more effectively with low-viscosity oil, while a thicker oil is better for lubrication of the bearings and the tooth system.

Software. In contrast to hardware, the reliability of software is structured in a fundamentally different way. The reliability of software is not a function of time, and there is no variance in the production process. Failure is always due to a development error.

Not every error causes a failure, so we cannot evaluate the reliability of software in the sense of a serial system whose overall reliability can be determined from the total of its individual parts. It is generally accepted that no software is completely free of errors – and in fact how many errors a software contains is one way of defining reliability. However, people tend to only perceive the properties they use.

The quality of software is determined by the development process, which requires a systematic approach during design, testing and system integration.

3.8.4   Reliability concepts during product development

The reliability of drive technology is the result of systematic and planned alignment of all processes involved in product development to meet reliability requirements. The methods/tools used in product development to assure reliability also have a decisive part to play here, as does production itself. Only products that have been produced under stable process conditions can guarantee reliable operation.

The methods and tools used have three key goals:

- ensuring that customer requirements are met by the drive,

- ensuring that the design meets the required reliability,
- ensuring that processes are in place that allow the product to be produced reliably without errors.

The following section deals with the most important measures employed in the product development process.

Quality planning. The goal of quality planning is to secure customer requirements for a product, including defined targets in terms of reliability. Among other things, the quality planning process specifies which methods are to be used at what stage in the product development and production process to ensure the reliability of the product. Here targets are set, implementation processes are defined and the necessary resources are provided.

Learning from mistakes. The first aspect at the start of product development is to look back on experience gained from previous projects. This can, for example, take the form of a review. The input variables are the evaluations of error databases, service reports, change procedures and CRM databases. At the end of a project, things that went well or not so well throughout the project should also be recorded. These summaries are valuable input variables. In addition, the personal experiences of service and sales staff should be collected and documented.

Calculating reliability. The calculation of reliability is already incorporated in the design process, where the failure rate and service life are considered. The key reliability figures of all components used can also be taken as a basis for calculating the failure rate of mechatronic drives in the design process. Assuming that every component is required for the function of the entire system, the key component figures can be added together. The utilisation and the ambient temperature then flow into the calculation via correction factors.

Failure mode and effect analysis, FMEA. The FMEA is a systematic tool which is used in the development process to improve the reliability of products. The risks of a design are evaluated by development teams. Purchasing, development/engineering, production, logistics and not least the perspective of users are taken into account. And FMEA analyses can also be incorporated once development is complete to make changes to the product or for reassessing production procedures. In practice, FMEA analyses are performed at different levels and with different goals. The most common forms of FMEA are:

- Design FMEA: The design FMEA is used during the design phase, among other things to ensure that the user's needs and opinions have been taken into account sufficiently.

- Circuitry FMEA: The circuitry FMEA is used when developing electronic circuits. It helps the designers check the design of electronic circuits in inverters. Corresponding methods are also available to the designers of electromechanical and mechanically drives.
- Process FMEA: The process FMEA is used in manufacturing to detect and eliminate any potential sources of errors in the production process.

The implementation of an FMEA takes place in five stages:

1. Defining the system elements and their structure, task and function.
2. Performing a systematic error analysis with regard to potential errors, as well as the possible causes and consequences of these errors.
3. Risk assessment, determining the risk potential and calculating a risk priority figure $RPF = S \times P \times D$ with
   S Significance of the effects of the error (1-10),
   P Probability that the error will occur (1-10),
   D Probability that the fault will be discovered (1-10).
4. Optimisation and implementing measures. Solutions are sought on the basis of the Pareto analysis, starting with the highest risks first. The following options are available here:
   - Changes to the concept to eliminate error causes
   - Increasing the reliability of the individual components to minimise the occurrence of the error cause and
   - Implementing checks to ensure that any potential errors are discovered before product delivery
5. Review and reassessment on the basis of the solutions found.

Reliability tests. Testing the reliability of new products is broken down into two areas. The first area involves ensuring that the components work properly, while the second area involves checking the protection from failure. A range of defined tests are carried out to check performance. The tests themselves become more comprehensive the more complex the component. Alongside separate tests of hardware and software, joint tests (so-called integration tests) are also performed. And system tests check the overall function and compatibility of the various components with one another, using specifications based on a test plan. All the errors detected here are systematic errors, which can be traced back to design (hardware or software) or the manufacturing process.

The actual reliability of electronic components is examined with the aid of accelerated reliability tests. These serve as verification of the reliability calculated in the design phase. The tests themselves use deliberate changes of environmental and ambient conditions to expose the devices to increased levels of stress. The parameters of temperature, humidity and operating voltage allow various, defined accelerations to be achieved. The influence of these factors can be described using the Arrhenius equation, which allows statements to be drawn on how many operating hours one hour in the increased ambient temperature represents.

$$k = A \cdot e^{-\frac{E}{R \cdot T}} \qquad (3.96)$$

This equation describes the relationship between the reaction speed k of the activation energy E (a component-specific variable), the absolute temperature T and a constant of proportionality A.

The test procedures employed attack the components the hardest directly in their weak areas. A plug connection (such as control terminals of an inverter or connection of a servo motor) can, for example, be checked using a test procedure in an aggressive environment. The result contains details which allow prognoses to be drawn on the failure probability in future. The limits of the permissible ambient conditions for the component being tested may not be exceeded, since any failure caused in this way would not be a random failure, but rather failure due to overload caused by going beyond the specified operating range.

Quality assurance measures in the production process. The concepts we have so far examined for ensuring reliability of drive products essentially concern the development phase. In production, manufacturers must ensure that the designs are safely manufactured and that deviations or errors which might be caused by the components and production processes are detected and corrected. Comprehensive tests are used here to guarantee the high quality of the products.

With electronic products, the following tests are performed after the individual manufacturing steps:

- An automated optical inspection after assembly of surface-mounted components (SMT).
- An initial electrical inspection (in-circuit test) once the PCBs have been fully assembled.
- A high voltage test and a final test after device assembly. This is often also combined with a burn-in process (stressing the device at an increased temperature). The final test is generally designed in such a way that it mimics the intended later use of the device as closely as possible. On inverters, the load via motors, load cycles at the current limit and both short circuit and earth-fault tests are included in the scope of the test.

Motors are also comprehensively tested after assembly. As standard, this testing includes measurement of the electrical values, inspection of the isolation and checking of the current consumption, which is an indicator for smooth running and therefore also correct assembly.

The focus when testing gearboxes is to examine the mechanical tolerances of the components (gears, shafts) as well as to check running after assembly. The current consumption of the motor and noise generation offer indications as to whether production has been performed correctly.

This chapter has described the elements of a drive system in detail. The next question is how these drive systems are used. The following chapter will examine how to structure this and illustrate the process systematically for twelve drive solutions.

4 Mechatronic drive solutions

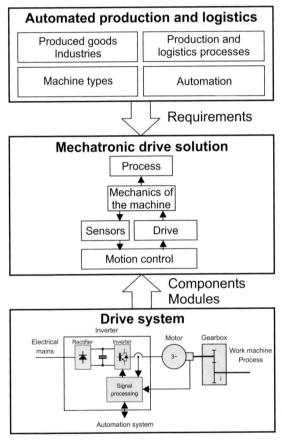

Fig. 4-1. Mechatronic drive solutions

Chapter 2 has dealt with the structure of highly automated production and logistics systems, in which most consumer goods are manufactured and distributed to customers. These production and material flow systems define the requirements for the drive solutions used.

In chapter 3 the different components of a drive system have been described. The tasks of the production and logistics systems are implemented with drives consisting of predefined modules and solutions.

Each of these drive solutions is a mechatronic subsystem of a production or a logistics facility (Fig. 4-1). They implement a process for which a machine with its mechanical parts is required. The drive supplies the machine with energy and controls the motions. Sensors recognise the state of the process. For operation of this process the crucial point is the motion control, i.e. the software which controls the process in such a way that it carries out the required function.

Based upon these interrelationships, mechatronic drive solutions can be divided into twelve groups [Le06, Ki06].

The first five of these drive solutions carry out material transport functions:

- *conveyor drives* are stationary systems for the transport of material,
- *travelling drives* move with the vehicle and the material to be conveyed,
- *hoist drives* move objects against the force of gravity,
- *positioning drives* place work pieces and tools at precisely defined positions,
- coordinated drives for robots and materials handling move objects in space using multi-axis kinematics.

The next four drive solutions can be found in production lines operated continuously or in fast cycles:

- *synchronised drives* move materials in a production line. Here many drives have to be synchronised exactly,
- *winding drives* feed the process with materials which have been wound up to form a coil in order to save space or rewind them at the end of the process,
- *intermittent drives* for cross cutters and flying saws single out the materials on a process line,
- *drives for electronic cams* are used for a large number of non-linear motions for machining steps.

The following two drive solutions execute production processes which change the work piece:

- *drives for forming processes* are used, for example, on presses and extruder,
- *main drives and tool drives* drive entire machines as principal drives or execute metal-cutting machining processes.

Finally, the conveyance of liquid and gaseous substances is presented:

- *drives for pumps and fans,* which are used for conveying liquids and gases and are often found in process engineering, in the infrastructure of buildings and factories, as well as in auxiliary units of production machines and systems.

To conclude chapter 4, taking four production lines and one logistics centre as examples, there is a description of to what extent the 12 drive solutions are used in those applications. Finally, trends are outlined for further development of the drive solutions.

4.1 Conveyor drives

Dr. Sven Hilfert

By conveyance we mean the transport or movement of people or goods over limited distances, usually on predefined paths.

The focus of this chapter is on stationary conveyor drives which mainly operate horizontally. If the drive moves with the transport, we refer to them as travelling drives, which will be dealt with in chapter 4.2. If the motion takes place against gravity, a hoist drive will be required, which is described in chapter 4.3.

4.1.1   Conveying process

The process of conveying is the most common process which is executed by electric drives. In all industrial production setups and logistics centres goods have to be moved, and in all highly automated processes this motion is controlled by conveyor drives which are actuated by the system control, usually a programmable logic controller (PLC). Conveyors can be found in industrial plants, to a large degree in storage systems for storing and distributing goods, in ports, in airports and in mining. Conveyors are also used within production machinery for feeding and discharging materials or work pieces.

Conveying and sorting materials can be either continuous or intermittent. Continuous conveyors operate constantly over a lengthy period and are used not only for transporting bulk materials and individual items but also for moving people (e.g. escalators). The material being conveyed is relatively evenly distributed on the conveyor. Generally speaking, it is fed to the conveyor and discharged from the conveyor while the load-bearing components are moving, and hence during operation of the drive motor.

With some designs the carrier can stop during the feed of materials and only the drive motor carries on running continuously. Intermittent conveyors (e.g. assembly lines) are normally classified as continuous conveyors. The operation of continuous conveyors is usually continuous duty (S1 duty). They are operated at a constant, fixed or variable speed. Starting and braking operations take relatively insignificant amounts of time in comparison to the length of operation and levels of acceleration are unimportant so such a system is usually ramped slowly. Only in some applications, e.g. with escalators or the handling of goods which can tip over, the demands for acceleration and braking times are important parameters when it comes to dimensioning.

Depending on the type of conveyor, transport can be horizontal, inclined or vertical over straight, curved or even completely three-dimensional sections. The classification of continuous conveyors and their accessory equipment is defined in DIN 15201.

By contrast, intermittent conveyors do not operate continuously but only at certain intervals, often only when materials have to be transported. In certain conveyor systems there can be units for ejection, building gaps or sorting. Intermittent conveyors usually contain travelling drives, hoist drives or positioning drives, which will be described in other chapters.

An overview of the various machine types of continuous conveyors is presented in Table 4-1.

Table 4-1. Overview of continuous conveyors

Continuous conveyors		
Unit loads, storage		**Bulk materials**
Bins and pallets		**Indoors and outdoors**
Roller conveyors	Power & free conveyors	Belt conveyors
Accumulating roller conveyors	Circular conveyors	Chain conveyors
Belt conveyors	Monorail overhead conveyors	Troughed chain conveyors
Segmented belt conveyors	Rotary tables	Scraper conveyors
Chain conveyors	Tilt tray conveyors	Chain bucket elevator
Chain ejectors	Circulating conveyors, paternosters	Screw conveyors
Belt ejectors	Z-conveyors	Vibratory conveyors
Pop-up ejectors		

For the conveying of unit loads often a combination of continuous and intermittent conveyors are required. A decision for one of these two types of conveyors particularly depends on the following criteria:

- desired quantity conveyed and distance,
- properties of the materials involved,
- investment and transport costs.

Table 4-2. Overview of typical parameters of conveyor elements

Machine type	Speed	Material conveyed	Throughput	Power per drive
Bulk materials (e.g. sand, coal)				
Belt conveyors	Up to 4 m/s	-	Up to 1,600 t/h	Up to 200 kW
Chain bucket elevator	Chains: 0.5–1.5 m/s Belt: 0.6–4 m/s	-	Up to 1,700 m³/h	Up to 100 kW
Screw conveyors	0.1–0.5 m/s	-	Up to 400 m³/h	Up to 25 kW
Scraper conveyors		-	Up to 1,000 t/h	Up to 150 kW
Chain conveyors		-	Up to 1,000 t/h	Up to 130 kW
Unit loads (bins b, pallets p)				
Chain conveyors	0.1–0.5 m/s	P: Up to 1,250 kg	Up to 1,200 p/h	Up to 2.2 kW
Roller conveyors	0.1–1.5 m/s	b: Up to 50 kg p: Up to 1,250 kg	Up to 1,000 b/h Up to 1,600 p/h	Up to 0.75 kW
Belt conveyors	0.5–2.0 (6.0) m/s	b: Up to 50 kg p: Up to 1,250 kg	Up to 1,000 b/h Up to 1,000 p/h	Up to 3 kW
Power & free conveyors	0.1–0.5 m/s	Up to 1,500 kg	-	Up to 5 kW
Chain ejectors	0.1–0.6 m/s	b: Up to 50 kg	Up to 1,200 b/h	Up to 0.25 kW
Diagonal chain ejectors	0.1–0.6 m/s	b: Up to 50 kg	Up to 2,900 b/h	Up to 0.55 kW
Belt ejectors	0.1–0.9 m/s	b: Up to 50 kg	Up to 2,000 b/h	Up to 0.25 kW
Pop-up ejectors	0.5–1.2 m/s	b: Up to 50 kg	Up to 3,500 b/h	Up to 0.25 kW
Rotary table	-	p: Up to 1,250 kg	Up to 180 p/h	Up to 0.37 kW
C-conveyors, Z-conveyors	0.1–0.5 m/s	b: Up to 50 kg	Up to 1,300 b/h	Up to 1.1 kW
Tilt tray conveyors	0.5–2.5 m/s	Up to 60 kg	Up to 12,000 pcs./h	Up to 3 kW

4.1.2 Mechanical design of conveyor systems

Below we deal with the particular machine types and the demands made of continuous conveyors. Intermittent conveyors such as cranes will be described in chapter 4.3 on hoist drives and monorail overhead conveyors and in chapter 4.2 on travelling drives.

Key parameters are indicated for common horizontal conveyor applications in Table 4-2. As can be seen from this table, there are various types of conveyor systems depending on the materials being conveyed, but also depending on the process parameters to be achieved, and they vary not only in terms of mechanical design but also according to the drive system used. These conveyor systems will be described below.

Belt conveyors. In the case of belt conveyors for unit loads the items (e.g. crates, pallets) are conveyed by means of a conveyor belt. Due to the smooth surface it is suitable for different sizes of items and for fragile packages. The main features are steady, quiet running as well as gentle transport of sensitive goods. With belt conveyors, inclines of up to 20° are typically manageable and specific designs are even feasible up to 60° in some configurations.

In the top strand the conveyor belt slides over a surface whilst in the bottom strand it can be guided by rollers. The bottom strand normally also handles belt tensioning by means of a tensioning device or tensioning unit with movable deflection pulleys. The transmission of power between the belt and the roller drives is effected by friction. Fig. 4-2 shows a schematic diagram of a belt conveyor and Fig. 4-3 shows a design which is used for unit loads in logistics centres.

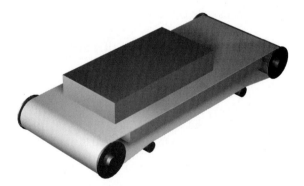

Fig. 4-2. Schematic diagram of a belt conveyor for unit loads

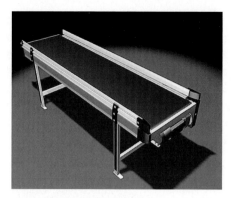

Fig. 4-3. Belt conveyor for unit loads

In terms of the position of drive mounting there is a head drive (positioned at the end, under the belt or at the side of the belt) and a centre drive (positioned at the centre, under the belt). In the case of the side head drive a bevel gearbox or helical bevel gearbox is typically used. The output shaft of the gearbox drives the roller drive direct. Where the head drive is positioned below the belt helical gearboxes are mainly used and the roller drive is driven with belts.

Belt conveyors for bulk materials are designed with circulating conveyor belts made of fabrics of textile or steel cord inlays with covers of rubber or plastic. The conveyor belts are supported by rollers and driven or decelerated via rollers by means of friction. The principle is illustrated in Fig. 4-4. These conveyors are often designed as troughed conveyors or pipe belt conveyors.

Fig. 4-4. Schematic diagram of a belt conveyor for bulk materials

Roller conveyors. Roller conveyors are highly versatile conveyors for transporting small unit loads such as bins, cardboard boxes, crates and tubs, or even large unit loads such as pallets and containers. They are suitable for conveying, buffering (accumulating roller conveyors) and in conjunction with additional units for distributing unit loads. The load is conveyed on transport rollers which are connected to one another by means of a chain or belt. Fig. 4-5 shows how they operate. A roller conveyor for unit loads weighing up to 50 kg with a centre drive is illustrated in Fig. 4-6. Roller conveyors can be designed like belt conveyors with geared motors mounted as head drives or a centre drives. Alternatively, it is also possible to drive individual rollers direct by means of a drum motor where the motor is actually integrated into the roller.

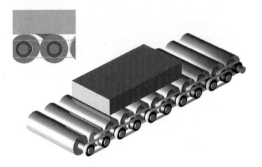

Fig. 4-5. Schematic diagram of a roller conveyor

Fig. 4-6. Roller conveyors for unit loads

One special design is the accumulating roller conveyor which is illustrated in Fig. 4-7. The intermediate roller transmits the driving motion from the circulating driven belt to the transport roller. The intermediate roller is spring-loaded so power is transferred by friction. When a container moves over a trigger lever, the downstream group of rollers is de-energised by swinging the intermediate roller aside. As a result the following bins come to a halt and only the bin at the stop is subjected to back pressure force. When the stop is released, the bin is able to move on and thereby releases the drive for the following bins. In this operation the accumulated bins are singled out as they approach, i.e. they are spaced in relation to one another.

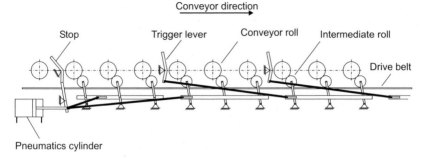

Fig. 4-7. Principle of a mechanical accumulating roller conveyor

As an alternative to the mechanical design illustrated the accumulating roller conveyor can also be designed electromechanically by providing it with accumulating rollers, which can be switched off. For this purpose an electromagnetic clutch and electromagnetic brake are integrated into individual rollers and these are actuated via a system of sensors installed at the conveyor. The accumulating roller is driven by the drive belt whilst all the other rollers in a group are connected to the accumulating roller via round belts. When the clutch is disengaged, the accumulating roller is switched off, and so too is the group of rollers.

Chain conveyors. In the case of chain conveyors the load (unit load or bulk material) is conveyed by means of a transport chain. Between the drive sprocket and the transport chain there is a positive connection. The principle is illustrated in Fig. 4-8. One frequent design for unit loads is a drag chain conveyors. These require dimensionally stable items with flat surfaces and they are mainly used for conveying, sorting and distributing unit loads with defined load carriers, e.g. pallets or bins. The drive system used and its integration into the conveyors are comparable with those of belt conveyors and roller conveyors. Depending on design, bevel gear, helical bevel or helical gear motors are used.

Fig. 4-8. Schematic diagram of a chain conveyor

Chain conveyors are used to convey bulk materials wherever the use of belt conveyors is not possible due to the condition of the material, e.g. hot material.

Scraper conveyors are of similar design to chain conveyors. They have cross-pieces fitted to the conveyor chain. The material is not transported on the chain, as in the case of the chain conveyor – the conveyor scrapes or pushes the material off a surface. Scraper conveyors are solely used for conveying bulk material such as overburden, coal, sand or in the vicinity of machine tools to take away swarf, die-cut parts, small components, wood chips, etc.

Chain bucket elevators. A chain bucket elevator is a closed conveyor system in which bulk material is transported vertically by elevator buckets without creating any dust. The elevator buckets, which are attached to chains or belts, rotate round two shafts, of which the upper shaft is the drive shaft (Fig. 4-9). At the bottom the bulk material is filled into the buckets by an infeed chute before being transported upwards to the conveyor head. By the circulation round the drive shaft the material falls into the outlet chute at the head of the chain bucket elevator as a result of the gravitational and centrifugal forces acting.

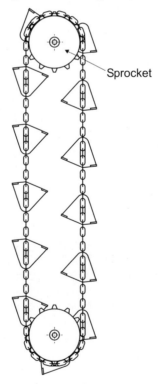

Sprocket

Fig. 4-9. Schematic diagram of a bucket chain conveyor

Depending on the type of pulling element there are chain bucket elevators and belt bucket elevators. In the case of chain bucket conveyors the power is transmitted by friction-type or a positive connection, depending on the sprocket, whilst in the case of belt bucket conveyors it is transmitted by friction, whereby the level of friction between the belt and the reel can be increased with a rubber lining.

The drive is often effected by helical bevel gear motors which are attached at the side of the chain bucket conveyor. If levels of power are high (> 15 kW) and the drive motor is powered from the mains direct, the drive unit is fitted with a hydrodynamic coupling to assist starting up and protect against overloads, as well as an additional auxiliary drive for maintenance purposes. When it is switched off, the bucket chain conveyor is prevented from moving back by means of a backstop on the gearbox or motor. If the motor is driven via a frequency inverter, there is no need for the auxiliary drive or coupling.

Screw conveyors. A screw conveyor is a typical bulk material conveyor for handling dusty, granular, small-size or even semi-liquid and fibrous materials. It is suitable both for horizontal conveyance and for steep to vertical conveyance with lengths of up to approx. 10 m. In process engineering it is used to convey, mix and meter. The principle is illustrated in Fig. 4-10 and an example is shown in Fig. 4-11.

Fig. 4-10. Schematic diagram of a screw conveyor

The drives used are helical geared motors, helical bevel geared motors or bevel geared motors which often drive the conveyor screw direct. The pipe or trough in which the conveyor screw is located is stationary. As the conveyor screw fitted with angled shovels rotates, the material is moved in an axial direction. This is also the basic principle of an extruder.

Fig. 4-11. Troughed screw conveyor

Tilt tray conveyors and sorters. The tilt tray conveyor is a typical unit load conveyor which combines the two functions of conveyance and sorting with one another. It is very often used, for example, in parcel distribution systems. The individual trays of the conveyor are linearly connected to one another and have a platform which can be tilted by up to approx. 30° at right angles to the direction of conveyance. The length of such platforms is typically 500 mm to 1,200 mm.

The conveyor usually picks up the material at a central point and distributes it to decentralised destinations. In the sorting process the material is discharged at the side by force of gravity and this can take place on one side or both sides depending on the design. It is necessary to ensure accurate coordination of the control system, the kinematics of conveyor tilt and the arrangement of the downstream conveyors.

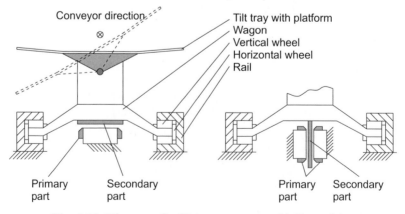

Fig. 4-12. Diagram of a tilt tray conveyor with linear drive

The conveyor can operate in a loop. For this, individual trays move along rails, which can comprise straights or curves. The wagons can be driven by a chain or screw with geared motor mounted as a centre drive or by means of linear motors using single-comb or double-comb technology (Fig. 4-12).

The platform is often tilted by a mechanism which is triggered by elements attached to the section. The individual platforms are then uprighted at an uprighting unit.

Distribution systems. The continuous conveyors described earlier, such as belt or chain conveyors for unit loads, are often combined with elements such as belt ejectors, chain ejectors, pop-up ejectors or pushers for sorting and distributing the items. These feed the materials onto the conveyor section, distribute them between individual sections and eject (sort) them from the section at the end.

Chain and belt ejectors are predominantly employed in conjunction with roller conveyors and chain conveyors, e.g. in bin and pallet handling systems. Basically they are designed for right-angled inward and outward transfer to and from the conveyor. To transfer the material two belt or chain strands are elevated by an electric motor or pneumatically within the conveyor. If the drive is an electric motor, eccentric or lever mechanisms are used. As soon as the material has been lifted, the belt or chain strand is started and the material is conveyed. The method is then similar to that of a belt conveyor or chain conveyors. One special design is the angled ejector, where ejection takes place at a typical angle of 30° or 45°. It is often used for sorting material on a conveyor section which is split into two sections further on.

Pop-up ejectors are used for ejecting bins or cardboard boxes from roller conveyors (Fig. 4-13). The material to be sorted is lifted slightly by the deflector rollers, which are recessed into the roller conveyor, diverted by rotation in a curve motion and thereby ejected. The pop-up ensures gentle ejection because there is no jerk or drawing action. The deflector rollers are lifted using eccentric or crank mechanisms, or they are pneumatic.

Fig. 4-13. Pop-up ejector

Pushers allow the simple transverse movement of light materials onto a parallel delivery section or one which is at an angle of 90°. By means of a pusher positioned at the side of the conveyor section the items are pushed off the conveyor. For this it is necessary to have smooth bottoms and spaces between the individual items due to the return motion of the pusher.

One special design is the rotary ejector fitted with a geared motor (Fig. 4-14). Here the arm is rotated by 180°, thus causing the items to be ejected. The disadvantage of this pusher is the sometimes relatively high impact required for sturdy goods. With special cam motion profiles for pushers with electric drives this impact can be considerably reduced in comparison to pushers with pneumatic drives. They are used in conjunction with roller conveyors and belt conveyors.

Fig. 4-14. Rotary ejector

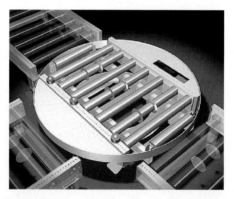

Fig. 4-15. Rotary table with roller conveyor

Rotary table. Rotary tables are used in bin and pallet handling systems for the inward transfer, ejection, storage and sorting of unit loads. In standard applications the rotation angle is 90° or 180°, thus making it possible to change the direction of the items or turn them over (Fig. 4-15). The round carrier of the rotary table has pivot bearings. It is driven by a geared motor, which acts either directly, via a traction wheel or via a pinion mounted on the motor shaft direct, which engages in a crown gear fitted to the rotary table. The most common motors used are bevel geared motors, helical bevel geared motors and helical geared motors with a hollow or solid shaft.

The base frame of the rotary table can be fitted with different conveyor systems such as roller conveyors, chain conveyors or belt conveyors.

Circular conveyors, power & free conveyors. Suspended conveyors are among the most important ceiling conveyor systems for the indoor logistics of unit loads. There are two basic categories: circular conveyors with a single chain and power & free conveyors (overhead twin rail chain conveyors).

Circular conveyors have a centre drive which drives a chain along a rail. The item is carried by suspension frames which run on rollers along a profile rail, which is suspended from the ceiling of the building (Fig. 4-16). The individual suspension frames are connected to one another, thus ensuring a continuous flow, as is the case in paint shops for example. The items are fed onto and off the conveyor with appropriate feeding and delivery units.

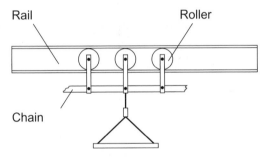

Fig. 4-16. Schematic diagram of a circular conveyor

Power & free conveyors (overhead twin rail chain conveyor) consist of a power chain (drive chain) and carriers which are guided along separate rails (Fig. 4-17). The power chain runs continuously whilst the carrier can be mechanically engaged and disengaged by driver cams. The carrier takes up the load. As opposed to the circular conveyor it is possible to install track switches, which allows for diverting and merging of the conveyor circuits, driveless buffer sections, and hence intermittent operation.

Like the circular conveyor, the power & free conveyor features a central caterpillar chain drive which engages with the (power) chain. Between the drive unit (geared motor) and the drive sprocket of the drag chain drive there is a safety clutch with a speed monitor which automatically switches off the conveyor chain if a set level of torque is exceeded (e.g. if the chain jams). The speed of the drive, and hence the speed of the conveyor chain, can be controlled as required using a frequency inverter.

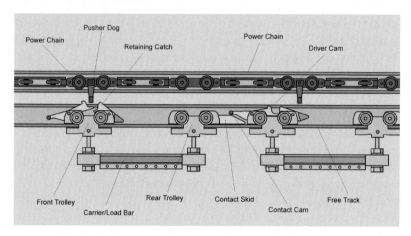

Fig. 4-17. Schematic diagram of a power & free conveyor

Circular conveyors and power & free conveyors are often used in mass production, for example:

- automotive production,
- production of large electrical equipment,
- production of domestic appliances,
- paint shops,
- refrigeration, and drying chambers.

A drive can propel circular conveyors with a length of up to 500 m and with a second drive the conveyor length can be increased to 2,000 m.

For long conveyors and a low frequency of items being conveyed it is advisable to install a monorail overhead conveyor, on which each conveyor is driven individually.

4.1.3   Drive systems for conveyors

Important requirements for the drive system of continuous conveyors include:

- long service life and a high level of reliability,
- ruggedness,
- little or no maintenance,
- fast, easy installation and assembly,
- high modularity,
- low energy consumption,
- low costs.

Operating conditions. Whilst many conveyor systems are located inside buildings, bulk material conveyors are mainly used outdoors. Consequently, special demands are made of the drives with regard to their degree of protection (typically IP 65). In order to prevent condensation inside the motor off-line heaters are used and in order to protect against corrosion suitable coatings are applied and the surfaces of bare metal parts are treated accordingly.

Drive dimensioning and drive power. The continuous conveyors in bin, pallet and bulk handling systems are mainly operated in S1 continuous duty and acceleration operations are relatively unimportant. The systems are usually ramped slowly so they do not place high demands on the dynamic response of the drive system being used.

In most cases the size of the drive is largely determined by the mechanical friction (losses) and possibly by a difference in height which has to be overcome (incline). Typical losses in continuous conveyors include:

- friction on a surface,
- roller friction,
- deformability and flexure of belt conveyors,
- sliding friction when the materials hit a stop,
- losses due to power transmission.

Geared motors. For the materials handling application it is mainly helical geared motors, helical bevel geared motors or bevel geared motors which are used in conjunction with three-phase asynchronous motors of two and four-pole design. Selection of the type of gear depends on the installation situation, i.e. the way in which the geared motor can best be integrated into the system and what type of gear is most suitable (output shaft coaxial or angled, solid shaft or hollow shaft). The technical differences with regard to torsional stiffness, backlash and noise emission are of minor importance.

For materials handling applications gear ratios up to $i = 50$ are used. These can be implemented with a single-stage design of worm gear, but worm gears have a low level of efficiency at high ratios so the S1 duty of the conveyor system will result in high energy costs. It is more economical to use bevel gears or helical bevel gears.

Brakes. Another important drive component is the brake. It is used, for example, to hold the position in distribution systems or to prevent a conveyor belt from running back. The most frequent application is use as a holding brake with an emergency stop function. In distribution systems such as chain ejectors, belt ejectors or pushers very high demands are placed on the brake switching frequency due to the high cycle rates, and hence on the sturdiness of the mechanical system of the brake. With these elements there can be up to 3,500 switching operations per hour so in three-shift operation there can be up to 25 million brake switching operations per year. It is also important to ensure that the brake is actuated properly, i.e. to ensure that when the drive starts up the brake is disengaged without any residual torque and the brake is only engaged when the drive is at a standstill – otherwise considerable wear can be caused to the friction linings on account of the high number of switching cycles.

Backstop. Bulk material conveyors are mainly fitted with a backstop instead of a brake because conveyance is often not only over a level surface but also over mounds and when the belt comes to a halt it must be prevented from running back. Due to the high inertia of the loaded conveyor belt the use of brakes is either not possible at all or only under certain conditions. When the conveyor belt comes to a standstill because a brake has been applied, high friction would be converted inside the brake and the friction linings would normally wear very soon. If the drive is switched off, the conveyor belt coasts until it comes to halt and the backstop then prevents it from running back, especially if the section is on an incline.

Actuation of the motor. Depending on whether the operation of the conveyor system is at constant speed or has to be varied, the drives are:

- operated with direct connection to the mains,
- connected to the mains via a soft starter,
- controlled by a frequency inverter.

Direct operation from the mains produces the least expensive alternative if no change in velocity or rotational speed is necessary. In the case of low-power drives, connection to the mains is direct and at higher levels of power it is via a star-delta starting circuit.

When the motor starts up, the rated torque is approximately doubled or even tripled with direct mains connection without star-delta starter which may cause jerks in the conveyor section or conveyor belt. This method of starting can only be implemented if such jerks do not have any detrimental effect on the materials being conveyed or on the mechanical system of the installation. In order to reduce the jerks when starting up asynchronous motors from the mains, motors with relatively low starting torque and stalling torque in relation to their rated torque are used. Alternatively the inertia of the motor is increased by the use of additional inertia or cooling fan with high inertia, so in effect a mechanical soft start is achieved.

If a change in the direction of rotation or a soft start has to be achieved electronically without any speed adjustment, e.g. to prevent items from tipping over, soft starters (motor starters) are used. Depending on design, these devices also include additional functions: integrated motor protection, brake actuation, the ability to evaluate sensors and communication via a bus system. In contrast to actuators they are wear-free thus help to increase operational reliability and system availability.

Frequency inverter. The rotational speed of an asynchronous motor is defined by the frequency of the supply voltage. For low-loss adjustment of the rotational speed of an asynchronous motor a variable frequency is necessary which is generated by a frequency inverter. The frequency inverter can be installed either in the control cabinet or decentrally on the motor itself or on the mechanical system of the conveyor. Since materials handling systems usually cover a large amount of space, the decentralised drive system is frequently used. As a result, the amount of installation work necessary can be reduced, and hence the cost as well.

Intermittent operation occurs, for example, when feeding and ejecting goods, forming gaps between items, and sorting. In simple applications drives are also operated from the mains direct or in conjunction with a motor starter. However, the proportion of inverter drives is much larger in comparison with conveyors in continuous operation. In these applications one reason for this is the necessary positioning function (see chapter 4.4). Servo drives are rarely used with continuous conveyors – they are mainly used on intermittent conveyors such as cranes and storage and retrieval units (chapters 4.2 and 4.3).

Linear drives. Linear drives are rare in materials handling because the low demands in conveyor systems in terms of positioning accuracy and dynamic response do not justify their high costs. One example of the use of linear motors in materials handling are the tilt tray sorters described above. They are controlled by servo inverters with feedback systems to detect the position of the moving part of the motor.

Connection to the control system. Communication between the drive components and a control (PLC) is handled via field bus systems, e.g. PROFIBUS or PROFINET. Via AS-i (actuator-sensor interface) sensors along the conveyor sections (e.g. photo electric sensors) are evaluated and actuators are operated such as decentralised drives, with their integrated switching, controlling and protection functionalities.

Conveyors often include sensors at the various conveyor sections which detect the materials or items. These are usually photo electric sensors. If decentralised frequency inverters are used, it is useful to initially have the sensor signals detected by the frequency inverters and then transmit them to the system control via the field bus system. This dispenses with the need for a further field bus interface for the sensors.

Safety technology in conveyor systems. The safety of conveyor systems is generally assured by ensuring that only trained staff has access to the systems. The speed is restricted so that the staff working there can recognise and evade any hazard. Emergency Off switches are fitted, with which the system can be brought to a halt.

There is one exception if staff has to work on the materials or items being conveyed. In this case care must be taken to ensure that the conveyor does not start up while staff is still in the vicinity of the materials or items. One example of this is rework stations in automotive production, but also workplaces where manual loading and unloading operations are performed. Here it is useful to use a safety system inside the drive, whereby the most common technique used is the "Safe Torque Off", STO. In individual cases it may be necessary to use "Safely Limited Speed", SLS.

4.2 Travelling drives

Frank Erbs

As with conveyor drives, travelling drives are used for the conveyance of goods. They move vehicles with their payload horizontally or over inclines. As opposed to conveyor drives, the drive is mounted on the vehicle, resulting in differences for the drive system.

4.2.1 Travelling process

Travelling drives are used on intermittent conveyors. With conveyor drives on intermittent conveyors, which were described in chapter 4.1, the drives are installed in the stationary part. In the case of the intermittent conveyors described here they are located on the vehicle, and hence on the moving part.

As opposed to continuous conveyors, intermittent conveyors provide more flexibility at higher costs. For this reason they are used wherever the volume of transport of the conveyor section is lower compared to intermittent conveyors.

There are track-bound and non-track-bound vehicles. Track-bound vehicles are easier to guide but they are less flexible with regard to routing. Since with many automated in-house transport operations the route is fixed, they are more frequently encountered in automated operations. For non-automated in-house transportation the manually operated fork-lift truck is still the most common industrial means of transport. This will not be dealt with in this book because it is not automated.

Travelling drives in intermittent conveyors are used in the following applications:

Table 4-3. Applications and typical characteristics of travelling drives

Application	Industry	Velocity (m/min)	Masses moved (t)	Power (kW)
Rail vehicles	Logistics, materials handling	100	100	Up to 150
Monorail overhead conveyors	Materials handling, logistics, transportation, automotive industry	130	5	5
Overhead cranes and gantry cranes	Materials handling	200	Up to 200	Up to 500
Storage and retrieval units (S/R machine)	Logistics	240	5–15	55
Industrial trucks	Materials handling, assembly, automotive industry	40	4	10

- in rail vehicles (e.g. transfer carriage),
- in monorail overhead conveyors,
- in overhead cranes and gantry cranes,
- in storage and retrieval units,
- in driverless transport systems.

Table 4-3 provides an overview of the applications with their typical characteristics.

4.2.2   Material flow systems with travelling drives

Rail vehicles. Rail vehicles are known for the transportation of passengers and goods over long distances (railway system). In in-house transportation rail vehicles are used whenever a route is used so rarely that the installation of a stationary materials handling system would not be economic. The vehicles generally also contain devices for fixing, moving, loading and unloading the materials or items (Fig. 4-18).

The rails on the route are normally recessed into the ground. In some rare cases the rails are elevated.

Rail vehicle drives are based on the force of friction acting between the wheel and the rail. In addition to steel-steel contacts, rubber wheels are also used. When accelerating and braking the transmission force of static friction must not be exceeded or else the drive will slip.

Fig. 4-18. Track vehicle system in the automotive industry

The travelling speeds are usually limited for safety reasons. For high travelling speeds the travelling area is often fenced off so that nobody can enter, in order to rule out any hazard.

The applications of travelling drives described below are basically track-bound vehicles.

Monorail overhead systems. A monorail overhead system (electrified monorail system) transports goods without making contact with the ground. It consists of rails which are attached to the ceiling of the building or to columns, as well as travelling gear which conveys the items on suspension frames. Since each vehicle is controlled individually, monorail overhead systems are highly flexible. The vehicles travel to their destinations automatically.

Monorail overhead systems are driven by a friction wheel running on the rail. The entire vehicle control system is installed on the vehicle, whereby in addition to drive functions other control and safety functions are also necessary. The entire control and drive system of an electrified monorail system comprises:

- signal and energy transfer to the vehicle,
- decentralised motor control,
- geared motor with friction wheel.

Overhead and gantry cranes. In addition to the swivel drive, tower cranes also have a travelling drive for the trolley. Overhead and gantry cranes have two trolleys with one travelling drive on each trolley. The drive can be located on the trolley or sometimes the trolley is moved by a stationary drive using cables.

Gantry cranes require two drives to move the gantry. With small spans these can be mechanically coupled (drive shafts). In the majority of gantry cranes, there will be two travelling drives which are actuated by one or two inverters.

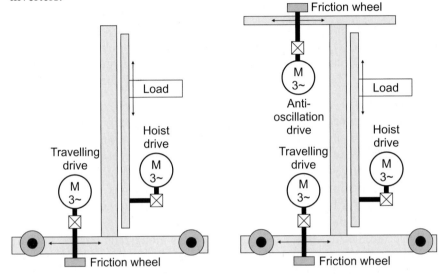

Fig. 4-19. Storage and retrieval units with and without anti-oscillation drive

Storage and retrieval units (S/R machine). The travelling drives of storage and retrieval units move the vehicle along the bay aisle to the rack slots or to the pick-up and delivery stations. Since there is normally only one vehicle in an aisle at the same time, high acceleration and high speed of travel are essential to achieve a high throughput.

For the travelling drive of storage and retrieval units wheel drives or belt drives are used. Toothed belts are tensioned over the travelling distance and these pass through the drive on the vehicle via an omega guide. The omega arrangements are able to transmit the acceleration forces arising. An additional effect of using belts is the increase in belt length during acceleration. This has a negative influence on positioning accuracy. Belt drives can therefore only be used up to an aisle length of approx. 60 m.

For these reasons it is almost only wheel-driven vehicles which are used nowadays, where two friction wheels are pressed onto the rail. The advantage is the lower elasticity, which thus leads to fewer vibrations during acceleration and braking. The two friction wheels are often driven by two asynchronous motors which are controlled by one inverter. For wheel-driven storage and retrieval units there is basically no limitation in terms of aisle length. Aisle lengths of several hundred metres are in use today.

Modern storage and retrieval units for bin handling systems operate with rack heights of 18 m and more. With travelling drive concepts implementing three to four drive motors (two motors for the bottom travelling drive and one to two motors for the upper "anti-oscillation unit") speeds of up to 6 m/s and accelerations of up to 4 m/s² are achieved (Fig. 4-19).

In pallet handling systems less dynamic travelling drives are used. Here it is a high materials handling capacity which is important. Issues such as wheel wear or oscillations therefore play a minor role. Characteristical values for pallet S/R machines are roughly 2 to 4 m/s travelling speed and 1 to 3 m/s² acceleration so they are much lower than the figures for bin handling systems. The multi-motor technology described above is employed here as well.

Curve-going storage and retrieval units. Curve-going storage and retrieval vehicles are required whenever several bay aisles in a high-bay warehouse have to be served by one vehicle. In such a case the vehicle has to negotiate curves in the transfer aisles so it has to be of curve-going design. The limitation to one vehicle is expedient whenever material flow in the warehouse is low because during negotiation of curves it is necessary to reduce travelling speed.

Since the S/R machines run on two wheels along rails but the radii of curves are very small in relation to the wheelbase, this means that the wheels have to travel at different speeds.

This change in speed must be actively controlled. As a result, transverse accelerations are minimised during the negotiation of curves.

When the front wheel enters the curve, the rear wheel is still travelling along the straight and slows down. When the rear wheel enters the curve, the front wheel is back on the straight. This brings about an increase in speed for the rear wheel, which can be more than double, depending on the radius of the curve and the wheelbase.

In such a case a separate inverter is required for each travelling drive which adjusts the necessary travelling speed. A speed controller determines the speeds of the front and rear drives during the negotiation of curves.

Industrial trucks. As opposed to the travelling drives already described industrial trucks travel without tracks. As a result they are much more flexible but they also require control and safety functions.

Automated driverless transport systems (DTS) (Fig. 4-20) move at a maximum speed of 1.67 m/s. The drive is effected by wheels. In addition to the travelling drive, a positioning drive is required for steering. For high mobility it is also possible to make all the wheels steerable and provide them with drives.

Fig. 4-20. Automated driverless transport system

Position sensors for travelling drives. The range of travel of travelling drives is normally much larger than that of positioning drives (chapter 4.4). For accurate travel to positions, it is necessary to determine the present position precisely. If there are only a few discrete positions to reach, proximity switches are often used which supply switching information.

The angle sensors built into the drive on the vehicle are often not very suitable for determining the positions to be reached because, for example, they are unable to output the exact position on the route due to wheel slip. For this reason laser distance sensors can be used if the position has to be determined very accurately. By this method accuracies of 2 mm are possible over distances of up to 300 m. Laser distance sensors can only be used on straight routes, like those in high-bay warehouses.

One alternative on routes with curves (e.g. on monorail overhead conveyors) are serial bar code strips positioned along the route, which are then read by an optical bar code reader on the vehicle. This method makes it possible to determine absolute position on the route with an accuracy of 1 mm. The measuring range with this method is practically unlimited.

4.2.3   Drive systems for travelling drives

Whilst the driving properties of travelling drives are not much different from those of conveyor drives (low dynamic response, see chapter 4.1) and positioning drives (high dynamic response, see chapter 4.4), it is the position of the drive on the vehicle which leads to different requirements and solutions.

Operating conditions. Travelling drives are used in buildings as well as outdoors. The degree of protection of drive motors and gearboxes has to take this into account. On relatively large vehicles such as storage and retrieval units there is a separate control cabinet on the vehicle for the drive electronics. Consequently, the same devices can be used as in the stationary control cabinet but travelling drives are subjected to higher mechanical loads so the components have to have a high level of shock and vibration resistance. In the case of smaller vehicles such as monorail overhead conveyors the drive electronics and the vehicle control system are normally accommodated in a housing with a high degree of protection (IP 54).

Dimensioning of travelling drives. The power of travelling drives which run constantly for a long period and only accelerate for a relatively short period is dependent on friction, driving resistance and any difference in height which has to be overcome. With dynamic applications the drive power is dependent on the velocity profile and the masses which have to be moved.

The input variables for dimensioning are as follows:

- mass of the vehicle including the drive train,
- mass of the payload,
- wheel diameter,
- wheel material and rail material,
- travelling speed, acceleration / deceleration,
- braking delay (Emergency Off),
- maximum incline of the route.

The necessary torque for travelling drives is generally determined by the acceleration and mass of the vehicle. Here too there are two different cases. Firstly, there are drives where the stationary levels of torque predominate and determine the selection of components, e.g. on vehicles on inclines or with very long routes, which thus accelerate and brake rarely. Secondly, in the case of dynamic travel applications it is the levels of acceleration which normally determine the size of the necessary motors and inverters. The thermal load on the components is then of minor importance.

Dimensioning of storage and retrieval units. Since especially with travelling drives on storage and retrieval units (Fig. 4-21) with short cycle times there are no system-specific travel profiles available, such drives are dimensioned according to a standardised profile, the so-called FEM cycle. FEM is the "European Federation of Materials Handling and Storage Equipment" (French: "Fédération Européenne de la Manutention"). The technical rules issued by the FEM are highly respected and enjoy a similar status to that of industrial standards.

The input variables of this profile are aisle length and aisle height. The cycle is so called because its profile describes one storage operation and one retrieval operation per journey.

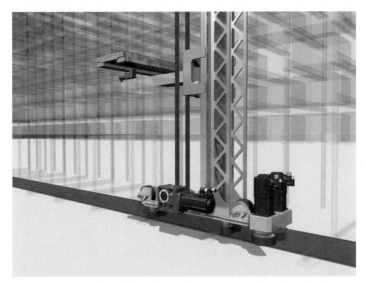

Fig. 4-21. Travelling drive of a storage and retrieval unit

Based on a standardised travel profile the capacity of a storage and re-trieval unit is determined by establishing the storage and retrieval times. The travel profile consists of a triangular journey $s_1 - s_2 - s_3$, whereby the points s_1 (0,0), s_2 (1/3 height, 2/3 length) and s_3 (2/3 height, 1/3 length) are defined (Fig. 4-22). Empirical practice confirms that this travel profile is matched both by the number of cycles and by the thermal load on the drives with maximum deviations of 5%. All in all, dimensioning using the FEM cycle is a good, practice-proved method of dimensioning storage and retrieval units without any accurate knowledge of the materials handling movements.

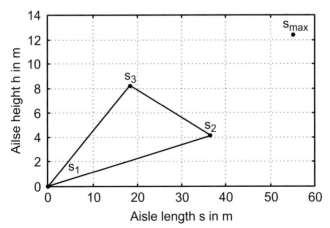

Fig. 4-22. FEM cycle

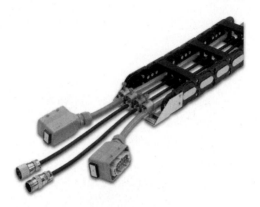

Fig. 4-23. Energy chain

Energy and signal transmission to the travelling drive. A travelling drive is mobile. Consequently electricity and signals have to be sent to the vehicle. For this purpose there are systems with and without sliding contacts.

If distances are limited, *energy chains* can be used (Fig. 4-23). These have to be dimensioned for the travelling cycles occurring over their lifetime. The cables used must also have sufficient fatigue strength. Types with a service life of 25 million motions are available.

Fig. 4-24. Cable festooned system

In cranes and hoists which move along the ceiling of the building *cable festooned systems* are often used (Fig. 4-24). Here the cables are suspended in several loops from mobile trolleys which run along profile rails. The advantage is the minimal amount of material used and the option of also laying hoses for compressed air or liquids. The disadvantage is the drawback due to cables hanging down. The system operates at travelling speeds of up to 4 m/s.

Fig. 4-25. Motorised cable reels

The transmission of energy to large gantry cranes is often performed using *motorised cable reels* which normally include a separate winding drive (Fig. 4-25). The transfer from the stationary to the mobile system is via rotating sliding contacts in the bearing of the cable reel.

Fig. 4-26. Conductor rail systems

If there are long travelling distances and curves, energy is often transmitted by means of *conductor rail systems* (Fig. 2-26). These have to be sealed so that they are protected against dust and dirt. Due to the sliding contact they are subjected to wear, which has to be taken into consideration during the dimensioning procedure and the maintenance concept. Normally, there are at least three poles per phase in order to achieve sufficient operational reliability.

Contactless power transfer to mobile systems can be performed using *inductive energy transfer system* (Fig. 4-27) [Bu06b]. The system consists of a medium-frequency transformer in which a conductor loop builds the primary winding and a pick-up system on the vehicle accommodates the secondary winding. Frequencies of 25 kHz are used. A stationary inverter is required to supply the power. The pick-ups contain rectifiers and provide the energy for the travelling drive. This inductive system is very suitable for transfer carriages because here the energy transfer has to take place on the floor so it is particularly prone to soiling. Due to a reduced frequency of failures and down times the additional cost of the contactless energy transfer system pays off.

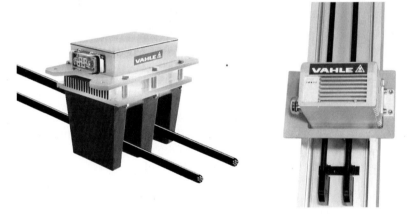

Fig. 4-27. Inductive energy transfer systems

For the transmission of signals the same systems can be used as for energy transfer. In order to reduce the number of cables a communication system is always used. To make the communication system independent of the energy transfer system and provide it with higher reliability, however, optical transmission systems (optical data links) and radio link systems (wireless communication) are used as well.

Geared motors. In travelling drives mainly geared motors with axial, angular or planetary gearboxes are used. The usual travelling velocity and wheel diameters require a reduction of the motor speed by means of a gearbox. Angular gearboxes are used if the geared motors are positioned in the direction of travel so that they cannot influence the width of the vehicle. Worm gears are only rarely used on account of their low level of efficiency. An angular gearbox with a hollow shaft can be coupled to the shaft of the drive axis direct.

Gearboxes for monorail overhead conveyors are dimensioned in such a way that their bearings can directly support the weight of the vehicle with its payload. Consequently, no special bearing is required for the drive shaft with the drive wheel. Gearboxes for monorail overhead conveyors have a disengageable clutch which allows manual movement of the vehicle if the drive system should fail, and hence with the brake applied.

Asynchronous motors are typically used in travelling drives because they meet the demands for high process reliability on account of their rugged design. For dynamic applications field weakening is often used because in conjunction with s-shaped ramps it allows better exploitation of the motor torque. Synchronous motors are used in applications with very high dynamic response (very fast storage and retrieval units).

Group drives and multi-axis systems. High motor outputs are implemented using two separate motors, e. g. with the two travelling drives of a storage and retrieval unit. The two travelling drives of a bridge or gantry crane also belong in this category. These two motors – as asynchronous motors – can be supplied very simply in parallel from one inverter. In this configuration the slip characteristic of the asynchronous system is utilised in order to implement load sharing between the two motors without any complicated control algorithms. Possible tolerances in the mechanical system (e.g. minimal differences in friction wheel diameters) or in motor production are compensated by automatic displacement of the operating point on the relevant motor characteristic.

However, there are also concepts with two inverters where the drive control ensures symmetrical torque splitting. These concepts are applied especially if two drives are far apart, e.g. between the two drives of a large gantry crane or the travelling and anti-oscillation drive of a storage and retrieval unit.

Brakes. Brakes are required in travelling drives in order to reliably hold the vehicle in the stopping positions and to brake the vehicle if there is a power failure. For dimensioning purposes the maximum stored kinetic energy and the probable numbers of emergency stop braking procedures have to be taken into account.

Actuation of drive motors. Simple travelling drives are still operated without drive control in some cases. However, this produces limitations if defined accelerations and braking ramps have to be implemented and stopping positions have to be reached accurately. For this reason travelling drives are normally equipped with controlled drives.

Inverters in travelling drives. Depending on the process requirements these applications use either frequency inverters or servo inverters. For vehicles which are typically controlled via limit switches frequency inverters are ideal (monorail overhead conveyors, gantry cranes).

If, on the other hand, high demands are made in terms of dynamic response and positioning accuracy, the use of servo inverters with an appropriate feedback system is necessary (storage and retrieval unit, industrial trucks). In most cases a resolver is used for the feedback system because at typical gear ratios it provides sufficient positioning accuracy.

Motion control in the higher-level control. With this concept the inverter only acts as a speed actuator (Fig. 4-28 a). The higher-level control transmits a speed setpoint to the inverter (usually via a communication system). The position of the vehicle is detected by the external position sensor and sent to the control. Normally, distance measuring devices with communication systems are used for this.

The control handles calculation of the positioning profile and thus generates the speed profile for the inverter. Consequently, position control is also inside the control unit.

One disadvantage of this system is the dead time in the control structure of the positioning control loop, which has to be compensated by standby factors, thus restricting the dynamic response of the system. In addition, calculation of positioning profiles requires computing power in the control, at the expense either of a powerful control (costs) or of higher cycle times (loss of dynamic response).

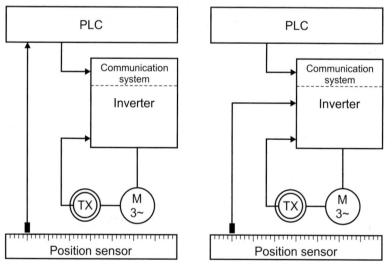

Fig. 4-28. Motion guidance by the control or in the drive

Motion control in the drive. With this concept the inverter is used with a built-in positioning control (Fig. 4-28 b). The higher-level control sends a position setpoint to the inverter. The inverter generates its motion profile on its own. The position of the vehicle is also detected via the external measuring system but it is no longer sent to the control but to the drive. A velocity reduction can also be specified via the control (override), making it possible to adapt velocity to the situation.

This system has no dead times in the control structure which is due to closing the positioning control loop via the communication systems. Furthermore, the software of the control is relieved because it does not have to compute positioning profiles. Generally speaking, a system PLC in this configuration can actuate many more drives than is the case when motion control is performed by the PLC.

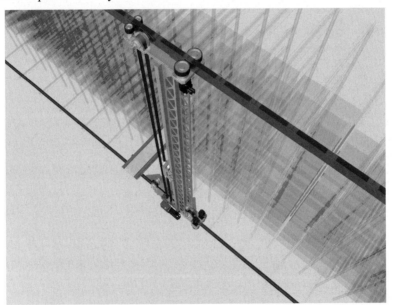

Fig. 4-29. Anti-oscillation drive

Anti-oscillation drives in storage and retrieval units. Due to the mast, storage and retrieval units have a high centre of gravity which, together with the limited stiffness of the mast, tends to generate oscillations. When accelerating and braking this creates forces which can reduce the contact pressure of the wheels on the rail. In the worst case a travelling drive can become derailed or cause the storage and retrieval unit to tip over. This situation must always be avoided by selecting suitable acceleration ramps or suitable control concepts.

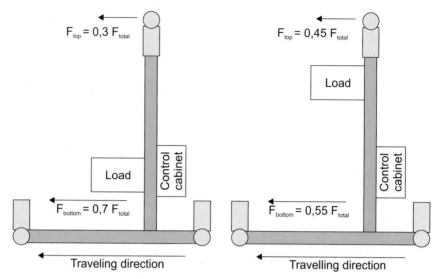

Fig. 4-30. Load distribution between travelling drive and anti-oscillation drive

In order to prevent the mast from oscillating, storage and retrieval units are sometimes provided with an additional travelling drive as an anti-oscillation drive (Fig. 4-29). No additional position sensor is normally used for this drive so the drive is only operated with speed control. Due to the high levels of acceleration, however, mast stresses and strains occur which the anti-oscillation drive is unable to compensate directly. This is illustrated in Fig. 4-30.

Depending on the position of the payload platform the effective mass is displaced towards the upper or lower drive. This displacement is approx. 15% to 20%. Only if the two drives were operated with stiff coupling, would the load distribution be constant. The oscillation depends on mast height, mast elasticity and acceleration. The same behaviour results when braking to halt. Here the aim is to stop without an overshooting of the mast as far as possible, firstly in order to reduce the additional loads on the mechanical system and secondly in order to prevent any loss of time resulting from settling time.

With suitable methods of torque splitting between the two drives, which may still be dependent on the current height of the load, it is possible to optimise force distribution which minimises mast oscillations.

Control functions in monorail overhead conveyors. The control units of monorail overhead conveyors comprise all the functions which are necessary for operating the vehicle (Fig. 4-31) [Ha04]. In addition to actuation of the travelling drive, either via a motor starter or via a frequency inverter, they include the entire motion management of the vehicle, including speed control, motor monitoring, distance monitoring, data transmission to the higher-level control, and possibly other diagnostic options. In order to prevent vehicles from running into one another, either special distance sensors are used so the vehicle can automatically detect when the minimum distance from a vehicle up front has been reached. Or a block section control enables sections for the vehicles and thus prevents collisions. The evaluation of position scales on the route is also handled by the control unit of the vehicle automatically so it can determine its actual position and thus travel to the destination positions.

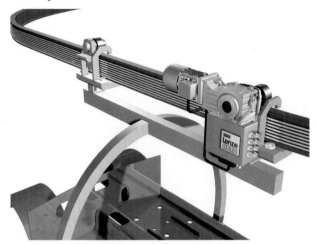

Fig. 4-31. Control and drive unit of an electrified monorail system

Emergency Stop. In Emergency Stop situations there are often deviations from the normal braking procedure of the vehicle. Usually shorter braking ramps are used in order to achieve shorter braking distances. The braking power of the brake transistor and brake resistor has to be dimensioned for this. If this braking procedure is assisted by the motor brake in order to achieve higher braking torque and hence a shorter braking distance, care must be taken to ensure that the motor brake is not subject to excessive wear and hence excessive maintenance. Therefore it is reasonable to execute such Emergency Stop scenarios solely with the inverter and motor in order to protect the brakes.

Safety technology. Travelling drives are predestined for the use of inverter-integrated safety functions because in situations where people are within their range of travel they have to be reliably prevented from moving or their velocity has to be restricted.

4.3 Hoist drives

Sabine Driehaus, Karl-Heinz Weber

Hoists and vertical conveyors have the task to move or position masses vertically under the influence of gravity.

Applications where loads have to be transported along an inclined plane and work therefore needs to be done against the influence of gravity can also be carried out with conveyor drives, as described in chapter 4.1. Load positioning, a job frequently demanded of lifting gear, is described in chapter 4.4. The emphasis of this chapter is on lifting applications where loads are moved vertically.

4.3.1 The lifting process

In the lifting process, persons or loads are transported and moved vertically. This very broad field of application includes the following areas:

- transport of persons by lifts, escalators and stairlifts,
- cable-operated lifts and cable cars,
- winches in theatre lifts,
- moving materials in the areas of logistics and production, vertical carousels,
- building site/construction and gantry cranes, loading equipment for containers,
- winches, capstans,
- Stackers, swing stackers, pallet lifts, windlasses, scaffolding shaft lifts,
- lifting platforms, lifting stations, scissor-type elevating platforms,
- storage and retrieval units,
- industrial doors as sectional or high-speed doors,
- tower drives of accumulators or material stores.

There is smooth transition from applications with low to moderate acceleration of 0.1 to 2 m/s² (< 0.2 g), which are frequently effected using rope systems, up to very high accelerations of 60 to 70 m/s² (6 to 7 g) for materials handling applications.

Important parameters are listed in Fig. 4-4 for the most frequently encountered hoisting and vertical conveying applications.

Table 4-4. Overview of typical parameters of hoists and vertical conveyors

Application	Industry	Velocity (m/min)	Mass/pay-load (t)	Power output (kW)
Storage and retrieval units	Logistics	120	10	5–150
Cranes/winches	Conveyors	200	150	1–120
Belt lifters	Assembly	110	2.2	10–37
Platforms	Conveyors	120	1.5	5–22
Elevating platforms	Conveyors, logistics, transport	10	10	2–15
Freight lifts	Conveyors, logistics, transport	120	5	2–50
Building lift	Construction	24–100	0.1–10	1–200
Passenger lifts	Technical building services	50–1,000	0.5–10	2–250
High-speed doors	Technical building services	180–240		2.2–3
Monorail overhead conveyors	Conveyors, logistics, transport	15	4	2–10

4.3.2 Mechanical design of hoists

As the areas of operation of hoists and vertical conveyors are very different, there are also different mechanical designs and force transmissions.

Winches for cranes, theatres and materials handling. In winches, power is transmitted from the drive to the load via a cable drum and a cable. This can be done with or without a vertical guide (rollers, slide rails). The cable can be run over one or more sheaves, or be reeved (principle of a pulley system). Force can only be transmitted in one direction using cables or belts. For this reason, they are only employed in applications with minor acceleration, well below the acceleration due to the Earth's gravity of 1 g (1 g = 9.81 m m/s²). Typical winch applications are cranes of all kinds, winches in theatre lifts, storage and retrieval units and manually-operated hoists. A cable traversing unit controlled from the winch speed is coupled mechanically or also electrically with the winch via a traversing drive, and ensures the cable is traversed properly in one or more layers. In single-layer systems grooves may be applied to the cable drum, in which the cable is laid when winding.

Fig. 4-32. Gantry crane

Cranes designed as Electric Overhead Travel (E.O.T.) cranes, gantry and construction cranes (Fig. 4-32) or simple hoists (Fig. 4-33) lift loads by means of cable winches and operate with various reevings. The drive units are often mounted on travelling construction elements just like the travelling drives.

Fig. 4-33. Hoist

In order to lift sets and scenery vertically in theatres, a number of parallel cables are often used per set, acting on the scenery via cross beams. The cables of the platform and set lifts are wound synchronously around mechanically-coupled winches arranged axially next to each other. The winches are powered from a central hoist drive.

Tangential nip rollers ensure that the cable is laid safely into the grooves in the drum. The cables can have a length of up to 50 m in length. Systems with counterweights are also used to lift the scenery in emergencies.

Instead of cables, in the field of materials handling, several layers of belts are wound on or off, e.g. to transport plates to a machining process. This principle has position-dependent variable diameters, as is also the case in winding drives.

Hoists with driving pulleys and counterweights. Hoists with drive pulleys have the advantage over winches that the cable length is practically limited only by the cable's own weight. Hoists with driving pulleys, such as e.g. passenger lifts, are invariably fitted with counterweights that run up the side of the shaft, parallel to the lift cage.

The counterweight is generally designed to be half the payload. This means that the hoist is in equilibrium with half the payload.

In addition to the two directions of rotation, two torque directions (motor and generator) can occur, depending on the load.

The cables connect the cage with the counterweight directly or via suspension mountings, running in V-grooves with a minimum angle of contact over the driving pulley. The driving pulleys around which the cables pass transmit the drive torque by means of friction against the cable and hence to the system consisting of the cage and counterweight.

For safety reasons, a number of parallel cables (6 to 8) with a very high safety factor (12) are used, so that a single cable will be able to carry the load. A redundant safety concept ensures extremely high safety, even in the event of power failure or cable breakage.

A new development to be mentioned here are belts (sheathed, steel-reinforced flat belts) as replacements for steel cables, which offer the advantage of increased running smoothness and a longer service life.

Reevings. Cables in lifting gear can be run over one or more pulleys, known as reevings (principle of pulleys). With the cable running over multiple pulleys, the circumferential speed of the cable drum is a whole-number multiple of the hoisting speed. This reduces the required torque on the cable drum by the reeving factor and increases the speed by the same factor. In the case of materials handling equipment, reevings can also take the form of guide pulleys for toothed belt. Reevings have no effect on the drive power (apart from highly-dynamic movements).

Fig. 4-34. Hoist with cable run over a reeving and driving pulley

Hoists with endless belts or chains. If high acceleration and/or positioning accuracy is required, or if there are compelling design reasons, endless belts or chains to which the hoisting equipment is coupled are used in hoisting applications. These can be shaft, belt or support lifts for pallets, for example, some of which have very high lift heights of up to 20 m. They are equipped with counterweights, generally set to half the payload.

Fig. 4-35. Hoists with endless belts

Fig. 4-36. Heavy-duty monorail overhead conveyor

Hoists for materials handling are fitted with endless toothed belts, with or without counterweights (Fig. 4-35). Alternatively, the hoist winds itself with an omega-type wheel set along a fixed tensioned belt, where the direction of travel can lie between the horizontal (X axis) and the vertical (Y axis). A further method of transmission is the use of chains, used with heavy duty monorail overhead conveyors, for example (Fig. 4-36). Vertical carousels for warehousing purposes are operated with chains or toothed belts.

Hoists with spindle drives. A number of parallel angle-synchronous spindles may be used for raising and lowering lifting platforms. The spindles are coupled electrically or mechanically. These systems are self-locking due to the generally small spindle pitch. This means that in the event of failure of the drive torque, the spindle cannot be turned by the load. The lifting platform therefore maintains its position in the event of a power failure.

Hoists with rack and pinions. Translation to linear motion can also be effected by a rack and pinion. In this system, a pinion rolls along a vertically-fixed rack, and the lifting platform slide blocks run on sliding or ball bearing bushes. The drive system is in the lifting cage.

Hydraulic hoists. Hydraulic drives are used for lifts in particular, acting on the lift cage either directly or via cables.

Eccentric or lever mechanisms. Eccentric or lever mechanisms are used to overcome small hoisting heights, e.g. in transfer units or ejectors. These non-linear mechanisms are used in conveyors to transfer pallets or bins from one conveyor line to another.

Hoists with scissor mechanisms. Scissor mechanisms are used in scissor-type elevating platforms in production and logistics processes to achieve short-travel lifting movements (Fig. 4-37). These mechanisms can be powered by hydraulic cylinders, electrical spindle drives, eccentric mechanisms or cables. Due to the geometry, an angle correlation exists between the motor speed and the lifting speed. Geared motors are used as the electric drive.

Fig. 4-37. Scissor-type elevating platform

Hoists for industrial doors. For industrial high-speed doors, a geared motor drives the door via a gear and chain. The door consists of individual articulated panels. For safety reasons, compensation springs acting against gravity ensure that the door remains in equilibrium in its partially-opened position if the drive should fail. Motor and generator operation in both directions is possible here as well.

Sensors on hoists. The following status variables are important in hoisting applications and need to be detected by suitable sensors:

- the position,
- the mass of the load to be lifted.

Position sensors. In order to be able to approach defined positions, in the simplest case an operator is needed who can give travel instructions by means of keys or a joystick. This kind of manual operation is standard with cranes and hoists. As soon as lifting equipment needs to be operated automatically, position detection is necessary. With fixed stopping positions (e.g. storeys in buildings, shelf levels), this can be done with proximity switches that first switch the drive from operating speed to inching, and then stop it at the desired stop position by means of an additional limit switch. For more demanding applications, relative or absolute position sensors are fitted to the motor or winch, or linear measuring systems (e.g. laser distance sensors) with or without reference sensors on the hoist are used. Evaluation and monitoring is done in the inverter or control unit. Redundant double encoder systems are frequently used for applications with higher safety standards and for highly-dynamic control. One encoder on the motor for speed detection, and a second encoder on the load for low-tolerance position measurement.

Load sensing equipment. To detect the actual load on the cable, the drive motor can be fitted with torque sensing shafts or force measuring bearings, or a load-measuring device can be fitted directly to the cable, acting on the inverter or control unit. It is also possible to determine the load torque indirectly from the motor control variables.

4.3.3   Hoist drive systems

High demands are generally placed on the drive systems of hoists, since dangerous motion caused by the load moment quickly occurs if the drive should fail. The use of hoist drives is therefore governed in many areas of application by regulations and standards intended to ensure safe operation. For this reason, integrated safety functions are generally included in hoist drives in addition to the drive functionality.

The majority of hoisting applications have to be operated with defined motion profiles and precise sequences, and are therefore fitted with controlled drives. Geared brake motors operated directly from the mains are only now used where there are minor requirements on the performance of the hoist drive.

Operating conditions. Since hoists are used not only in industrial enterprises but also in buildings, there is a wide range of operating conditions, which are set out in chapter 3.2. Drive motors and gearboxes must sometimes comply with high enclosures (IP 54 or IP 65), e.g. for outdoor or wet areas. Appropriate surface treatments of bare metal parts, e.g. brake discs, are sometimes used for corrosion protection.

Noise emission. If lifting applications such as passenger lifts or theatre platforms are used in acoustically sensitive surroundings, low-noise operation of the drive system is necessary. This also applies to inverters, motors, gearboxes, brakes and brake resistors. The inverter must be operated with a switching frequency of ≥ 16 kHz, and the gearbox, brakes and brake resistors must be designed in a low-noise design. Worm gearboxes with low-noise running properties are suitable here, for example. Alternatively, direct drives may be used instead of gearboxes.

Dimensioning of hoist drives. Hoist drives build up potential energy when lifting, which is released again when the load is lowered. A cable-operated drive system without a counterweight operates in two quadrants (positive and negative speed with positive torque). Hoist drives with counterweights operate in all four quadrants.

The power P_{Hoist} to be applied is calculated from the mass m that has to be lifted, the velocity of travel v and the acceleration a:

$$P_{\text{Hoist}} = \frac{\left(\Delta m \cdot g + m_{ges} \cdot a\right) \cdot v}{\eta_{mech}} \tag{4.1}$$

Δm here is the difference between the mass of the lifting gear with the payload, and a counterweight. Apart from the static torque, additional dynamic torque must be provided by the drive for accelerating and braking. Fig. 4-38 shows the changes over time of speed, power and torque for a hoist with a winch.

Hoist drives with cables are generally operated with accelerations from about 0.5 to 2 m/s². This produces a dynamic load component of about 5 to 20 % of the stationary torque. The dynamic components in the acceleration only become significant with large accelerations. Materials handling equipment with vertical axes is sometimes operated with very high accelerations of up to approx. 60 m/s². In these cases, the dynamic load components are dominant when designing the drive.

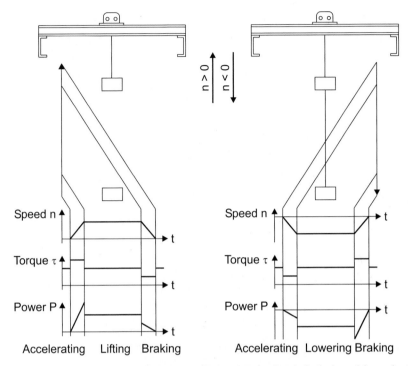

Fig. 4-38. Diagram of speed, torque and power over time of a hoist with a winch.

In *hoists with counterweights*, the stationary and dynamic torque proportions change quite considerably. The counterweight reduces the stationary torque while increasing the dynamic torque, since the counterweight also has to be accelerated. When using compensation springs (weight compensation, safety function) in door drives, the dimensioning must take into account the compensation effect on the load side. A characteristic curve or graph of the resulting torque requirements for the load as a function of position is necessary here.

The design of hoist drives must be carried out very carefully and with sufficient static and dynamic reserves. For passenger lifts, for example, the drive must be able to move a mechanically jammed cage out of the safety wedges.

Geared motors. Geared motors used in hoisting applications are predominantly those with axial, angular or planetary gearboxes. If the gearbox acts directly on the winch, it can do so either axially through a shaft journal and a coupling, or directly on the winch shaft journal, in which case a torque plate takes up the reaction torque. Here, the gearbox acts as a shaft-mounted gearbox. If a motor is intended to drive two cable drums, angular gearboxes with two drive journals are used. If a hoist is operated with large loads at low speeds and small loads at high speeds, this can be done by using a gear change mechanism or by operating the motor with field weakening in order to make better use of the output power.

Worm gearboxes are used in applications in acoustically sensitive areas. They offer the advantage of low-noise operation, and beyond a certain transmission ratio the useful effect of being self-locking for torqueless phases of the drive. When using worm gearboxes, the significantly lower efficiency must be taken into account in the dimensioning process.

Hoist drive gearboxes are generally combined with asynchronous motors. Synchronous motors are used when the compactness of these motors, or their low moment of inertia on materials handling axes, offer advantages. With very large power outputs (e.g. container cranes), DC drives continue to be used.

Fig. 4-39. Direct drive with drive pulley

Gearless direct drives. Direct drives are increasingly used for passenger lifts and in theatres. They have the advantage with passenger lifts that the resulting compact drive units can be installed directly in the lift shaft without the need for separate machinery room (Fig. 4-39). For hoist drives in theatres, the main advantage of direct drives lies in their extremely low acoustic emissions. In order to suppress operating noise as much as possible, the inverter must be operated at an inaudibly high switching frequency of 16 Hz.

Another advantage of direct drives is the significant energy saving compared to conventional geared motors, and the ease of service.

Direct drives predominantly use synchronous motors, but sometimes asynchronous as well.

Group drives. With large motor power outputs, two separate motors are used, one on each side of the winch. As asynchronous motors, these two motors can be very easily supplied in parallel from one inverter (Fig. 4-40 b). If two synchronous motors are used on a winch or winch system, one inverter is allocated to each motor. The symmetrical distribution of the torque is then effected by the inverters (Fig. 4-40 c). This can similarly also be done with two asynchronous motors. If redundancy is called for, this can also be achieved with multi-axis systems designed so that one drive can take the entire load.

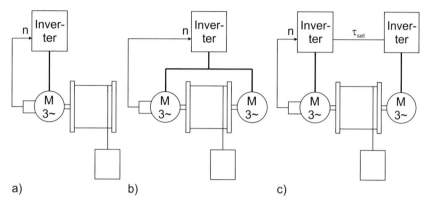

Fig. 4-40. Winch with single-axis (a), group (b) and multi-axis (c) drive.

Brakes. Electromechanical brakes are necessary to hold loads reliably and safely, particularly in the event of a power failure, and are subject to high demands in terms of reserves, reliability, redundancy and monitoring. A number of systems are used, depending on the required safety class, e.g. double brakes on the motor or one motor brake and an additional electro-mechanical disc brake on the winch, which can be fitted with several brake callipers and circuits. The brakes are generally dimensioned for at least double the rated torque of the drive motor.

For hoist drives in acoustically sensitive areas, special brakes in terms of safety and noise are sometimes used.

Mechanical emergency brakes are particularly used with lifts, where they wedge the cage with mechanical acceleration equipment in the guide rails (arrester hook) so that it is securely braked in the event of a cable breakage.

Sliding rotor motors. Sliding rotor motors offer a particular integration of brake and motor for hoist drives. The surface of the rotor is conical, which means that when it is energized an axial force is produced. When the motor is not energized, a spring pushes the rotor against the brake surface. As soon as the motor generates torque, the rotor is pulled towards the stator and the brake is released.

The advantage of the sliding rotor motor is intrinsic safety: if the motor is unable to generate any torque, the brake is automatically applied. No external switching elements are needed for this action. Disadvantages are the high manufacturing resources required, which lead to higher costs, and the increased current requirement due to a larger air gap. These reasons result in the increasing replacement of sliding rotor motors by standard motors fitted with spring-operated brakes.

When sliding rotor motors are operated with frequency inverters, attention must be paid to the very different excitation current requirements between the start-up and running phases.

Handwheels and manually-released brakes for emergency operation. In order to be able to operate the lift or hoist in the event of a power failure or the absence of an energy supply, handwheels and manual releases are provided on the brakes. This is only possible within a limited power range and with geared motors.

Actuation of the motors. Simple hoists are operated with asynchronous braking motors actuated by a reversing contactor circuit. The motor can be equipped with a separate, integral electromechanical brake, or is designed as a sliding rotor motor with a combined rotor braking system. With frequent switching operations, the service life of the brake and switchgear and the energy losses of the motor and brake must be taken into account in the design phase.

Somewhat more comfort is offered by pole-changing asynchronous motors with two separate windings (4/16 or 4/12 pole), where the higher-poled winding is used for inching or DC braking to move up to the destination position. A brake is compulsory here as well.

The best operating behaviour of hoist drives is achieved by using an inverter. Gentle acceleration and braking ramps and a variety of travel speeds can be set. Approaching to specific positions is also significantly easier to achieve and more accurate compared to uncontrolled drives. The spectrum of inverters ranges from frequency inverters with V/f-control to servo drives with speed or position control. Due to their superior characteristics, controlled drives with inverters are finding increased use in hoists.

Use of field weakening with frequency inverters. The use of field weakening in asynchronous motors offers a major advantage in terms of reducing the drive power to be installed for the scenarios "heavy load at low speed" and "light load at high speed". Furthermore, with highly-dynamic applications and changing loads, field weakening can have a positive effect on the control and acceleration behaviour by reducing the load matching factor.

There is a danger with field weakening that the drive can stall and the motor can no longer follow the magnetic field. If the hoist is operated in the field weakening range, the field weakening factor must be matched to the stall torque changes and the motor characteristics. A load-dependent set point limitation for the field-weakening range is an essential function to prevent inadmissible speeds under heavy loads. The actual load can be identified by the inverter or detected by a sensor.

Dissipation of regenerative power. Since regenerative power is generated in descending under load, the inverter must be able to dissipate this. The different methods of dealing with this power are set out in chapter 3.4.1. In hoists with controlled drives, brake choppers with brake transistors or regenerative power supply units are used to dissipate the regenerative power.

Inverters with brake choppers offer the advantage that even in the case of mains failure they can maintain the controlled braking operation by using the regenerated power from the motor.

Braking resistors for hoisting applications must able to convert power for prolonged periods. For this reason, other designs are used than those for short-term conversion of kinetic energy.

With regenerative power of more than 5 to 10 kW, the use of regenerative power supply units is expedient for energy reasons. This limit will move down to smaller power magnitudes in future in order to obtain better energy efficiency.

Drive control. A hoist drive must be able to control the motor smoothly because of the active load in all operation quadrants of the torque-speed characteristic, especially in the transitions between quadrants and at standstill. This requirement exists irrespective of whether the system in question is or is not equipped with speed feedback. In sophisticated systems a far higher speed operating range (e.g. 1 : 2,000) and operation at standstill under full load must be possible in order to hold the load without activating the brake.

Different drive control methods are employed for hoist applications, differing from each other both in terms of drive performance on the one hand, and costs on the other [Ti99]. The areas of application of these methods are explained below.

V/f-control without speed feedback. For simple freight lifts or lifting equipment in the power range up to 7.5 kW, V/f-operation of a frequency inverter in conjunction with an asynchronous motor is often sufficient. The fundamentals of V/f-control are described in chapter 3.4.4.

When using this mode for hoists, it must be ensured that the voltage boost (V_{min}) is set to a value that the motor current does not fall below 1.67-times the rated motor current. There is no correction of the V_{min} setting as a function of motor temperature and load.

In most cases, this constant V_{min} setting is used for lifting and descending. This causes an excessive current at small negative speeds (descending) which has to be provided by the inverter and further heats up the motor. Specifying a reduced V_{min} setting for descending avoids this effect, provided that the inverter can change over the V/f-control parameters during operation (e.g. by a parameter set change).

In order to avoid thermal problems with the motor due to the high V_{min} setting, the motor should only be used up to 75 % of the rated motor torque. The inverter must generally be selected one power output step bigger than the motor to be able to provide the high current for the lower speed range.

As soon as the set-point speed of the drive is less than one to two times the rated slip speed of the motor, the brake must be activated to ensure safe load handling on the transition between brake and motor. A transition between lifting and descending without activating the holding brake is not possible. This load transfer between motor and brake is subject to jerking. Due to this limitation, the speed setting range of a V/f-controlled hoist drive is between 1 to 10 and 1 to 20.

Despite the disadvantages stated, this method is frequently employed in the lower power range, since up to 2.2 kW it offers a very good price-performance ratio.

Additional speed feedback can be used for slip compensation and standstill monitoring.

Due to their imperfect motor excitation in the lower speed range, drive systems with inverters using V/f-control should only be used for power outputs up to 7.5 kW, ensuring that they have sufficient power reserves.

Servo control with speed feedback. Servo inverters with field-oriented current-impressing motor control (chapter 3.4.4) ensure an excellent highly dynamic torque control, a high speed setting range and high overload capacity. Continuous transitions between lifting and descending are possible, as is holding the load at a standstill without the support of the brake. Using an integrated angle controller, the load stiffness at standstill can be increased. This control method can be used with both synchronous and asynchronous motors with speed feedback. The load is taken on smoothly.

The field-oriented control provides the motor with the current actually needed to control the load. With partial-loads in the lower speed range, this results in substantially less current consumption than is necessary with V/f-control. There is no need for the inverter to be overdimensioned, and no unnecessary heat is generated in the motor.

Table 4-5 shows the properties and applications of both methods of drive control for hoists.

Drive functions for hoists. In addition to controlling the drive, further functions must be provided for hoisting applications. These can be implement in higher-level control systems or in the inverter or distributed over both. The following functions are needed for hoisting applications:

Table 4-5. Control methods for hoist drives

Control method	Effectiveness	Applications
V/f-control without speed feedback	O	Simple hoisting applications, e.g. high-speed doors, freight lifts
Servo/vector control with speed feedback	++	Precision hoisting applications in automatic or manual operation, e.g. storage and retrieval units, passenger lifts, stacker cranes

- travel profiles for motion control,
- positioning at the holding positions,
- braking management functions,
- load-dependent speed limitation,
- active load monitoring.

Travel profiles. Motion control must ensure that the natural frequency of the spring-mass system is not excited by the travel profile. This means that s-shaped ramps and limited acceleration times are used. For increased ride comfort in passenger lifts and occasionally for transporting sensitive goods, low-jerk motion profiles are used (s-shaped, square sinusoidal, specific mathematical polynomials). These offer the advantage that the change in acceleration (the jerk), and therefore the influence of force on the mass to be moved, is finite or continuous. Mechanical vibrations are only slightly excited by this. In fact, a long rope with a load represents a spring-mass system with very little damping.

Positioning at the holding position. A simple positioning functionality can be provided in the form of switch-off positioning with inching and stopping triggered by a limit switch. For more sophisticated hoists with positioning functionality, a position controller is necessary that can evaluate absolute or relative position sensors with different interfaces and protocols. If different winch diameters occur when winding, this should be compensated electronically by a diameter calculation.

Brake management. The control of the holding brake on a hoist is so closely connected with the motor control that it can easily be done by the inverter.

Ideally the switching element for actuating the brake will be located directly in the inverter. Alternatively, the brake can be actuated by an external relay from a digital output from the inverter.

The objective of a good brake management is to achieve as smooth as possible a torque transfer from the brake to the inverter. Since the brake is not released suddenly, but the braking torque continually decreases during the brake disengaging time, the inverter must have already built up a torque when the brake is disengaged.

With simple frequency inverters operating a hoist with V/f-control, there is generally no specific brake management. Instead, the brake is actuated as a function of the speed set-point at one or two times the rated slipping speed of the motor. This results in the load being taken up with a jerk.

A continuous increase or decrease of the motor torque during the load commutation between the motor and brake, which can be implemented in the drive control by torque limitation, prevents unwanted noise. If the load is detected and processed by the motor control, this contributes to low-jerk transitions between the motor and the brake.

Load detection. Load detection is necessary to prevent inadmissible loads from being lifted and for low-jerk load transfers between the motor and brake. Overload monitoring often includes monitoring the correct release of the brake. If the inverter tries to run against the incorrectly-applied brake, this will be detected as an overload status.

Load-dependent speed limitation. This function enables a higher drive speed in the partial-load range by utilizing the field weakening range. The advantages of using field weakening on hoists have already been discussed. The inverter limits the maximum permissible speed as a function of the actual load. It is assumed that the load does not increase during a lifting action. Without this function, an inverter in descending operation with an excessive load could operate with field weakening and no longer be able to brake the load on its own. When braking the load in descending operation, a higher torque is generally needed than for a constant descending speed.

Multi-axis applications. If multi-axis systems (one inverter per motor) are part of the same lifting process and if the drive motors are not mechanically coupled, the two drives must operate synchronised at the same angle. The software functions necessary for this are described in chapter 4.6 on synchronised drives. This function can be used on a bridge crane with two trolleys for transporting long goods (Fig. 4-41).

Such multi-axis systems with friction-type or positive-type couplings between the motor and load are operated in torque ratio control. If this is done with group drive systems (several asynchronous motors parallel to one inverter), the motor control must support this mode of operation.

Installation and automation. Inverters for hoists are frequently controlled by a PLC, lift or crane control units. Separate monitoring units are also used to meet high safety requirements.

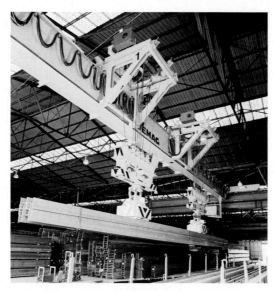

Fig. 4-41. Two hoists operating in tandem for loading long goods

Demanding requirements on the function are generally implemented in special lift or crane control units that contain the logic functions, set-point generation and monitoring. Crane control units are often fitted with wireless remote control. The very extensive regulations and standards are not dealt with here.

Safety equipment for hoists. Lifts are subject to highly specific safety requirements and must be checked regularly by the technical supervisory authorities. The technical regulations are set out in European directives.
In the event of a failure of the electrical power supply, it is necessary for passenger transport and also sensitive areas of goods transport for the lifting equipment to be capable of continuing operation from a UPS or batteries with a direct connection to the DC bus of the inverter.

In emergencies the motor is separated safely and partly redundant from the inverter by a motor contactor, or two contactors connected in series. Alternatively, this can be done by a multichannel safety function in the inverter ("Safe Torque Off", STO).

In drive systems with synchronous motors, the winding is short-circuited in an emergency. This causes a dynamic braking, and is used also in the event of a power failure. The braking effect depends on the speed.

In normal operation, electromagnetic brakes are switched on the AC or DC side; switching on the AC side takes longer but is quieter and the switchgear is not so heavily stressed (chapter 3.3.7). When switching on the DC side the response time of the brake is markedly less, so in emergencies the brake is switched on the DC side. In the event of a fault, in safety-oriented systems the AC and DC sides are separated into two channels. In safety-relevant areas, the brakes are fitted with manual operation by means of a lever or cable mechanism. A two-channel brake actuation system ensures redundancy and safety. Monitoring the brake for wire breakage, current and actuation (microswitch) increases operational safety. If the brakes are only rarely actuated, the brake disc can corrode. Consequently the effectiveness of the brake is tested at regular intervals and if necessary conditioned by briefly running against the applied brake.

Sequential monitoring functions such as mains and motor phase failure, detection of mains undervoltages, over- and undervoltage in the DC bus, mains voltage asymmetry, braking and monitoring the position sensor are elementary components of inverters for hoists.

4.4 Positioning drives

Andreas Diekmann, Karsten Lüchtefeld

Positioning drives are used to move conveyed material or work pieces to a certain target position. The issue of positioning objects to a target position has already played an important role in chapter 4.2 Travelling drives and chapter 4.3 Hoist drives, but it has not been explained in detail. Hence, in this chapter the general architecture of positioning drives will be described in detail.

In this chapter point-to-point positioning tasks will be introduced. For these kind of tasks the crucial point is the target position and not the entire continuous path which is pursued to reach those positions. The more complex and multidimensional movements will be explained in chapter 4.5 (Coordinated drives), chapter 4.8 (Cross cutters and flying saws) and in chapter 4.9 (Cam drives). These drive solutions also perform positioning tasks, however, they are more difficult to describe and to implement than the point-to-point positioning tasks specified in this chapter.

4.4.1   The positioning process

Positioning is defined as the action of moving one ore several movable machine parts from a starting position to a predefined target position. During positioning a mass m has to be moved along a certain distance s in a certain time t. The target position must be reached maintaining the predefined accuracy Δs. Mass m and distance s depend on the particular application. The distance s and mass m can vary within fixed limits (maximum mass of the conveyed material, longest path within traversing range).

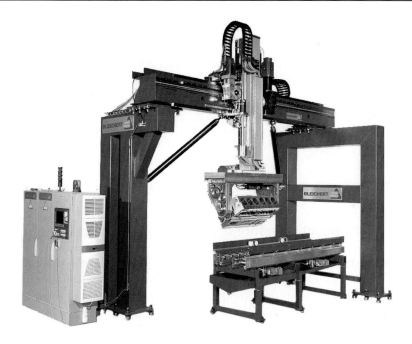

Fig. 4-42. Handling machine with servo positioning drive

Positioning tasks are often non-productive times in production machines and don't add any value to a process. Such a positioning task could be for example approaching different stations to load and unload machines. In this case the positioning task has to be carried out as fast as possible. To achieve this, a short time t for the positioning task and therefore also a highly dynamic positioning drive is required.

The quality of the positioning task is defined by the accuracy Δs maintained while moving an object to its target position. The required accuracy depends on the particular application. This means that e.g. in conveyor or travelling drives an accuracy in the range of several millimeters is sufficient while material handling requires an accuracy in the range of 0.1 to 1 mm. In production processes accuracy requirements can even be a lot higher. Precise machining requires an accuracy in the range of micrometers. In general, higher positioning accuracy means increased efforts regarding the mechanical parts, sensors and drive control.

Positioning drives that move conveyed material or work pieces to a predefined target position are to be found in the following applications (Fig. 4-42):

- loading and unloading of work pieces in production machines,
- pick-and-place-machines, e. g. for placing electronic components,
- assembly machines,

- rotary indexing tables,
- tool changers,
- adjustment of limit stops in production machines,
- travelling and hoist drives in storage and retrieval units,
- palletiser,
- stopping of conveying processes at an exact position in continuous and intermittent conveyors (conveyor belts, monorail overhead conveyors, industrial trucks),
- travelling to a stop position in hoists,
- rotating rollers to predefined maintenance positions.

4.4.2   Mechanical structure of positioning systems

Positioning tasks can be performed by means of linear or rotary movements, whereas the majority of positioning tasks is performed by linear movements. If a rotating drive is used, the rotary movement of the motor must be converted into a linear movement. The linear mechanisms available for this purpose and their properties have already been described in chapter 3.6.6. The particular mechanism notably determines the achievable speed and therefore the dynamics of the positioning task as well as the repeatability meaning the quality of the task.

The different types of linear mechanisms for use in positioning drives can be grouped together in the following way:

- Spindles are used for high positioning accuracy and low speeds. The positioning path is limited.
- Toothed belts allow for higher speeds, but provide lower positioning accuracy of about 0.1 mm. Compared to spindles the traversing path is longer, but still limited.
- In contrast, the traversing path of rack and pinions is unlimited, whereas the accuracy is lower and backlash occurs.
- Besides the mechanisms mentioned above, transmission elements such as ropes and chains can also be used. Positioning tasks can also be performed by travelling drives, whereas the achievable accuracy and dynamics are limited.
- All applications requiring high speeds as well as high accuracies can be equipped with linear motors (see chapter 3.3.5). Here, the conversion of the rotary movement of a rotating motor into a linear movement can be dispensed with.

Toothed belt drives are widely used for material handling applications, since they provide high speed and an accuracy which is sufficient for this kind of application.

In general, spindles are used for precise adjustment of limit stops and precise positioning of work pieces in production machines.

Linear drives are increasingly used for very precise and fast positioning of electronic components.

Rotary positioning tasks are performed e.g. by rotary indexing tables. In this case direct drives are used for high-precision applications in order to avoid backlash.

4.4.3   Drive systems for positioning

The main requirements for positioning drive systems are the following:

- High dynamics in order to achieve short positioning times,
- sufficient accuracy according to the particular application, and
- high reliability.

Dimensioning of positioning drives. As opposed to continuous processes, positioning tasks are highly dynamic and require the drive to constantly change between acceleration, constant travelling and deceleration operations. For the dimensioning of the drive the positioning profiles, the required torque and the required speed have to be determined. In general, the required peak power is taken into consideration for selecting the drive components. In a few cases it can also be the required continuous power.

Gearboxes. In some applications gearboxes are not necessary, because the permissible input speed of a mechanism can be high enough to allow for the use of a motor without speed reduction. This can be the case in spindles or toothed belt drives. However, in many other cases positioning drives require a speed reduction by means of a gearbox.

The most important criterion for selecting a gearbox for positioning applications is a low gear backlash. For this reason, planetary gearboxes with low backlash are often used. But also helical gearboxes and bevel gearboxes are used, as long as their backlash doesn't affect the required positioning accuracy.

In positioning drives, gearboxes are exposed to constantly changing load conditions, because positioning is characterised by fast acceleration and deceleration as well as the reversal of the rotational direction. Gearboxes and in particular the connecting elements to the motor and to the mechanical parts must be designed to resist these kinds of loads in order to guarantee a high reliability. For the connection between gearbox and motor, friction-type press-fit connections and direct mounting have proven in use, whereas key joints can be worn out too early.

Motors for positioning drives. The selection of the motor type has a crucial influence on the achievable dynamics of the system. Standard motors are not optimised for high dynamics and can therefore only be used in applications with low dynamics requirements. In general, servo motors are used for positioning applications. Especially synchronous servo motors provide low mass inertia and high overload capability. This allows for fast acceleration and deceleration operations.

In order to avoid the negative influences of gearboxes and mechanical parts on a system, direct drives are increasingly used for highly precise positioning applications. In rotary applications high-pole torque motors which are generally set up as synchronous motors are used as direct drives. Linear motors are used for fast and precise linear movements.

Position sensors. In applications with low positioning accuracy requirements the position is determined by means of the built-in incremental encoder of the motor. The various encoder systems and their characteristics have been described in chapter 3.3.6. For use in positioning applications they can be grouped together as follows:

- A resolver provides high robustness and an accuracy which is generally sufficient for material handling applications. It measures the absolute angle of one revolution and is therefore well suited for the use in synchronous motors. However, it requires the homing of the entire positioning drive.
- Encoders with sin-cos signals provide a considerably higher accuracy and are therefore used in applications with high positioning accuracy requirements.
- If absolute encoders and especially multi-turn absolute encoders are used, homing of the drive is generally not necessary anymore after power-on.

In applications with high positioning accuracy requirements the actual position must be measured directly at the mechanical parts. Through this, accuracy restrictions due to mechanical transmission elements (e.g gear backlash, errors in the pitch of the spindle) can be avoided. In this case encoders are applied directly at the load. In linear mechanisms electromagnetic or optical linear scales are used. Optical linear scales can achieve an accuracy of less than $1\,\mu m$.

Actuation of the motor in positioning applications. In applications with very low positioning accuracy requirements a motor can be switched off as soon as the target position has been reached. In this case an inverter is not needed. But the achievable accuracy is extremely low so that these applications cannot really be called positioning drives. However, this set up is still widely used especially together with geared motors, due to the low costs.

In order to achieve a higher positioning accuracy without using an inverter, pole-changing motors have been commonly used in the past. When the first proximity switch has been reached the motor switches to low speed. The second proximity switch causes the motor to stop. Due to the low speed, a higher positioning accuracy can be achieved without having to reduce the total travelling speed to that low level. These days the cost efficiency of this solution is no longer relevant, since the additional costs of the pole-changing motor including its control can as well be invested in inverters which offer many more advantages.

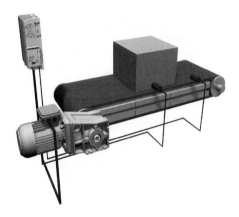

Fig. 4-43. Positioning system with frequency inverter / switch-off positioning

Switch-off positioning with frequency inverter. If only a few discrete positions are to be approached with low accuracy (> 1mm), then frequency inverters in combination with proximity switches can be used (Fig. 4-43). This type of positioning corresponds to the configuration with pole-changing motors except for the fact that here the frequency inverter controls the speed. The proximity switches have to be interpreted directly by the frequency inverter in order to reduce positioning inaccuracies resulting from signal propagation delay and jitter. Frequency inverters provide fixed frequencies (JOG-values) for speed changes that can be controlled by means of digital control terminals. The digital inputs are sampled by the inverter software whereas the sampling rate causes a jitter that is generally in the range of 1 or 2 ms which influences the repeatability of the positioning sequence.

Positioning systems with frequency inverters are commonly used for conveyor, travelling and hoist drives with low positioning accuracy requirements (approx. 1 mm). These systems offer the advantage of low costs, because proximity switches are only required along the traversing path and position sensors at the motor can be dispensed with.

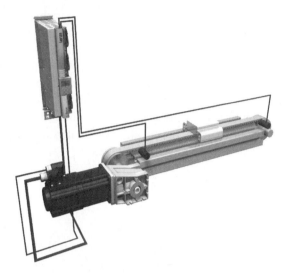

Fig. 4-44. Servo positioning system

Servo positioning drives. Servo drives are used for all positioning drives where high dynamics and high positioning accuracies are required (Fig. 4-44). Therefore they will be described in the following sections.

The positioning sequence of servo drives. The positioning sequence is divided into three steps (Fig. 4-45):

- the acceleration up to positioning speed,
- travelling with constant speed,
- deceleration to standstill including the arrival at the target position.

Acceleration and travelling at constant speed in principle corresponds to the start-up of any drive. Due to jerk-limiting profiles (see chapter 3.7.4) oscillations of the mechanism can be avoided.

The crucial point of the positioning sequence is the transition from travelling with constant speed to deceleration. What's important here is that the drive stops at the desired position. To achieve this the inverter calculates the residual path for the deceleration ramp, in order to determine the right deceleration point.

The dimensioning of the components according to power requirements, which at the same time means to define the costs of the drive system, depends on the required peak power. With the maximum speed and acceleration selected freely, the minimum peak power for a certain positioning task (traversing path s and traversing time t) can be achieved if the total positioning time is segmented equally into three parts which are acceleration, travelling at constant speed and deceleration.

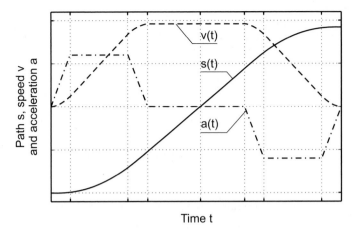

Fig. 4-45. Servo positioning sequence

If the travelling speed which is required therefore can't be achieved due to limitations, shorter positioning times can only be achieved by increasing the acceleration, which results in a higher peak power for the drive.

It is also important for the achievable positioning time and accuracy, to how much jerk the drive is exposed when it reaches the target position. Mechanical transmission elements have limited stiffness and therefore show natural vibrations. For this reason it is necessary to use jerk-limiting motion profiles in order to avoid vibrations in the spring-mass system of the drive train. The corresponding rules of dimensioning have been described in chapter 3.7.4. In a particular application it is important to know the lowest natural frequency of vibration of the mechanical parts and to configure the acceleration ramps for acceleration changes in a way that they are below the resonance frequency.

Additional software functions in positioning drives. In order to simplify the use of positioning drives, some additional software functions are required:

- configuration of the coordinate system,
- limit position monitoring,
- positioning profiles,
- positioning modes,
- homing,
- positioning sequence control.

The coordinate system. The user of a positioning drive is interested in the position of an object. Hence, the software should take into account all the influences and conversions that result from the mechanical parts, the gearbox and the position sensor. In particular, therefore the following parameters must be known for input into the system:

- the feed constant which determines the traversing path per revolution (pitch of the spindle, circumference of the toothed belt sprocket or the drive wheel),
- the reduction ratio of the gearbox,
- the resolution of the position sensor.

These parameters depend on the architecture and the particular system components and therefore have to be determined and input into the system, only once during development and commissioning of the positioning drive. After commissioning, inputs concerning positions of an object can be made by means of the coordinates of the mechanical parts.

Limit position monitoring. All positioning drives with limited traversing range (e.g. spindles) require limit position monitoring in order to avoid damage of machines due to crossing of the limit positions. In practice, hardware limit switches and software monitoring of the permissible traversing range are used for this purpose.

Hardware limit switches. Limit switches are installed prior to the limit positions of a machine in positive and negative direction respectively. As soon as the positioning unit activates one of the limit switches, a fast stop of the machine is initiated. The drive decelerates to stop at the quick-stop deceleration ramp. Furthermore, an error message is generated in order to enable the control or an operator to correct the error. If a limit switch has been activated the drive can be relieved in manual mode. The limit position monitoring is constantly active during automatic operation. In practice, it is desirable to avoid tripping of the limit switches which represent the final safety measure before machine damage occurs.

Software monitoring of the traversing range. In order to avoid tripping of the hardware limit switches, positioning drives provide software limit positions. These are two position marks used by the software in order to limit the traversing range of the positioning unit. Before a positioning sequence is initiated, the software checks whether the target position is located within the permissible traversing range. If the target position is located outside the permissible range, the positioning sequence won't be started and an error message will be initiated.

Positioning profiles. In order to group the different parameters for a positioning sequence together, it is useful to create positioning profiles. The following parameters are contained therein:

- the desired target position,

- the acceleration, the travelling speed and the deceleration of the positioning sequence,
- if applicable, the final speed at arrival at the target position (if the speed for the transition between several profiles is other than 0),
- the maximum jerk which is generally specified as the time needed for increasing and decreasing the acceleration.

The software of a positioning drive should be capable of storing and selecting between several positioning profiles in order to minimise efforts for the setup of parameters by an operator or a higher-level control.

Positioning modes. The positioning mode specifies, in which way a target position is to be interpreted. The following positioning modes are to be distinguished.

Absolute positioning. In this positioning mode the target position is specified as the absolute position in the coordinate system (Fig. 4-46). The drive system compares the actual position with the defined target position, calculates the difference between those and travels to the target position whereas the set profile parameters and the software limit positions are taken into consideration.

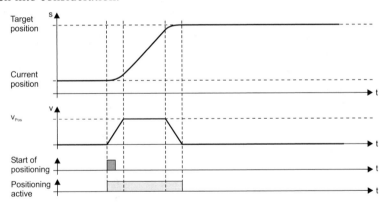

Fig. 4-46. Absolute positioning

The absolute positioning is used when the drive is supposed to move its axis from a random position to a new predetermined position.

Relative positioning. In this positioning mode the target position is specified as the distance to the actual position (relative position) (Fig. 4-47). Here the actual position is not important. The drive travels from the actual position to the target position according to the set profile parameters. The software limit positions are also taken into consideration. If the relative positioning would cause the drive to cross the software limit positions, the relative positioning would not be executed and a warning message would

be generated. Furthermore it has to be noted, that during relative positioning the internal distance integrators which detect the actual position might
overflow. This occurs frequently in positioning applications where the
traversing path is not limited (e.g. conveyor belts, rotary indexing tables).
In such applications the modulo positioning must be used.

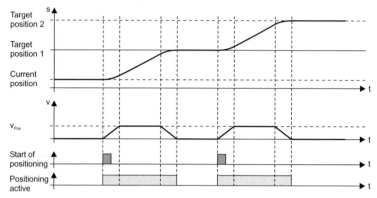

Fig. 4-47. Relative Positioning

Modulo positioning. In this positioning mode relative movements can be
performed endlessly, e.g. rotary indexing table applications (Fig. 4-48).
For this purpose the internal distance integrators are reset before every
start of a positioning sequence in order to prevent them from overflowing.
However, by resetting the distance integrators the absolute reference point
of the axis is deleted. In addition, the software limit positions are not monitored during modulo positioning, because the traversing path is considered
to be unlimited.

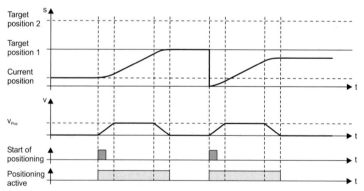

Fig. 4-48. Modulo positioning

Sensor based positioning. Sensor based positioning is used if conveyed
material whose exact position on a conveyor is not known has to be moved
very precisely to a target position (Fig. 4-49). A sensor, such as a photo

electric sensor, detects the conveyed material and stores the actual position of the drive at this moment via a touch probe input (see chapter 3.4.3) and then initiates a relative positioning sequence. Actually, sensor based positioning is quite the same like switch-off positioning with frequency inverters, except for the fact that here the positioning accuracy is considerably higher. The stopping of a travelling drive at an exact position after it has passed a sensor can as well be implemented that way.

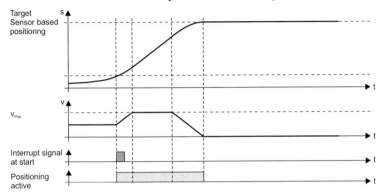

Fig. 4-49. Sensor based positioning

Homing. Many encoder systems such as incremental encoders, resolvers or single-turn absolute encoders only measure relative movements or the absolute angular position of one revolution of the motor. However, the total traversing path of a positioning application generally requires several revolutions of the motor. If the inverter is switched off there is no reference between the revolutions of the motor and the position of the mechanical parts anymore. In order to re-establish this reference, homing is required after power-on. Only if a position sensor which measures several revolutions or the entire traversing path absolutely is used, homing is not necessary.

For precise homing the machine provides a home position switch. Mostly, it is not precise enough to determine the home position by means of the home switch alone. The zero pulse of the encoder is notably more precise. Hence, the drive first travels until the home position switch is activated. Then the position of the next zero pulse of the encoder is stored as the home position for the absolute positions of the mechanical parts of a machine. Here, the home position switch is used to determine the traversing path of one revolution of a motor which is used for the homing of the absolute positions by means of the zero pulse of the encoder (Fig. 4-50).

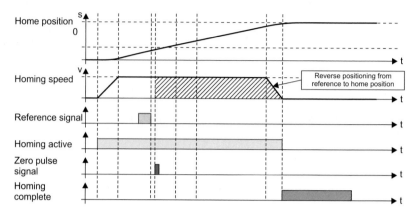

Fig. 4-50. Homing with home position switch

The software enables the zero point of a position to be different from the machine home position by means of a certain parameter.

In order to avoid the installation of an additional proximity switch a limit switch can be used as a home position switch (Fig. 4-51). In this case the drive first travels, until the limit switch is activated. Then the rotational direction is inverted and the position of the first zero pulse is used as the home position.

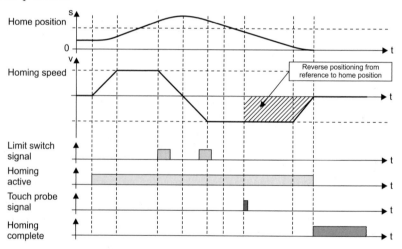

Fig. 4-51. Homing with limit switch

If the mechanical parts provide precise sensors, their signals can also be used for determining the home position by means of a touch probe input (Fig. 4-52).

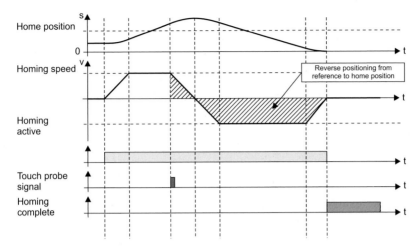

Fig. 4-52. Homing with touch probe sensor

Homing with torque limitation. Due to the increasing cost pressure in machine construction, developers are increasingly trying to avoid the use of sensor systems wherever it is possible. Hence, a homing procedure based on torque limitation of the drive has been developed for this case. Here the system travels in positive or negative direction until it reaches the mechanical limit stops of a machine. The drive then keeps on generating torque until a preset torque limit has been reached. By means of this position the machine home position then is defined.

All in all, positioning drives provide many homing procedures for many different kinds of constellations in machines. All procedures share the same disadvantage which is the time effort for homing. Furthermore, the machine concept has to allow the drive to first travel to the home position.

Setting the home position. If the positioning drive provides a position sensor which measures absolute values (e.g. a multi-turn absolute encoder), then the home position has to be set only once during commissioning or after the exchange of components. The system then works with the home position that has once been defined.

Manual positioning. Positioning drives generally provide functions to move the drive by means of manual commands (Fig. 4-53). This is useful for operations like the setup of positions (teach-in), for relieving the drive after failure or for maintenance. In manual positioning the limit switches are also interpreted and the software limit positions of the traversing range are maintained.

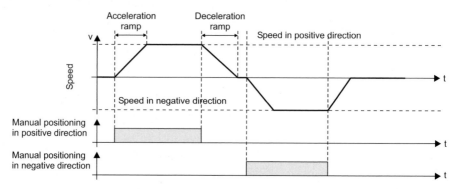

Fig. 4-53. Manual positioning

Positioning sequence control. Usually, positioning drives do not only perform a single positioning movement, but rather an entire positioning sequence. For programming of such sequences, positioning drives provide the possibility to link several positioning profiles with each other. At the respective profile transitions, delay times can be inserted or the drive can be programmed to wait for external signals. Using this feature, no additional control is required to implement a positioning sequence.

Connection to the higher-level control. In general, the positioning drive is connected to the higher-level control by means of logical signals which initiate positioning tasks and report the termination of the tasks to the higher-level control. The parameters for the positioning profiles and in particular for the target position are transferred via the parameter channel of the communications system. Due to this interface it is very simple to connect the positioning drive to the higher-level control.

4.5 Coordinated drives for robots

Ralf Scharrenbach

Specialised machines perform movements which are defined by the machine's geometry and design. The drive functions of such machines are simple and not coordinated.

In contrast, industrial robots either consist of a kinematic chain or a parallel kinematic mechanism. These kinematic systems have a simple mechanical structure. Industrial robots can be used for a wide range of handling purposes. This is reflected in the complexity of the robot control which coordinates three to six axes and thus enables the robot to perform the desired movements. In order to execute a multi-dimensional movement in space, several drives have to travel simultaneously in a coordinated way.

This makes robots very flexible and enables them to move to variable positions in predetermined paths. Hence, in comparison with the drives for point-to-point positioning which were introduced in chapter 4.4, they can perform tasks which are considerably more demanding.

4.5.1   The technological process of handling

Robots and handling systems are used in various applications in modern factory and logistics automation. Typically, the movements are executed with definite and often alternating target coordinates (Fig. 4-54).

According to the German standard VDI 2860 handling is defined as the spatial orientation of any geometrically defined body (e.g. the gripper of a robot or transported material) to a reference coordinate system (Fig. 4-55).

This definition points out the difference between handling and conveying or storing goods, because here either the body is not geometrically defined (e.g. gases or liquids) or the orientation of the body to a reference coordinate system is not clearly defined (e.g. conveyed items on a belt conveyor).

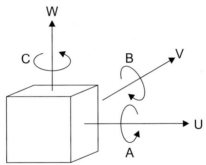

Fig. 4-54. Degrees of freedom according to VDI90

In addition to the logical processing of control commands the control of a handling system (central control, robot control) has to generate the coordinates for the drive motion. Which reference coordinate system is appropriate depends for example on the design of the robot or the user interface (for programming).

Figure 4-56 shows coordinate transformations where Cartesian XYZ coordinates are transformed into axis coordinates for the six axes of an articulated robot and vice versa.

Compared to the point-to-point positioning where only the target position of the respective drive can be stored and approached, a trajectory control allows for travelling in space according to a mathematically defined motion function.

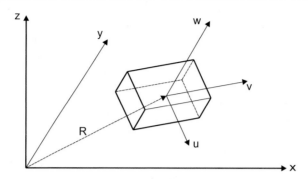

Fig. 4-55. Coordinate systems in handling according to VDI90

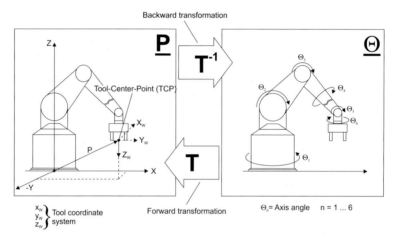

Fig. 4-56. Transformation of Cartesian coordinates into axis coordinates

In trajectory control, speed set points are calculated and transmitted with chronological synchronism for the motion control of the particular drives. The function of motion control has been described in chapter 2.3.1.

4.5.2   Application fields for robots

Handling functions can be categorised into "holding a quantity", "changing quantities", "moving" and "securing or monitoring". Thus, for example:

- metal sheets and welding guns are held in place during spot or laser welding,
- palettes are loaded for order-picking,
- adhesives are dispensed,
- measuring instruments and sensors are approached to work pieces,
- car bodies are handled,

- plastic parts are retrieved from injection moulding machines,
- pipes are bent,
- tools are used for milling, grinding and polishing,
- mechanical long-term stress is imposed on products in an automated way for quality assurance.

The weight of the tools, the work pieces or the conveyed material has a crucial influence on the design of handling machines. Payloads range from a few grams (e.g. chocolates which are placed from a belt conveyor into a box) up to more than 1,000 kg (e.g. train wheels that are stacked).

The material to be moved, the application field as well as the work envelope or workspace determine which machine type is used. The respective machine type then has to achieve the required accuracies (mostly repeatability) and the dynamic requirements regarding cycle times. Table 4-6 shows application fields with typical parameters.

Table 4-6. Application fields based on the classification of the VDMA (German Engineering Federation)

Application field	Accuracy [mm]	Cycles [rpm]	Power requirement [kW]
Painting and coating	0.1	< 20	< 5
Gluing and sealing	< 0.1	< 20	< 15
Spot welding	0.1	< 60	< 15
Continuous welding	< 0.1	< 20	< 15
Machining	< 0.1	< 60	< 10
Cutting	< 0.1	< 120	< 5
Assembly of small parts	< 0.01	< 120	< 5
Microassembly	0.001	< 120	< 1
Sorting	< 0.1	< 120	< 5
Measuring and testing	< 0.1	< 120	< 5
Picking and palletising	< 0.5	< 30	< 15
Handling in presses and forges	< 0.2	< 10	< 10
Handling in die casting machines	< 0.2	< 20	< 10
Handling in injection moulding machines	< 0.1	< 20	< 5
Research and development	< 0.1	< 120	< 5
Service robots	< 0.1	< 20	< 2

4.5.3   Mechanical structure of robots

Robots, which are also referred to as assembly machines or handling machines, are defined according to ISO 8373 as automatically controlled, reprogrammable, multipurpose manipulators with three or more axes. The link members of a robot are connected together via linear guides and rotary joints to form a kinematic chain. The link members, joints and their respective drives form the axes of the robot.

The main axes, also called base axes, are responsible for positioning the end effector and also the tool or work piece within the work envelope.

The hand axes, head axes or secondary axes are mainly responsible for orientating of the tool. Hence, they generally consist of a series of rotary joints.

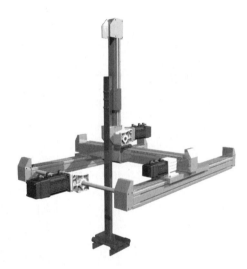

Fig. 4-57. Schematic diagram of a gantry robot or Cartesian robot

Gantry robots or Cartesian robots . This type of robot allows for movements according to Cartesian space coordinates so no coordinate transformation needs to be executed by the control (Fig. 4-57).

Hence, the implementation of the control is relatively simple. Due to the Cartesian space coordinates, in some applications gantry robots can be controlled by means of the motion control functions of an inverter without the use of a separate higher-level motion control. Due to the high stiffness of the construction, gantry robots cover large work envelopes. Typical applications are picking and palletising as well as to loading of assembly and manufacturing machines. Gantry robots share the disadvantage of potential collisions in the work envelope and a relatively low operating speed.

Fig. 4-58. Schematic diagram of a SCARA robot

SCARA robots or horizontal articulated robot. The design of the SCARA robot (Selective Compliance Assembly Robot Arm) which is sometimes called horizontal articulated robot is quite similar to the human arm (Fig. 4-58).

Typical applications are joining processes or placing components onto a printed circuit board or into magazines where high cycle rates and high positioning accuracies are required. The tare weight of the robot does not act on the drives. Due to the high stiffness, high speeds can be achieved in the vertical direction (Z-direction) and at the same time high ranges in the XY-plane.

The SCARA robot is preferably used for assembly and handling. The disadvantage of this robot type is its limited work envelope.

Vertical articulated robot. The vertical articulated robot has six axes of movements and thus provides all six degrees of freedom. It thus can be used for a variety of different purposes. The structure of the robot enables it to reach behind obstacles (Fig. 4-59). The motors for the secondary axis or the hand axis which are mounted at the rear of the robot arm can be used as a counterweight. Yet, due to the tare weight of the robot additional weight balancing is sometimes required. The end effector usually is an in-line wrist so that the three rotational axes cross each other at one point. As is the case with most of the robot types the bottom of the robot can be either mounted stationary on the floor, at the ceiling or the wall or it can be moved by means of a linear drive.

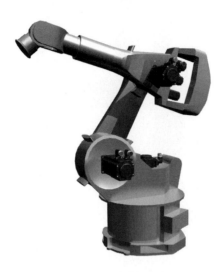

Fig. 4-59. Schematic diagram of a vertical articulated robot

The vertical articulated robot is a classical industrial robot and is used in numerous manufacturing applications.

Fig. 4-60. Schematic diagram of a cylindrical robot

Fig. 4-61. Schematic diagram of a spherical robot

Cylindrical robot, spherical robot. These types of robots result from replacing two of the three linear main axes of a Cartesian robot by rotational
axes (Fig. 4-60 und Fig. 4-61). The work envelope is approximately cylindrical or spherical.

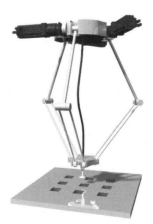

Fig. 4-62. Schematic diagram of a parallel robot (e.g. hexapod)

Parallel robot. In contrast to the robot types mentioned above (serial kinematic chain) the principle of parallel kinematics is based on the parallel
arrangement of the drives so that no drive is burdened with the weight of
another drive and only low masses have to be moved at the driven side.
(Fig. 4-62). Since tensile and compressive forces act evenly over the entire

cross section of the particular rods a high total stiffness of the system can be achieved.

Since the drive power is distributed over several independent units the particular portions of inaccuracies due to the inevitable part deviations might neutralise each other. This results in higher dynamics and a higher positioning accuracy.

A disadvantage of this technology is the limited work envelope which results from the independently working drives whose movements don't add. Compared to serial kinematic chains the kinematic and dynamic conditions are more complex which must also be considered. Parallel robots are used for handling low masses (e.g. sorting bulk material on a conveyor belt in packaging machines) but also for example in flight simulators.

Mechanical transmission elements. The mechanical transmission elements with their specific characteristics which are used for the various robot designs are the same which were described in chapter 3.6. They are also used for the positioning drives described in chapter 4.4. In particular, these elements are spindles, toothed belts, universal shafts and couplings.

There is a special feature about the joints of SCARA robots and articulated robots. They have a work envelope of $\leq 360°$ resulting in low speeds (< 10 rpm). This requires special transmission ratios between the motor and the respective joint.

Sensors. In general, robots have average accuracies (approx. 0.1 mm) and high transmission ratios between the motor and the mechanical parts. Hence, the angular resolution of a resolver in a servo motor is usually sufficient. Because resolvers can only determine the absolute angular position of one revolution of the motor, it is necessary to store the absolute position of the mechanical parts in order to avoid repeated homing of the robot after every power-on. This is generally done electronically by counting the revolutions and storing the last position of the motor at the time when the power supply is disconnected. Since in this case the brakes are engaged, the robot is expected to not move when turned off.

If these procedures are not sufficient, multi-turn absolute encoders can be used for detecting the absolute position over many revolutions.

In addition to determining the position, some tasks performed by robots require the measurement of the force. Joining processes performed by robots are only one example. For this purpose the robot joints are equipped with torque sensors since usually – due to the characteristics of the subsequent gearbox – it is not sufficient to only measure the torque of the motor. In some applications, the tool is equipped with force sensors.

Robots are often handling or machining materials with no geometrically defined position or increased tolerances. Here, additional sensors are used for determining the position of the work pieces. These can be measuring probes or image processing systems which are increasingly used. The robot control uses the detected positions for corrections of the subsequent movements.

In, general robots work in secured work envelopes. But sometimes robots and human operators work closely together. Consequently the hazardous area of the robot's work envelope has to be monitored, for example by laser scanners. Sometimes the work envelope to be monitored is derived from the actual position of the robot. In some applications it can be useful to limit the work envelope in order to protect the human operator who works close to the robot.

4.5.4 Drive systems for robots

Robots are general-purpose handling systems which are used in numerous applications. In general, the requirements regarding dynamics and accuracy are in the medium range. The cost pressure is very high.

Fig. 4-63. Articulated robot

In some manufacturing processes a high number of robots are used (e.g. automotive body-in-white applications) (Fig. 4-63 and 4-64). The failure of one robot would then cause the standstill of a whole production line which should of course be avoided and should, if necessary, be corrected as quickly as possible.

According to these general conditions, drive systems for robots have the following requirements:

- high reliability (high MTBF = „Mean Time Between Failures"),
- easy handling for service (low MTTR = „Mean Time To Repair"),
- low costs,
- close cooperation with the robot control system,
- integration of safety functions in order to design safe robot systems.

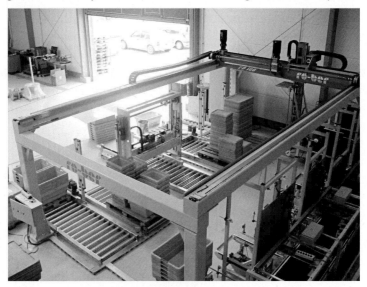

Fig. 4-64. Gantry robot

Dimensioning of robot drives. The precise dimensioning of robot drives is very difficult, because robots perform complex movements and the movements of the particular drives are dependent on each other. The payload and the actual movements required also have a big influence on the system. The actual load on the drives is determined by the following parameters:

- acceleration forces,
- weight,
- masses and thus the moment of inertia of the movable parts,
- size of the payload,
- actual position of the robot joints,
- the actual motion with the resulting acceleration of the particular drives.

The basis for the dimensioning of a drive is the required torque and the resulting inverter current. Short acceleration and deceleration ramps typically result in an overload factor of ≥ 3 between peak current and rated current. Other crucial parameters are the holding torque which is required to hold the payload and the respective continuous current of the inverter.

In practice, the drive power of robots is dimensioned in a way that referred to known sequences which can be executed with the maximum payload without generating a thermal overload in the drives are derived from typical applications. In particular applications where high performance and high dynamics are to be achieved, the resulting load on the corresponding drives (speed-torque characteristics and the resulting temperatures of the inverter, motor and gearbox) is determined and the maximum speeds then are defined accordingly.

Gearboxes. The gearboxes in robots must have low backlash in order to not influence the positioning accuracy in a negative way. For transmission ratios of up to $i=100$ planetary gearboxes are preferably used. Partly, helical gearboxes and bevel gearboxes with low backlash are used as well.

Robot joints with high transmission ratios (typically $i > 200$) require special gearboxes. For example the harmonic drive gearbox or the cyclo gearbox both provide high stiffness and high transmission ratios in just one gear stage. They are very compact, have low weight and thus a low moment of inertia.

Motors. Due to the high transmission ratio between the motor shaft and the mechanical parts, the inertia of the motor plays the most important role. Hence, in robots synchronous servo motors are predominantly used. They provide low inertia and high overload capability which makes them well suited for dynamic applications.

Linear drives are only used exceptionally. In some robot joints high-pole torque motors can also be used.

Brakes. In order to prevent the robot arm from dropping when it is turned off, the robot drives are equipped with holding brakes. In Cartesian designs this is at least required for the z-axis. In articulated robots all drives generally have holding brakes.

Depending on the particular application the holding brakes are controlled either all together or separately.

Inverter system for multi-axis applications. Robots perform multi-dimensional movements which always require several drives. Thus, it is practicable to implement a drive concept with a common power supply, a common DC bus and power inverters for the particular motors. Since the multi-axis construction executes movements in motor operation and in generator operation simultaneously the power exchange by means of a DC bus is useful. The central power supply unit also contains the brake transistor and the brake resistor. The brake resistor takes up the energy which is regenerated during deceleration. Sometimes regenerative power supply units are used.

The functions of the inverters and the power supply modules are coordinated. Together as a system they are a cost-efficient alternative for single-axis applications with redundant functions [WaBo00]. Thus two or three inverters can also be grouped together in order to form multi-axis units.

Due to the power exchange between the axes and to the dynamic characteristics of the acceleration and deceleration ramps of the motion profiles, the average power consumption from the mains is relatively low. However, during acceleration for a short time the power consumption is very high.

Servo control of the drives. Robot drives are always controlled by means of a servo control which must provide a good dynamic response and a short rise time for torque control. In order to extend the operating range of synchronous servo motors, field weakening as described in chapter 3.3.3 can be applied.

Under certain circumstances narrowband filters can be required for the servo control in order to avoid resonances (digital notch filter in software).

Distributed functions between control and inverter. Robots execute movements using multiple axes so that it is most practicable to implement the motion control and the calculation of the desired positions by means of the robot control. Therefore industrial PCs are often used, as described in chapter 2.3.1. Due to the complex kinematics, high computing power is required for the transformation of the Cartesian coordinates into the axis coordinates.

The motion control is able to derive the torque requirements from information about the trajectory path, the kinematics of the robot and the payload. By means of these parameters the torque controller or current controller of the servo inverters can be feedforward-controlled in order to achieve a precise motion along the trajectory path.

The speed and angle controllers can be integrated into either the servo inverters or the control. In addition, the evaluation of the angular encoders can also be executed by the servo inverter or additional central components. In general, the particular design is developed in accordance with cost considerations. Six-axes articulated robots especially allow for very cost-effective designs by centralising functions such as speed control or the evaluation of encoders for groups of axes. As a result, the number of interfaces and the computing power requirements for the servo inverters decrease.

For cost reasons the control and the servo inverters are generally connected by means of a communication system. These must guarantee synchronous operation of the system so that all drives can cooperate synchronously and thus perform precise multi-dimensional movements.

Emergency stop and power failure. In case of emergency stop or power failure robots must stop very quickly. After recovery of the power supply or reset of the emergency stop the robots must also be able to restart very quickly and without any homing procedure in order to immediately resume the production. In general, special drive functions are provided for these cases.

As an example, the servo motor is short-circuited via the inverter in case of an emergency stop in order to achieve fast deceleration of the drive. The produced regenerative power is converted into heat inside the motor.

In case of power failure a buffered 24V-power supply enables the signal processing system to continue operation so that position detection and diagnostics can still be executed. The brakes are not activated before standstill of the drive. At the moment of standstill the actual positions are stored in order to allow for a restart without the need for homing. It is expected that in the meantime, the drives have not been moved by external forces.

Safety functions of robot drives. In the past, robots had to be stopped and disconnected from the power supply in order to achieve an emergency stop. By integrating safety functions into the drive, as described in chapter 3.4.7, a safe shut-off of the power electronics can be achieved faster and with less effort.

Additional safety concepts such as the safe limiting of the work envelope or the operating speeds are executed by the robot control in cooperation with position sensors and other sensor types. If safety rules are violated in the control the inverters, power supply units and mechanical brakes build different shutdown paths which are then to be activated.

4.6 Synchronised drives

Karl-Heinz Weber, Uwe Tielker

Synchronised drives are the heart of continuous production processes in which continuous material is produced or refined at high speed. They are often combined with winding drives, which are operating at the beginning and end of these processes (chapter 4.7). Alternatively, cross cutters and flying saws can be used at the end of a production line to cut the material and thus form the transition from continuous to fast cycled discrete part manufacturing. These are presented in chapter 4.8. Synchronised drives are the most important drives in processing lines, as they control material flow and, therefore, the machining process.

4.6.1   Technological processes

Synchronised drives are involved in the production process of continuous materials. This technological task consists of the following processes:

- transport,
- stretching, shrinking, tensioning, drawing,
- bracing, feeding,
- rolling,
- cross-section reduction,
- alignment,
- slitting,
- coating, impregnating,
- squeezing,
- refinement, twisting, spinning, twining,
- machining,
- printing and embossing.

Printing units with single drives are a special type of synchronised drive. They function as adjustable electronic gearboxes operated with precise angular synchronicity at high resolution and very good smooth running so that the individual colours of the individual printing units lie on top of one another in precise register. The trimming of individual printing units can be trimmed manually or automatically via registration mark controls or camera systems. Precision up to 0.01 mm on the product is required to achieve a pin-sharp print image. Table 4-7 provides an overview of the most important applications.

Table 4-7. Application examples with typical parameters

Application	Industry	Speed (m/min)	Power (kW)
Manufacturing and refining of paper or film webs	Paper Plastic	1 – 2,000	0.37 – 100
Manufacturing and refinement of textile webs	Textile	5 – 150	0.37 – 55
Slab and thick-plate rolling machines	Metal		Up to 10,000
Rolling, annealing and refinement of sheet metal	Metal	5 – 300	0.37 – 200
Wire-drawing machines	Wire	10 – 2,000	0.75 – 250
Printing of paper or film webs, textiles	Printing	5 – 1,000	0.75 – 110

Other synchronised drives that are primarily used for forming are described in chapter 4.10.

Materials. The spectrum of materials to be processed and their cross-section profiles is very broad. At the same time, the tensile forces and transport speeds also vary. The consistency of materials with regard to elasticity, surface, hardness, bending forces, tensile strength, width, thickness and thickness tolerance, density, grammage and cross-section profile also varies greatly. For paper, the dimension for material thickness is specified as the grammage in g/m^2. Micrometers are used for films, and millimetres for other materials. Therefore the following list of materials that are processed is not exhaustive.

- base paper, coated and printed paper, tissue paper,
- films of every type, from lorry tarpaulins to packaging, covering, household, adhesive, labelling, protective, peeling and technical films,
- rubber and caoutchouc webs,
- sheet metal, metal films and bands made of steel, non-ferrous metals and aluminium,
- composite materials (mixture of several layers of different materials, e.g. paper, plastic and metal films),
- packaging materials made of paper and composite materials for many consumer goods,
- textiles, woven fabric and knitware in yarn and twine production as well as in the weaving, dying, coating, printing and many other processes,
- plastic, cord, glass-fibre and carbon threads,
- felts, fleeces (non-woven)
- wires made of steel, copper, aluminium, brass (some also coated or enamelled),
- electrical cables for communication and power transfer,
- ropes and strands used for wire ropes or copper leads,
- wire mesh made of different materials for industrial or medical purposes,
- fabrics, e.g. textile fabrics,
- floor coverings, e.g. carpeting or plastic coverings.

4.6.2   Machine types in processing lines

Due to the variety of different machine types, only a few typical machines with a high proportion of synchronised drives are described here. The main focus here is on the synchronised drive modules in these machines.

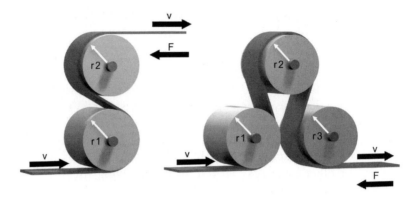

Fig. 4-65. S roll unit (left) and omega roll unit (right)

Braking units and fixed speed points. Braking units implement fixed speed points in a production process and thus absorb the tensile forces generated by other production equipment. To couple high tensile forces into the material web, roll arrangements are generally designed with large angles of contact in the form of S or omega roll units (Fig. 4-65).

The following equation applies for the forces and torques (r_n: radii of the rolls, τ_n: torques of the drives, F_n: tensile forces of the material web):

$$F_1 + F_2 + \frac{\tau_1}{r_1} + \frac{\tau_2}{r_2} + \frac{\tau_3}{r_3} = 0 \tag{4.2}$$

The individual rolls can be mechanically coupled and equipped with single drives. The transferable brake torque of the braking unit depends – besides factors due to the mechanics of the machine – on the static friction coefficients between the material and roll surfaces. In applications in which this coefficient can be significantly reduced, e.g. due to soiling, appropriate measures are to be taken, for example the provision of anti-slip control in the control structure to prevent slipping between the circumferential roll speed and the material web.

Heat and cooling rolls. Heat and cooling rolls are used to supply thermal energy to a material web or to dissipate it, so as to make other processing steps possible, e.g. the adhesion of two material webs. A heat transfer medium (e.g. water steam) flows through the heat and cooling rolls in order to generate the corresponding temperature on the surface of the roll. To vary the energy that is to be supplied to or dissipated from the material web, the angle of contact of the heat or cooling roll is adjusted (Fig. 4-66). Depending on the requirements of the process regarding temperature constancy, e.g. with varying production speeds, the angle of contact can be adjusted based on the material web speed. This can be done manually or automatically via material web temperature control.

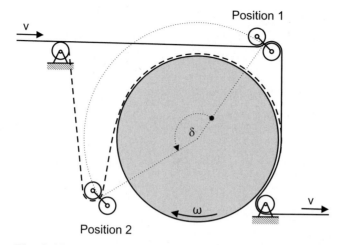

Fig. 4-66. Heat or cooling roll with variable angle of contact

Roll formers. In roll formers, continuous sheets are rolled out into trapezoidal sheets (for industrial facades) or guard rails (for driving lanes) in multiple steps with parallel, axially-arranged forming rolls. The rolls plunge deeper and deeper into the material from step to step to form its final profile. The drives involved in this process are operated as an electrical shaft or are speed-controlled.

Rolling mills. Steel or non-ferrous metal sheets and bands are rolled to reduced thicknesses in multiple stations in rolling trains for wide and flat products. For this purpose, the materials to be rolled are fed between two parallel rolls with an adjustable gap. Several such stations are arranged in succession in order to achieve bigger reductions in thickness. Since the material becomes brittle during the rolling procedure, it must be subjected to an annealing process before further processing. There are many other roll techniques for a wide variety of products, but they will not be outlined here. The speed of the drives used here must be very precisely controlled, and load surges must be compensated for quickly.

Wire-drawing machines. In wire-drawing machines, metal wires are unwound, drawn to reduced cross sections via drawing dies and then rewound. Like in rolling mills, in wire drawing machines several stations are arranged in succession.

Straightening machines. Straightening machines are used to straighten unwound, non-planar wire or web material (e.g. thick wires, sheet metal, metal bands) via targeted bending so that it can then be processed further. The material to be straightened passes through rolls arranged with an offset and nested into each other.

Fig. 4-67. Continuous annealing system

Continuous annealing lines. Steel and non-ferrous metal sheets (steel and steel alloys, copper, copper alloys, bronze, nickel silver and brass) are given the desired thickness by rolling processes. Since the material changes structure and becomes brittle during the rolling procedure, it must be subjected to an annealing process after rolling to gain its original structure for further processing. This occurs in annealing lines, which consist of winding drives at the respective input and output stations and numerous synchronised drives (Fig. 4-67).

The sheet metal coils are unwound via winding drives and led over S roll units with synchronised drives. To ensure continuous operation when changing the coils (travel through the oven must be continuously without stopping), the material web is led through storage towers, with sufficient capacity to bridge a coil change at production speed.

The web then runs through degreasers and additional S roll units and is then led into the annealing furnace via a dancer with a low and constant tension. The oven is heated via gas or electricity and is the core process of the system. There are horizontal and vertical continuous annealing furnaces where the thermal treatment of the metal web is performed.

Annealing in the furnace is followed by a cooling section involving driven cooling rolls. The web is then led through an etchant via S roll units and nip rolls. There are annealing units with and without integrated etching station. After the etching, the surface of the web is cleaned with rotating brushes in the cleaning sections (washing machines). Various nip rolls remove any remaining water. Synchronised drives are found at all processing stations.

The web is then lead into the storage tower of the winding station via S roll units. It then continues on to the winding reel via S roll units and conveyor belts. Separation and attachment equipment is found in the input and output groups for automated coil changing. The synchronised drives are operated as electrical shafts with dancer and tensile-force controllers and torque ratio controllers.

Altogether, such a continuous process requires a large number of drives that can be operated at exact speeds and torque ratios.

Stranding machines and armouring machines. Stranding machines twist (as helicoids) individual fibres or wires into a cable (Fig. 4-68). Typical applications include:

- wire ropes made of steel or stainless-steel for lifting equipment or tensioning ropes,
- catenary wires for cable cars,
- aluminium ropes with steel cores for high-voltage overhead lines,
- copper leads for communication cables, glass-fibre cables, copper wires for power cables,
- plastic or hemp ropes for lifting equipment,
- steel cord leads or Bowden tensioning leads,
- armouring of power cables.

Fig. 4-68. Stranding machine for steel ropes

Many different types of stranding machines are used, depending on the end product: e.g. basket stranding machines with and without reverse rotation, single- and double-strike stranding machines, armouring machines and skip stranding machines. The basic principle of them all is similar, however. The reels with the individual wires are braked and unwound mechanically or via a winding drive. They are overhung and located in a rotating flier or basket that is driven via the main drive of the machine.

The rotating flier causes the individual wires to be twisted into a rope that is fed to the rewinder via a pull-off disc (fixed speed point). The speed ratio between the fixed speed point and the flier determines the length of twist (number of rotations per cable length).

With very thin wires, the wire spools can also be arranged outside the flier, feeding it via an opening in the rotation centre. Gimbal-mounted pull-off discs and rewinders can be located inside the flier here.

The range of diameters of wires to be twisted is very wide and extends from several micrometers to some millimetres. The drives located in the flier are supplied with power via slip rings. The synchronised drives used are operated via electrical shafts, whereby the speed ratio influences the length of twist.

Laminating lines. In laminating lines, several layers of different material webs are joined (e.g. plastic films, paper and aluminium foil) and thus refined. A laminating line for the fasteners of baby nappies (production speed: 300 to 400 m/min) is described here as an example.

In this system, a total of three layers are joined. A middle web made of elastic material (a highly stretchy film with rubber band-like properties) is enclosed by two "non-woven carrier material" outer webs. All the individual webs are unwound from "contact winders" which are followed by "peel-off rolls" and "splice rolls". Since the material must be supplied via an automatic roll change, the previously mentioned units are doubled. At the winders the material is changed at a standstill.

After unwinding, the material web is lead through a material accumulator that ensures continuous material supply of the laminating mechanism when changing the bales. It supplies the laminating mechanism while the winder stops for the changing phase. The accumulator drive is tension-controlled and positions itself indirectly using the differential speeds.

There are three of these machine modules in all, one for elastic and two for non-woven material. The web with elastic is cut lengthwise in parallel webs in a cutting station. The webs are twisted in a "twister" and moved to the actual laminating station. The laminating mechanism is comprised of two parallel rolls with an adjustable gap, the laminating roll and the pressure roll. The non-woven webs coated with hot adhesive on one side enclose the elastic web like a sandwich in the gap and adhere it. The adhesive is applied via rubber-coated rolls.

The laminated composite fabric is then cooled down via cooling rolls. So-called activators, which then break the fibres of the outer non-woven webs near the enclosed elastic film, are applied so that the elastic stretch effect is applied to the non-woven carrier material. The fabric then runs to a contact winder via an S roll unit. The contact winder stores it temporarily for further processing.

The synchronised drives used here are operated as electrical shafts with tensile-force controllers and torque ratio controllers.

Stenter frames. Stenter frames are frequently used in the textile industry to secure woven goods after a refinement process. The material web, which is fixed to chains by means of pin adapters, is transported (Fig. 4-69) via two speed- or angle-controlled chain drives.

The position of the chain drives can be adjusted laterally so that a shearing force in feeding direction can be applied or the width of the web (e.g. in a thermal drying process) can be set to a defined value.

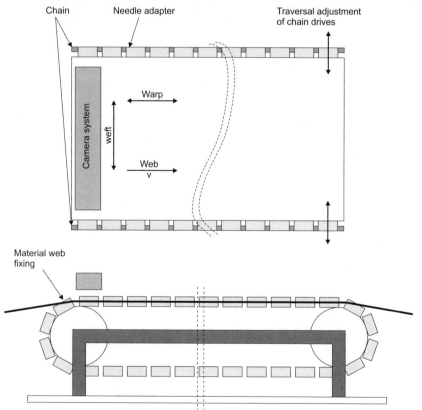

Fig. 4-69. Schematic diagram of a stenter frame

Fig. 4-70. Entire stenter frame system

Another task of a stenter frame can be to determine the alignment of the warp and weft threads of the material web after an upstream production process. For this purpose, the position of the corresponding threads is detected by optical sensors (e.g. camera systems). A correction signal from a corresponding electronic control unit ensures that the threads are brought together at the desired angle (generally 90°) via the speed adjustment of one of the chain drives. Fig. 4-70 shows an entire stenter frame system.

Main drives of printing units. While the individual printing units of a multi-colour printing machine were coupled mechanically in the past (generally via gear wheels), rotary printing machines have been increasingly equipped with single drives since approx. the mid-1990s. These drives must be operated in angular synchronism and are therefore also to be classified in the synchronised drive category.

A printing unit like the one shown here (Fig. 4-71) is designed for printing flat materials, such as corrugated cardboard. The actual printing process occurs between the print roll and an impression cylinder. The printing plates with the printing image are fixed to the print roll and supplied with ink by one or two ink application rolls (for different qualities).

All three to four print rolls involved in the printing procedure must be precisely synchronised in order to achieve a good overall print image with the printing units for other colours.

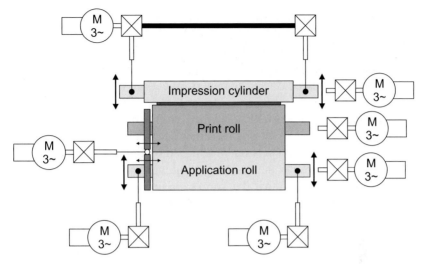

Fig. 4-71. Schematic diagram of print roll arrangement

Due to the high requirements to all drive elements involved in the printing procedure with regard to smooth running, stiffness and adjustability, only high-quality components are suitable. In addition to minimal backlash, gearboxes must provide high stiffness. Some requirements can only be fulfilled with appropriate special designs or by the use of direct drives.

Encoders with very high resolution are used so that high precision is achieved with the product.

Auxiliary drives of the printing unit. In addition to drives that are directly involved in printing, there are a variety of auxiliary drives in the positioning drive category (Sect. 4.4).

To ensure the adaptation to the material thickness, the impression cylinder must be adjusted. Depending on the degree of automation of the machine, this can be carried out manually or as motor-driven positioning. If a homing procedure is not acceptable after power up, an absolute encoder providing the current position must be used for this drive, as well as for the auxiliary drives described next. The quantity of ink that is applied to the material in the printing unit is defined by the position of the ink application roll. Variants with manual or motor-driven positioning are possible, depending on the degree of automation.

An additional auxiliary drive is used to define the axial printing position on the material. Several printing units can thus be positioned in reference to the incoming material in such a way that the different colours produce the entire image true to size. Again this positioning can be carried out manually or via motor-driven adjustment, depending on the degree of automation of the machine.

Slitting machines. Slitting machines are used in a variety of industries whenever it is necessary to cut a material web, e.g. film, paper, corrugated cardboard, hard or soft rubber, plastic panels, metal or foam, into a useful size for production. Slitting machines are also used for trimming winding coils that were damaged during transport by cutting off a thin strip at both edges.

Slitting machines are usually designed with a number (between 2 and 50, depending on application and material) of circular knives attached to a knife roll. The circular knives can be adjusted on the knife roll manually or with positioning drives to define the cutting widths. The knife roll is generally speed-controlled so that the knife's circumferential speed can be adjusted in order to optimise cut quality. It is adjusted within a speed range from synchronous to the material web to a considerably over-synchronous speed.

4.6.3   Mechanical transmission elements in processing lines

With synchronised drives, a stiff connection with low-backlash between drives and feed rolls is important. This implies special demands to the mechanical transmission elements that are generally required to connect drive elements to one another.

Drive elements. Axial *couplings* connecting the drive journal of the roll and the motor or gearbox must have a low backlash. If the radial misalignment changes during operation where a connection with the proper angle is absolutely imperative, a Schmidt coupling is used. This is a short, torsionally stiff power coupling for a large, changing radial misalignment. The transfer of the angular speed between the input and the output side is flawless here. It is used in systems with an adjustable roll gap (e.g. calender units, chill-roll units and calender stacks).

Toothed-belt transmissions are also occasionally used with synchronised drives. They are characterised by minimal friction and good damping characteristics. They are primarily single-stage transmissions with transmission ratios of approx. 0.3 to 3.

If parallel rolls must be adjusted (e.g. variable roll gap) during operation, *universal joints* with low backlash are used.

Decoupling elements. In multi-motor systems with synchronised drives, often particular parts of the machine must be decoupled so that an undesired interaction between the individual process steps is avoided. Pathless and path-bound actuators and sensors are available for this purpose.

Loop accumulators without tensile force. Loop accumulators are sections with synchronised drives in which a material web enters and exits with almost no tensile force (except gravitational force caused by the weight of the material web). In the case of long sections recesses in the foundation are sometimes used. They are often implemented in the transition from constant to intermittent material feeds (punches, guillotines etc.). The loop position is detected without contact with analog or digital optical sensors. Control can be of the two-point, three-point or continuous type.

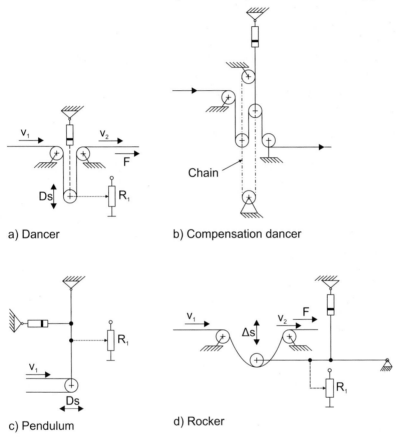

a) Dancer

b) Compensation dancer

c) Pendulum

d) Rocker

Fig. 4-72. Dancers, rockers, and pendulums

Dancers, rockers, and pendulums. Dancers, rockers, and pendulums are material accumulators. Dancers (Fig. 4-72 a and b) redirect a material by 180 degrees, whereby the material can also be redirected several times in a meandering fashion. Accumulators, with the material being redirected only once at an angle of less than 180 degrees are referred to as rockers (Fig. 4-72 c) or pendulums (Fig. 4-72 d).

These devices are used to set the material tensile force and ensure synchronous operation of the drives. The material tension can be applied via fixed or adjustable weights or – more conveniently – via pneumatic cylinders. Pneumatic cylinders offer the advantage of an electrically actuated adjustment of the tensile force via proportional valves. For wide tension setting ranges ($F_{max} / F_{min} \geq 150$), asymmetrical double-cylinder systems are used.

Good dancer systems are characterised by ease of mechanical movement, minimal masses, good weight compensation of the rolls, minimal friction of the pneumatic cylinders, and good damping characteristics. Pendulums with magnetically-controllable force are also used for very sensitive tensile forces (e.g. with thin wires or threads). These systems are characterised by very low friction. During the design phase, particular attention should be paid to low-mass systems and minimal cylinder friction in the sleeves of the pneumatic cylinders.

Position sensors on dancers or rockers, which are preferably designed to function without contact, provide position values for the control loops of the drive.

Other sensors in processing lines. Beside the described dancers, rockers and pendulums, which provide measured values to the drive control in addition to their function as material accumulators and an adjustment element for the material tensile force, processing lines also contain other sensors that are required for control of the process. These sensors must detect the material web tension, the material thickness and the material speed.

In *tension measuring stations*, the material web tension is measured as an actual value for a superimposed tension control. With this pathless measurement, the material web is led over redirection rolls at a specific angle. These redirection rolls are equipped with inductive or strain-gauge sensors in the bearings. This allows for high tensile-force precision and wide tensile-force setting ranges.

To measure the transport speed precisely and without slipping, encoders with measuring wheels that move along with the material web are used.

Thickness measuring systems are installed to monitor the thickness of material webs and to influence the upstream production process via the corresponding actuators (e.g. spindle adjustment of roll gap, gap opening of flat nozzles and metering of the material applied). A variety of measurement methods are used, e.g. inductive or capacitive sensors and optical methods.

4.6.4   Drive systems for synchronised drives

Synchronised drives are quasi-stationary drives (i.e. with a steady circumferential speed) that can be operated at a precise and constant speed, angle or torque. Changes in speed are generally only made when the material is changed or when the system is stopped. To achieve a high refinement quality of the product to be processed, good smooth running and the ability to make precise adjustments to the drive torque are important drive characteristics.

Drive dimensioning. In addition to using a suitable drive, its proper physical dimensioning and suitable mechanical transmission elements (gearbox, toothed belt, coupling etc.) are important for optimum machine operation. The physical dimensioning is determined by the process parameters of tensile force, speed, moment of inertia and acceleration for emergency stop. The rated power of the motor and inverter is generally designed for stationary needs. Additional dynamic power required for acceleration and braking phases can be provided by the overload capacity of the drive components.

Geared motors. For high transmission ratios, axial or angular (helical, bevel or planetary) gearboxes are used and sometimes operated in combination with belt transmissions. If the gearbox acts directly on the shaft of the roll, this can be done axially with a coupling via a shaft journal; or the gearbox acts directly on the shaft journal of the roll as a shaft-mounted gearbox. Gearboxes are used very frequently with synchronised drives. It is preferable to use geared motors with standard three-phase AC motors and integrated encoders with synchronised drives, as the demands for dynamic performance in quasi-stationary operation are minimal.

AC or DC drives. In the past, the classic synchronised drive was a DC drive because of its especially good and simple control characteristics and the excellent smooth running. In the meantime, the control of modern AC drives has caught up to or exceeded that of DC drives. The operational reliability on less stable mains is better with AC drives. In terms of maintenance and mains load, the AC drive is clearly superior. For these reasons, AC drives are used for synchronised drives almost exclusively.

Direct drives. If the speeds of the process and motor match, the drive motors can act on the rolls of the machine axially without a gearbox via low-backlash couplings. Alternatively, they are designed as hollow-shaft motors and coupled without backlash to the input shaft ends of the rolls with clamping elements. The advantage here is that the gearbox and any transmission faults are eliminated, so that highly precise and very constant transport speeds can be achieved. At very low speeds, multipole three-phase motors are used – also referred to as torque motors – with a rated speed of, for example, 100 to 300 rpm.

DC bus connection. With three-phase systems, power coupling can be executed very easily and effectively via the DC bus.

In processing lines, the motor of the unwinder generates electrical energy. The synchronised drives of the processing line are predominantly operated in motor mode and thus consume electrical energy. By connecting the DC buses of the inverters, it is possible to exchange power between the drives of the machine. If all drives of the production machine are operated with DC-bus connection, the mains is only loaded with the resulting process power and minimum idle power.

This reduces operating costs by saving energy and the investment costs for switchgear necessary for the power supply of the machine.

Generator operation with brake units or regenerative power modules. Short term generated power during deceleration under normal operating conditions, in a quick stop or emergency stop, is dissipated via a central brake chopper or is returned to the mains via a regenerative power unit. Continuously generated power by individual drives that are not connected to drives in motor mode via the DC bus, should be fitted with regenerative power modules for reasons of energy efficiency.

Field weakening. Field weakening is only used for synchronised drives in certain situations, especially those where high speeds are required with low web tensions and low speeds with high web tensions (but not high speeds and high web tensions at the same time). When using field weakening, the rated power of the inverter and motor can be reduced by the field weakening factor.

Group drives. Drives in roll arrangements (e.g. S roll units or omega roll units) with constant speed, can be operated with one inverter controlling several asynchronous motors in parallel. The torque is distributed nearly symmetrical here. Deviations of approx. 10 to 20% can arise due to motor tolerances.

Motion control for synchronised drives. With synchronised drives, motion control can reasonably be implemented locally in the inverter [BrGeKl02, BrGe04]. This relieves the higher-level control and supports modularisation of the entire structure of the system, where the individual drives already execute local tasks. Examples of this include:

- Torque-set drives that automatically limit the speed to a permissible value in case of a loss of traction between the drive wheel and the material web.
- Setpoint calculation for subsystems comprised of several drives (e.g. S for roll unit drives operated at oversynchronous or subsynchronous speed).
- Controllers for process parameters implemented in the inverters.
- The adaptation of controllers to make them capable of automatically adjusting themselves to changing moments of inertia.

Process monitoring by the drives. Another requirement of the drive system is the monitoring of the process via the sensors in the inverter. Speed- or angle-controlled drives measure speeds and angular ratios very precisely and can relay this information to a control centre via a communication system. On the other hand, the measurement of tensile forces and torques of the process only via the sensors of the inverters is clearly less precise, as existing mechanical faults, e.g. bearing friction, cannot be detected.

Drive control for synchronised drives. A variety of different control methods are used in synchronised drives, depending on the process.

Torque control. In many processes, the drive must set the tensile force and, therefore, a torque. This can be implemented with open loop or closed loop control.

Tension control. Open loop tension control incorporates the fact that a drive system comprised of a gearbox, motor and inverter can adjust the torque, and thus the tensile force, repeatedly with a certain precision and resolution. The quality of this torque adjustment that can be achieved is limited by perturbations in the individual components, which cannot be fully compensated for (Table 4-8).

Table 4-8. Perturbations for the torque setting of drives

Drive component	Perturbations	Measure for minimising the influence
Gearbox	Temperature of the gearbox oil $\tau = f(\vartheta)$	Use of synthetic oil with the most temperature-independent properties possible. Use of toothed belt transmissions instead of gearboxes, for example.
Motor	Temperature of the motor $\tau = f(\vartheta)$	Measurement of the winding temperature.
Inverter	Offset in the current measurement leads to smooth running faults	Measurement of the offset values.

The typical setting ranges for the torque at the motor shaft shown in Table 4-9 can be achieved depending on the selected control method.

Speed control. Many synchronised drives are operated speed-controlled. Depending on the required accuracy the speed control can be implemented without a speed sensor (open loop) or with a speed sensor (closed loop). The different methods and their properties are described in chapter 3.4.4.

Table 4-9. Torque setting ranges of different control methods

Control method	Torque setting range
Frequency inverter with vector control without speed feedback	$\tau_{max} / \tau_{min} < 5 - 10$
Frequency inverter with vector control with speed feedback	$\tau_{max} / \tau_{min} < 10 - 20$
Servo control	$\tau_{max} / \tau_{min} < 30 - 50$

Tension and dancer position control. In a production line with several drives arranged in succession the speed, angle or torque of synchronised drives are in a fixed proportion to one another. To decouple individual processes within a system, dancer or tensile force control is used.

Tensile force control. In a drive system with a wide torque setting range, static faults can be corrected by an implemented torque or tensile force control. Here, the torque setting drive is superimposed by the tensile force control loop, which receives its actual value via a suitable torque shaft or tensile force sensor.

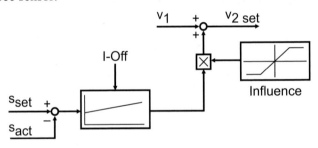

Fig. 4-73. Principle of dancer position control

Dancer position control. With dancer position control, a speed-controlled system is usually superimposed by a position control loop for a dancer roll (Fig. 4-73). As long as the dancer does not hit its mechanical limit positions, the static tensile force is only defined by the mass of the dancer roll In order to vary the tensile forces, it is possible to load the dancer roll as described above, e.g. via a pneumatic cylinder.

Angular control, electric shafts. An angular control is implemented if the process requires the various synchronised drives to operate precisely in register. Measurement of the angular position at the motor shaft and processing of the information in an angular control loop allows it to be implemented with high precision. Multiple angular-controlled drives can be connected in parallel or cascaded structures, depending on the requirements of the respective application.

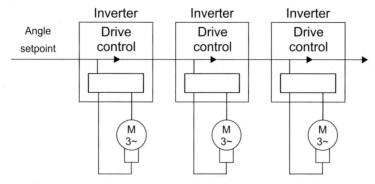

Fig. 4-74. Parallel structure of synchronised drives

Parallel structure. The parallel structure of several angular-controlled drives (inverters 1 to n) represents the replacement of a conventional mechanical shaft (Fig. 4-74). All drives receive the same angle setpoint to which the position of the motor shaft is adjusted. The advantages of such a parallel structure in comparison to a mechanical shaft are the absence of a reaction between the individual stations with pulsating loads and the easy modularisation of the entire system. When using synchronised drives, this parallel structure is used, for example, when a material web is to be transported without being influenced by external forces.

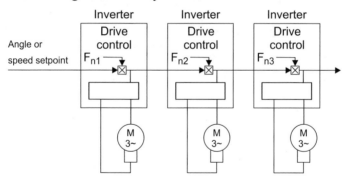

Fig. 4-75. Cascaded structure of synchronised drives

Cascaded structure. The second option for interconnecting several speed or angular controlled drives is a setpoint cascade (Fig. 4-75). In each drive it is possible to evaluate the received setpoint with a factor F_n before passing it on to the next drive.

The advantage of this structure becomes evident when it is necessary to stretch a material web in succeeding stations in a defined way. By changing a factor, the stretching between two stations can be changed without influencing the previously defined stretching between the other stations.

Special operating conditions and scenarios. Processing lines are designed for executing production processes for very long periods and at high levels of productivity. The set-up of such lines is generally complex since special operating conditions, such as a mains failure or malfunctions in individual drives, can be compensated for so that the material is not damaged, the process is shut down in a controlled manner and restarting is possible without great effort. For this purpose, the drives of such processing lines often have special functions, which are shown in the following example.

Failure of the mains power supply. If the mains power supply fails, there are various options for ensuring a controlled shut down of a process.

A redundant *separate AC mains supply* can be installed and switched to if a malfunction occurs. This often requires disproportionately high effort, however. Furthermore, such a separate mains network must be continuously maintained so that it may be switched to without delay. All components, especially line-commutated inverters such as regenerative power units and DC speed controllers must also be suitable for the separate power supply.

In an *uninterruptible power supply (UPS)*, the supplying AC mains is rectified and buffered by an accumulator. Charging and monitoring electronics ensure that there is sufficient power in the accumulator for bridging a mains failure. The connected equipment is supplied by an inverter both during normal, fault-free operation and during a mains malfunction. This very sophisticated method of safeguarding the power supply for a production line is advantageous since it continually supplies all equipment with power (from three-phase AC motors connected directly to the mains through to control units).

With *accumulator buffering of the DC bus*, all relevant inverters of a machine are coupled on the DC bus and connected to an accumulator. If the supplying mains is now to fail, the power supply of the bus would be supplied by the power stored in the accumulator via relatively simple electronics. This bridges the interruption in the mains supply.

This is a cost-effective alternative to an uninterruptible power supply for the drive power. The fact is utilised that the inverters for the drives are already present so that an additional power conversion by the UPS is not necessary. This UPS need only be dimensioned to handle the relatively small power requirement of control, sensors and other equipment.

Utilisation of kinetic energy. The rotational energy stored in quickly rotating machine elements that feature sufficiently high moments of inertia can be utilised for a controlled shutdown should the mains supply fail.

Scenario in which a component of the drive train fails. If the failure of a component of the drive train may not be allowed to interrupt the process, a redundant design can be a viable solution.

If the drive trains essential for the process are only half utilised, the remaining trains can continue to control the process should one train fail. As the components are not as heavily loaded during normal operation, this results in a further reduction of the probability of failure. With this approach, the control structures and the controller must be prepared for a corresponding failure scenario to be able to continue controlling the process despite the failure of a component.

An alternative concept is to use a redundant drive train to which the system switches should a fault occur. This is simpler from a technical standpoint, but leads to a considerably more influence on the process.

4.7 Winding drives

Markus Toeberg, Karl-Heinz Weber

Winding drives are used to store continuous material before or after processing. Together with the synchronised drives described in chapter 4.6, they form the heart of continuous manufacturing processes. They are also often used to provide continuous material to intermittent manufacturing processes.

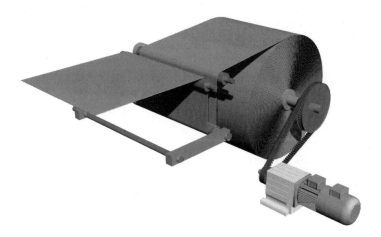

Fig. 4-76. Dancer-controlled centre winder with drive

4.7.1 Applications of winding drives

The winding process. Winding drives are used in a variety of production processes. The materials processed can be homogeneous material webs (e.g. a paper web), fabric, mesh, wire, yarn or thread. From a mathematical standpoint, a wound-up material (wound-up bale) is a spiral. The material is wound up with a defined tensile force that is dependent on the consistency of the material and the diameter.

Unwinder drives are found at the beginning of refinement processes, and rewinder drives at the end (Fig. 4-76). Unwinder and rewinder drives are often – but not necessarily – used in the same process. The material can also be cut and separated after the refinement process.

Every material to be wound makes different demands of the winder drive, and in some cases these demands are very high. The material to be stored must not be negatively impacted by the winding process. Winding drives are quasi-stationary drives, i.e. they are operated at a steady circumferential speed. Other drives that refine the wound material (often synchronised drives) are situated between the winding drives. This refinement could be the printing of a paper web or the coating of fibreboard. When transferring a material from large rolls to small rolls, an unwinding procedure and a rewinding procedure are coupled directly.

Table 4-10. Materials and typical parameters

Application	Material	Speed (m/min)	Power (kW)
Winding base and printed paper and film webs	paper, film	30 – 2,000	0.37 – 400
Winding of textile webs	textile fabric	5 – 150	0.37 – 55
Winding of sheet metal or metal foil	metal	5 – 300	0.37 – 200
Winding of cables	cable	5 – 250	0.37 – 75
Winding of wires	wire	5 – 2,000	0.75 – 200

The wound material is designated differently in the various areas of application and industries, e.g. wound bales, rolls, coils, reels, spools, trees and batches. Table 4-10 provides an overview of materials and process values for winding applications.

Materials to be wound. The spectrum of materials to be wound and their cross-section profiles is very broad. In addition, the tensile forces involved can also vary from 0.5 to 30,000 N. There are tables available for determining the required tensile forces. The materials vary greatly with regard to the following characteristics:

- elasticity,
- surface,
- hardness,
- bending forces,
- tensile strength,
- width,
- thickness and thickness tolerance,
- density,
- grammage,
- cross-section profile

With paper, for example, the dimension for material thickness is specified as the grammage in g/m^2. Micrometers are used for film, and millimetres for other materials. The following list of materials that are wound is by no means exhaustive.

Paper. Base paper is produced with paper machines and wound onto winders for large spools of up to 10 m in width and with diameters over 2 m at a speed of up to 2,000 m/min at the end of the process. Quite often, paper webs are first wound up to then be printed, cut to length or laminated.

The raw material for corrugated cardboard used for packaging cartons is lead to a corrugated cardboard machine via rolled up paper webs. This machine works at production speeds from 250 to 300 m/min. At the end of production, the material is cut to defined lengths and formats with cross cutters.

Large paper bales are cut lengthwise and crosswise for further processing via winding and slitting machines before they are wound up as smaller units. Examples of this might include wallpaper, which is unrolled from larger rolls to smaller rolls after printing and embossing, and paper rolls that are processed further in packaging or printing machines.

Paper is processed at high production speeds both during manufacture and when undergoing further processing. This defines the requirements of the winders, including the switching from one roll to the next.

Film. Films of all types, including packaging, covering, household and adhesive films, are produced via extrusion and calender processes. These raw films are wound up after production for subsequent storage. In further processes, they are unwound, refined, separated, cut into smaller bunches or rolled onto other rolls. Films are very sensitive materials, so the winders must work with high precision.

In the tyre industry, rubber and caoutchouk webs are processed continuously, and winders are very often involved.

Sheet metal, metal films and metal bands. Fine sheet metal made of a variety of materials (steel, non-ferrous metals, aluminium etc.) are produced in roll processes and wound up into unfinished sheet metal coils. Winding drives are also often used as rewinders and unwinders in further processes (e.g. annealing, rolling, coating etc.). Large masses are involved when metals are wound; bending forces that affect the winding process also arise.

Composite materials. Composite materials are comprised of several layers of different materials, e.g. paper, plastic and metal films. Packaging materials made of paper or composite materials for many consumer goods are produced with the aid of winding drives. The composite material used in beverage boxes, for example, is made from joined paper and plastic rolls.

Textiles, woven and knitware. There are many winding applications in the textiles industry, both in yarn and twine manufacture and in weaving, dying, coating and printing processes. The wound rolls are called spools, trees or batches here. The production of yarns in the textiles industry involves very high production speeds, which then also lead to high speeds in the winding process.

Plastic thread. Plastic thread is produced via extruders and spinning pumps, stretched to the determined size in heating zones and finally wound up with winding drives. Winding drives are involved in the production of tear-proof cord threads, e.g. for vehicle tyres. Glass and carbon-fibre threads are characterised by extreme resistance to tearing and minimal elasticity.

Wire. Steel, copper and brass wire is drawn in drawing machines to the required cross section and wound onto reels or spools at the end of the drawing process. The further processing of, for example, enamelled copper wires, cables and ropes occur with the aid of winding drives.

Cables, ropes, leads. Electrical cables and wire ropes are produced from twisted wires in stranding machines. Winding drives are used as reels, spools and cable drums both at the beginning and at the end of production. In the case of gantry cranes covering large distances, supply cables are wound and unwound via winders depending on the crane position.

Floor coverings. Various types of floor covering, such as carpeting, linoleum and plastic coverings are wound onto rolls for production, storage and transport up to delivery to the end customer.

Billboards. Billboards are often positioned into the desired viewing area via intermittently-operated winders and unwinders.

4.7.2   Machine types and drive elements for winders

Winding machine types. Various different winding machines and winding principles are used, depending on and matched to the materials to be wound. There are two basic principles:

- centre winder,
- contact or surface winder.

Centre winders. The centre winder is the most frequently used winder. With the centre winder, the drive acts on the centre of the winding bale (Fig. 4-77). The torque is transferred via the winding shaft and sleeve to the material web. This passes on the torque from the inner layers to the outer layers. There is a technological advantage here, since the refined surface of the material web is not contacted by the rolls, as is the case with contact winders.

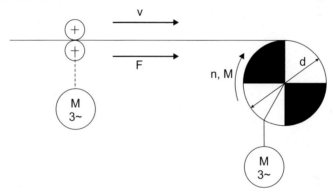

Fig. 4-77. Principle of a single centre winder

With a small diameter, the winding drive requires a low torque and a high speed. With a large diameter, on the other hand, a high torque and a low speed are needed. Since the torque determines the size of an electric motor, electric motors with high drive power that is difficult to control are used in centre winders. Inverters can expand the speed range of the winding drive by using the motor in the field weakening range. With large diameters and material widths, this principle is subject to technological limits that can be better solved with a contact winder.

Quick roll change. The way a roll change is carried out is decisive for the productivity of machines. A distinction is made between single and multiple winders here.

Single winder. Single winders are used as economical solutions when production permits the entire system to be stopped to change a roll (Fig. 4-78).

Fig. 4-78. Single centre winder

Single winders are used in conjunction with material accumulators when the roll change is to occur at full or somewhat reduced production speed and with the winding drive stopped. This principle can only be implemented at speeds up to a moderate production speed, as otherwise the accumulator must take up an excessively large amount of material.

Multiple winders with flying splice. If higher production speeds are involved and stopping of the entire system during a bale change is not permitted or desirable, multiple winders are used. With these machines, there are two or three winding points on a pivot-mounted turret or turning disc. The turning frame is rotated by a drive that receives speed and position setpoints via special algorithms so that the minimum possible changes in tensile force arise in the web during the turning phase. Thus inactive winding points can be loaded and unloaded while the other winding point is active.

Fig. 4-79 illustrates the principle of a double winder with automatic roll change. Inverters 1 and 2 drive the two winders; inverter 3 moves the turning frame during a roll change. Inverter 4 operates the synchronised drive for the material web.

Roll change and cutting systems ensure a flying splice from one winding point to another at full production speed. Automatic loading and removal systems can handle a fully-automatic roll change, including transport of the rolls to and from the machine.

Double winders are generally sufficient (Fig. 4-80). With a higher changing frequency, triple winders are also used.

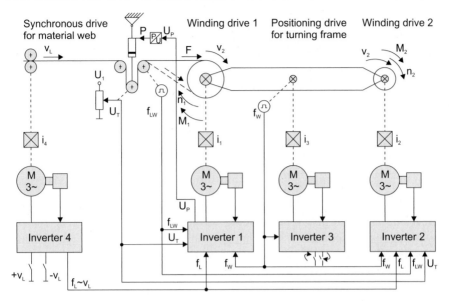

Fig. 4-79. Principle of a double winding turret machine as the centre winder

Fig. 4-80. Double winding turret machine as a centre winder

Moving or fixed motors with multiple winders. There are two different mechanical concepts for the arrangement of the motors and winding shafts in multiple winders. In the one concept, the drive motors are located in a fixed machine component and act on the winding shaft from the motor shaft via belt transmissions in a planetary arrangements. In the other concept, the motors are located on the rotating component of the winding machine (the turret), whereby the inverters can also be found on or outside the rotating component in a control cabinet. The motors or inverters are supplied with power via rotating slip rings.

Pressure rolls. To avoid blisters when rewinding very thin films at high speeds, pneumatic pressure rolls exert a defined contact force at the perimeter of the winding bale. Pressure rolls may also be required to prevent folding, constrictions and telescoping (lateral shifting of the wound material) during the flying splice of a thin film. They are positioned via linear units when the splice occurs.

Reel change systems. With winding turret machines, special reel change systems ensure separation and changing of the material web with a flying change from one winding point to another. With rewinders, the empty cores are prepared with adhesive tape or alternatively the reel change system wraps around the empty cores and handles the change to cores which therefore need not to be prepared.

Traversing drives. If wound-up materials with a round or rectangular profile, such as wire, rope, cable, yarn and thread, are rewound, the winder is equipped with an upstream traversing drive that guides the material, laying layers next to and on top of one another and creates the desired pattern. It is controlled by the winding shaft speed and a desired laying pattern.

Other centre winders. Special designs of centre winders include non-circular cross-section profiles, e.g. rectangular or square profiles. Common centre winders include "board winders", which feature a board-like cross-section profile, and winders for wrapping square packaging lashings, e.g. europallets. Special control algorithms may be required here.

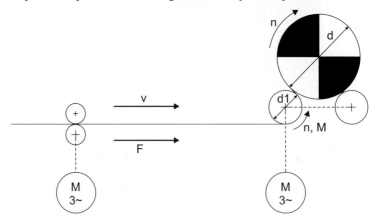

Fig. 4-81. Principle of a contact or surface winder

Contact or surface winders. Contact winders are only used for homogenous material webs, preferably as rewinders. With contact winders, also referred to as surface winders, the drive acts on one or more contact rolls at the circumference of the winding bale, whereby the contact rolls are mechanically coupled or driven via individual drives (Fig. 4-81 and 4-82). There is a technological advantage here, since the required drive torques and speeds do not change based on a varying diameter. This results in considerably reduced drive power in comparison to centre winders.

The circumferential speeds of the contact rolls can vary. Contact winders featuring a support drive in the centre for rewinding the inner layers tightly are also used. The effect of the centre drive is considerably reduced with larger diameters. In addition, a mechanism that counters the force of gravity in order to relieve the load on the shaft can act on the centre to avoid damage to the material at high speeds. The tensile force transferred to the circumference of the winder arises between the contact rolls below the winding bale and an upstream drive with a fixed speed with a defined speed ratio. Winders of this type are used with paper machines, among others. Automatic roll changes are possible while at a standstill or at full production speed.

Control methods of contact winders. There are two main methods for controlling contact winders.

Fixed speed ratio between the contact rolls. In this principle, the second contact roll runs at a faster defined speed via a fixed mechanical coupling or a separate drive, whereby the winding tightness characteristic is determined.

Fig. 4-82. Contact driven or surface winder

Variable torque ratio between the contact rolls. The drive of the feeding roll is operated with an adjustable speed ratio to the upstream machine. The second roll drive supports the first by means of a diameter-dependent torque ratio. The winding tightness characteristic can be determined with this ratio. Measuring of the diameter can occur optically or acoustically without contact.

Mechanical transmission elements in winding systems. Mechanical transmission elements are sometimes located between the drive and the winder. Toothed-belt transmissions, for example, are used frequently. They are characterised by minimal friction and good damping characteristics.

If larger centre winder coils and their pedestals must be moved to load the machine, an axial coupling is made using universal joints with low backlash. Axial couplings between the winding shaft and motor or gearbox output side must also have low backlash.

Sensors, actuators and assemblies involved in the winding process. As the speed, torque and, in some cases, the tensile force change during the winding process due to the spiral principle, certain decoupling components or sensors are required between the winding drives and the other synchronised drives of a production machine. In addition, monitoring sensors are used to gauge the material web when high speeds are involved.

Dancers, rockers and pendulums. Dancers, rockers and pendulums are material accumulators that can be located downstream or upstream of the unwinders and rewinders. The design and functional principle are described with the synchronised drives in chapter 4.6.3. Dancers, rockers and pendulums are used with both single and multiple unwinders and rewinders, but in particular with unwinders, which must often process eccentric winding bales. Position sensors (preferably contactless) at dancers and rockers provide actual position values for drive control loops.

Tension measuring stations. In tension measuring stations, the material web tension is measured as an actual value for a superimposed control of the tensile force. They also are described in chapter 4.6.3. This allows for high tensile-force precision and wide setting ranges – but they cannot compensate for eccentricities of winding bales.

Accumulators, loop accumulators and towers. These assemblies are used to supply or take up the material from unwinders or rewinders, respectively, during a roll change with stopped winders at full or partially reduced production speed. They serve for two tasks, a storage function in the roll change phase and a dancer function during normal operation. The principle is the same as that of a dancer with multiple redirections of the material web. The tension is generally applied by separate drives via spindles, chains or belts. The material is predominantly stored vertically here, but is sometimes stored horizontally as well. This equipment can also be operated in conjunction with dancers or tension measuring stations.

Loop accumulators without tensile force. Loop accumulators are sections in which a material web enters or exits upstream of rewinders or downstream of unwinders with almost no tensile force (except gravitational force caused by the weight of the material web). In the case of long sections recesses in the foundation are also possible. In conjunction with winders, they are often used to decouple intermittent material feeds. In band-splitting systems, in which sheet metal webs are split lengthwise, loop accumulators upstream of the winder are also used to take up material webs of varying lengths resulting from thickness tolerances. The loop position is detected without contact with analog or digital optical sensors. Control can be of the two-point, three-point or continuous type.

Diameter sensors. Analog sensors (optical, acoustic, potentiometers etc.) are used in certain applications to measure the winding diameter in the start-up and operating phases. This is predominantly carried out without contact.

Material web sensors. At medium and high production speeds, a tear in the material web must be detected directly and very quickly to minimise the damage. Contactless sensors that gauge the material are used for this purpose.

Edge controllers. To produce winding bales with straight edges with rewinders, the lateral guidance of the material web must be perfect. For this purpose, edge controllers that detect one or both edges of the material via sensors are required. The sensors affect the lateral guidance of the material via the adjustment of moving rolls by a controller. Alternatively, the entire winding station can be moved onto rails or sleds to ensure production with straight edges.

4.7.3 Drive systems for winding drives

Drive system requirements. The special requirements of winding drives arise from the wide speed and torque setting ranges that result from the winding concept and the necessity for excellent smooth running to ensure a good winding result with sensitive materials to be wound up. Each winding drive must be operable in two quadrants and for re-winders with directional changes in four quadrants. The transition from one quadrant to another must occur constantly. These requirements for the various winder types are examined in more detail below.

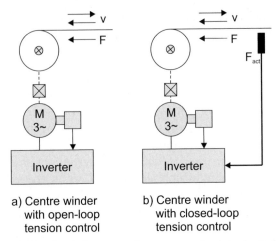

a) Centre winder
with open-loop
tension control

b) Centre winder
with closed-loop
tension control

Fig. 4-83. Control concept for centre and contact winders

Tensile force-controlled centre winders. In this concept, tensile force is controlled indirectly via the variable drive torque (Fig. 4-83 a). The torque must be changed depending on the diameter and tensile force. With some rewinders, the tensile force is also changed depending on the diameter and the surface consistency of the material (the so-called soft or taper tension characteristics). The required torque setting range of the drive system is derived from the product of the maximum diameter ratio and the variance in tensile force. The winding quality is thus essentially determined by the torque setting range of the drive and by disturbances that cannot be compensated for.

Tensile force-controlled centre winders. If a very wide torque setting range is required, a torque-controlled system may be pushed to its limits. These limits can be considerably expanded when a sensor measures the tensile force of the material (Fig. 4-83 b). The drive properties listed for torque-controlled centre winders also apply here. The winding quality is now essentially determined by the precision of the tensile force measurement. Tension measuring systems can also be equipped with double sensors (for the lower and upper areas of winding) for wide setting ranges. Due to the low dynamics of the tension-control loop, precise disturbance compensation (acceleration, friction etc.) is required, as is the case with torque control. Disturbances that cannot be compensated for negatively affect the winding result in situations where control is not steady (e.g. when winding begins).

The generation of a suitable speed profile is vitally important for the start-up of the winding drive, which must occur in synchronism with the acceleration of the synchronised drives of the production machine.

Torque setting ranges of centre winders. In practice, torque setting ranges from approx. 1:2 to 1:100 may be necessary. To assess a system's suitability for this purpose, the requirements of the winding process (including the permissible tolerances) must be compared to the capability of the drive system. Experience plays an important role here. It would not make sense, for example, to make demand for a tensile force tolerance of 1% (with regard to the actual value) with a very robust material. With wide torque setting ranges, the drive must be capable of a repeatable performance quality under any environmental and operational conditions that may arise.

Wide torque setting ranges can be solved easily with servo drive systems with encoders, whereby the top technical limit of servo systems is shifted upward because of ever more precise motor controls. It is currently 1:50 to 1:100. However, the specification of the torque setting range is directly linked to the precision that can be achieved. A wide torque setting range with high precision alone is not sufficient. The disturbances described in the following must also be kept well under control.

Mechanical disturbances. The mechanical friction is a disturbance that cannot be eliminated. A good winding result depends on the ability to keep this friction repeatable and low. If wide torque setting ranges are required, gearboxes with minimal friction (with a low number of stages and synthetic oil lubrication), toothed-belt transmissions or direct drives should be used. The precise and repeatable compensation of the friction (gearbox, bearings, winding shaft) by the inverter is also important. This can be executed by an adaptable, high performance process control.

Asymmetrical centres of gravity. In some winding applications, there are other operating conditions that must be taken into account. For textile dying processes with so-called "jiggers", major changes in the density of the winding bales can occur as a result of waterlogged wet batches. These changes in density are linked with large cyclical torque fluctuations. Both the control structure and the measurement of the centre winder must be able to compensate for this.

Dimensioning of a centre winder. The motor frame size is determined by its torque. The maximum torque of a centre winder is achieved with the largest reel diameter and thus the lowest speed. The power of the drive system to be installed is therefore considerably higher than the process power. Although high speeds and high torques do not occur simultaneously, the drive must be able to provide both.

Centre winders with defined material tensions are mainly dimensioned to continuous operation. The dynamic reserves of the inverter are mostly sufficient to brake the drive even in emergency situations. For very high inertias and short brake paths the brake torque may become the relevant factor for the dimensioning. For winders with intermittent operating modes the dynamic drive torque determines dimensioning.

The required rated motor power P_{Type} of centre winder drives using field weakening is calculated as follows:

$$P_{Type,rewind} = \frac{F_{max}(d_{max}) \cdot v_{max} \cdot d_{max}}{\eta_{mech} \cdot k_{Field} \cdot d_{dim}} \quad (4.3)$$

$$P_{Type,unwind} = \frac{F_{max}(d_{max}) \cdot v_{max} \cdot d_{max} \cdot \eta_{mech}}{k_{Field} \cdot d_{dim}} \quad (4.4)$$

F_{max} is the tensile force, v_{max} is the web speed, d_{max} and d_{min} the diameter limits, k_{Field} is the field weakening factor and η_{mech} the mechanical efficiency. The rated motor power P_{Type} can be considerably reduced with a smart combination of the dimensioning diameter d_{dim} and the field weakening factor k_{Field}. The field weakening of an asynchronous motor acts like an electronic gearbox.

The power P_{Supply} to be provided from the mains or from the DC bus is only the active power (process power), including all losses.

$$P_{Supply,rewind} = F_{max} \cdot \frac{v_{max}}{\eta_{mech}} \quad (4.5)$$

$$P_{Supply,unwind} = F_{max} \cdot v_{max} \cdot \eta_{mech} \quad (4.6)$$

For thick metal sheets the bending torque is also to be considered.

An overdimensioning of the drive (motor, controller, gearbox) is to be avoided for torque-controlled systems. Otherwise, the torque resolution and the torque setting range are restricted.

If there are two winding drives on one shaft, the required power is divided symmetrically. When using indexing gearboxes, the process parameters must be considered per stage.

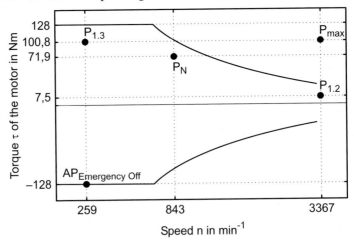

Fig. 4-84. Speed-torque diagram of a winding drive

Fig. 4-84 shows the dimensioning of a typical winding drive in the speed-torque diagram of the motor with the most important powers. The winding ratio results from the process parameters: $q = d_{max} / d_{min} = 1,350$ mm / 104 mm = 13. The motor only uses field weakening with the factor $k_{Field} = 4$. A drive power of 6.35 kW is to be installed although a process power of 2.73 kW is required only. The drive power would have to amount to 35.5 kW without field weakening.

Dimensioning of a contact winder. The diameter change does not affect motor speed and motor torque when a contact winder is used. This means that only the process power must be considered for the dimensioning of contact winders. Field weakening is useful only if low tensile forces are required at high speeds. The power requirement is as follows:

$$P_{Type,rewind} = F_{max} \cdot \frac{v_{max}}{\eta_{mech}} \tag{4.7}$$

$$P_{Type,unwind} = F_{max} \cdot v_{max} \cdot \eta_{mech} \tag{4.8}$$

If the drive power is to be split for two rolls the diameter-dependent torque characteristic for the second roll must be considered.

Gearbox selection. Winding applications generally require a transmission of the fast motor speed to a slower process speed. For this, low-friction and low-backlash gearboxes are preferred.

Here, positive-type belt transmissions are suitable up to a ratio of approx. 1 : 3 and motor speeds of $n_{Motor} \leq 6,000$ rpm. Worm gearboxes are not used since their non-linear and very high losses cannot be compensated for. Chain transmissions have backlash and thus the system is much more difficult to control.

Table 4-11. Gearbox suitability for winding drives

Axial gearboxes and other transmissions							Right-angle gearbox		
			Belt						
Helical gear-box	shaft mounted gear-box	Plane tary gear-box	Posi-tive con-nection	fric-tion-type con-nection	Chain	Direct drive	Helical bevel gear-box	Worm gearbox	Bevel gear-box
++	++	+	++	--	--	o	+	--	++

--	Not suitable
o	Conditionally suitable
+	Suitable
++	Very suitable

Direct drives. If the speeds of process and motor match, the drive motors can act directly on the winding shaft via low-backlash couplings. High-pole three-phase motors with a rated speed of 300 to 500 rpm are used for low speeds. Torque motors (high-pole synchronous motors) are not often used for winding drives because of their very small field weakening range as opposed to asynchronous motors.

Decelerating winding drives. A considerable kinetic energy is stored in rotating reels because of high moments of inertia and high speeds. This kinetic energy must be dissipated quickly in emergency stops, which requires efficient electrical, electromechanical and mechanical brakes. Combinations of different systems are also possible.

Motor brakes are part of the standard equipment of a winding drive. Additional mechanical brakes such as disc brakes can be used for very high inertias and speeds to support the drive in the deceleration phase. By this, a redundancy is achieved also.

Selection of the motor type. In the past, winding drives were mainly DC drives as this technology is excellently suitable for very precise torque settings. Today, AC drives with frequency inverters or servo inverters are increasingly used. Particularly in the small to medium power range, the three-phase technique is not only less expensive but also offers advantages such as a low mains load, electrically maintenance-free motors and unlimited motor standstill under full load.

Three-phase standard AC motors are not only sufficient for most winding applications but also to be preferred against servo motors. They are inexpensive, can be used for operation in the field weakening range and are suited to move large inertias in a controlled way because of their high intrinsic inertia. Particularly asynchronous motors with low rated speeds together with belt transmissions are excellently suited for many winding applications if their field weakening range is optimally used.

Separately driven fans are the best choice for motor cooling, since the highest losses occur at low speeds and the motor can be operated with high field weakening.

Asynchronous servo motors are used when compact designs or high maximum speeds are required.

Synchronous servo motors are less suitable for torque-controlled drives. The magnetisation strongly depends on the temperature which cannot be compensated for. Also detent torqueses are disturbing. In addition, the field weakening that can be achieved is clearly lower as for asynchronous motors.

Torque motors as direct drives can present an innovative solution when very compact and short designs are necessary, when e.g. mechanical brakes are replaced by winding drives in unwinders.

The fact that no gearbox is used eliminates friction as disturbing quantity. The achievable smooth running is up to ten times higher than for conventional servo motors.

Use of field weakening. As described above, the rated drive power to be installed is considerably reduced when field weakening is used. To achieve high field weakening factors, motors with a rated speed as low as possible should be selected. For this, the 29Hz technique with four-pole standard three-phase AC motors provides a good solution.

When field weakening is used, the speed above the rated motor speed is increased by reducing the magnetisation reciprocally to the speed increase. This means that the motor can be operated clearly above the rated speed with maximum voltage. Up to a field weakening factor of approx. 1 : 1.5 to 1 : 1.2 (depending on the stall torque) the output power at the motor shaft is almost constant. If the factor is higher the power decreases. In practical applications, maximum field weakening factors of up to 1 : 4, in special cases even up to 1 : 5 are used.

Group drives. When the motor power is high or the material is very wide, a motor is used on either side. In doing so, the drive torque is divided symmetrically and material torsions are avoided. Both motors as asynchronous motors can easily be supplied in parallel by one inverter. It is also possible to use two inverters where the torque is split symmetrically using the control system.

DC bus connection. The connection of several inverters via the DC bus allows a very simple and efficient solution where generator power and motor power of multi-axis systems are exchanged in the DC bus (and not via the AC mains as for DC drives). If an entire production machine is designed this way, the mains only has to provide the losses and the resulting motor process power. This results in a very low mains load and reduces operating costs by saving energy as well as investments in power supply installations.

Generator mode by brake choppers or regenerative power supply modules. Short-term generator powers which occur during normal stops, quick stops or emergency stops are dissipated by a central brake chopper or are fed back to the mains using a regenerative power supply module.

For reasons of energy efficiency, regenerative power supply modules should be used for a continuous operation in generator mode of unwinders which are not connected to drives operating in the motor mode via the DC bus.

Inverter selection. The inverter selection is determined largely by the required performance of the winding drive. Thus, simple speed-controlled winders can be achieved with frequency inverters and V/f control.

Servo inverters with a precise current control are used for powerful winders and processes with torque control. This is the only way to achieve high torque setting ranges and speed accuracies.

Winding process controls. For centre winders the following three control systems are mainly used (Fig. 4-85):

- open-loop tension control,
- closed-loop tension control,
- dancer position control.

In applications where there is no master drive between the unwinder and rewinder, speed-controlled rewinders can also be used (Fig. 4-85 d).

Open-loop tension control. Expensive and complicated sensors for web tension control are not necessary for open-loop tension controlled winders. The motor control of the drive is feedforward-controlled with a speed setpoint n_{Winder}. The torque τ_{Winder} is controlled according to the required tensile force F_{Web} (Fig. 4-85 a)

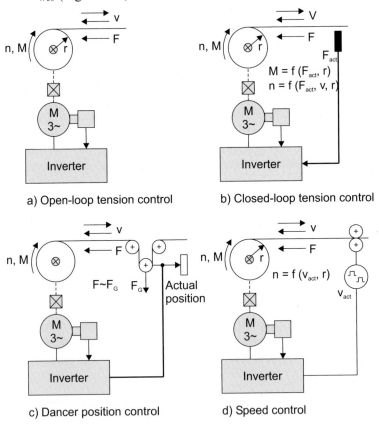

Fig. 4-85. Control methods for winding processes

$$n = \frac{v}{2\pi \cdot r} \qquad (4.9)$$

$$\tau = F_{web} \cdot r \qquad (4.10)$$

In this procedure, the winding quality and accuracy of the tensile force are largely determined by the accuracy of the diameter calculation and the disturbance variable compensation. The inverter can calculate the diameter precisely from the winding speed n_{Winder} and the material web speed v_{Web}.

Additional set torqueses are added to compensate the disturbance variable. These are necessary for acceleration and to overcome friction torques. The limits of the system are reached if the required friction or acceleration torqueses are higher than the smallest load torque. Friction is mainly produced in the gearbox. Speed and temperature of the gearbox oil can be compensated for, if necessary.

Table 4-12. Applications of open-loop tension control

Typical materials	Metal sheets, paper, textiles
Material web speed	Up to 600 m/min
Maximum torque setting range	1 : 60

Closed-loop tension control. Closed-loop tension control is used where the accuracy of an open-loop tension-controlled system is no longer sufficient, e.g. where large torque setting ranges are required or imponderable disturbance variables act on the winder.

In static operation, the quality of the tension measurement determines the accuracy of the tension control. The actual tension is compared to the setpoint via a process controller. In simple cases, the output signal is the reference of a lower-level speed or torque controller (Fig. 4-85 b).

Normally, the closed-loop tension controller uses a very small proportional gain ($V_p < 1$) and long reset time ($T_n > 0.5$ s) for reasons of stability. This means that this control system is not suitable to control very dynamic disturbance variables occurring during acceleration and unbalanced material.

Table 4-13. Applications of closed-loop tension control

Typical materials	Thin films, paper
Material web speed	Up to 1,000 m/min
Torque setting range	1 : 120

Dancer position control. In a dancer position control system the material web tension is exclusively generated by a dancer unit which is mounted upstream of the winder (Fig. 4-85 c). The advantage of this type of control is the ability to compensate for dynamic disturbance variables that can occur e.g. due to acceleration or unbalanced material web. Although the dancer position varies, the tensile force remains constant. The dancer position control is preferably used for applications with high tensile force accuracy and constancy requirements.

The winding motor is speed-controlled. The feedforward-control value for the motor speed is calculated from dividing the material speed by the actual radius. The dancer position control corrects the feedforward-control when the detected dancer position differs from the desired value.

The acceleration torque can be compensated for by means of feedforward-controlling the speed controller output. In general, measures for friction compensation are not taken for this control procedure.

Table 4-14. Applications of dancer position control

Typical materials	Cables, wire, foils, paper, textiles
Material web speed	Up to 1,000 m/min
Torque setting range	determined by dancer design

In dancer position control the limits are not determined by the torque setting range but rather by the mechanical structure of the dancer, the dynamics (weight, friction etc.) and the speed setting range of the drive.

Contact winder. Contact winders are speed-controlled and the winding hardness is determined by the differential speed of the contact rolls. If the contact rolls have separate drives, the winding characteristics can be determined by the torque distribution between the drives. The tensile force is generated either by dancer units or tension measuring stations which are upstream of the winder. These act on the winder drive by means of control loops.

Winding characteristics. During the winding process the motor torque rises proportionally with the reel diameter until the maximum diameter has been reached. If the reels have large diameters or the material web has a very smooth surface it occurs that the static friction between the layers exceeds its maximum value and the material on the reel starts telescoping. In this case, the tensile force F is permanently reduced with the increasing diameter. This is also called taper tension (Fig. 4-86). In the winding core the material is wound hard, whereas the material is wound increasingly soft with increasing diameters. The reduction of the tensile force is implemented inversely proportional to the diameter and is initiated at a certain diameter d_C.

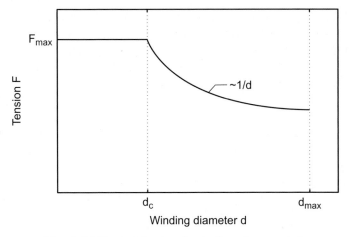

Fig. 4-86. Typical hard-soft-winding characteristic

Winding characteristics are used for all control systems mentioned above. In a dancer position controlled winder the tensile force can only be adjusted by respective actuators at the dancer units. This could e.g. be a pneumatic cylinder.

Disturbance variables compensation. The winding result is notably influenced by two disturbance variables especially in purely open-loop controlled systems. These are on the one hand mechanical losses due to friction in bearings and gearboxes and on the other hand acceleration torque. The impact of disturbance variables on the system can be minimised by means of a feedforward control of the motor torque, using certain set point values for a respective compensation.

The lower the torque is for the smallest tensile force set point value, the more important is the disturbance values compensation.

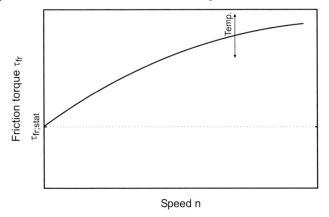

Fig. 4-87. Typical friction torque in geared motors

The *friction* is composed of a static component and a component which increases with increasing speed. The speed-dependant friction component varies with the different types of gearboxes as well as with the gearbox oil temperature. In order to compensate for this disturbance variable, additional torque set point values are used by means of a table function (Fig. 4-87).

The *acceleration torque* results from changing motor speeds and the actual moment of inertia. The inertia of the reel depends on the diameter, the width and the specific weight of the material and has to be permanently calculated. In addition, the moment of inertia of the motor also influences the acceleration torque.

Safety technology in winding drives. Winders are always just a part of an entire production process. Thus, safety precautions generally have to be made for the entire machine.

Winders with lower-level torque control must be prevented from exceeding the permissible speed range in case of *disruption of the material web*. An unwinder must not continue rotating in the unwinding direction after disruption of the material web, since material is then conveyed into the machine without any control. Hence, unwinders are generally feedforward controlled by the rewinding speed.

A safe and simple method for detecting the disruption of the material web is the monitoring of the diameter change. If the diameter of a rewinder or unwinder changes into the respective opposite direction, a disruption of the material web has occurred. The measures that are taken in case of such a failure depend on the particular process and the machine. Usually the entire production line is stopped by means of a fast stop.

Calculation of the residual length with automatic deceleration. The calculation of the residual length before a preset diameter is reached can be implemented by monitoring the actual diameter and the material thickness. This allows for an automatic deceleration of the machine before e.g. in case of an unwinder the material web is completely unwound.

4.8 Intermittent drives for cross cutters and flying saws

Andreas Diekmann, Karsten Lüchtefeld

Cross cutters and flying saws are used in continuous production processes to machine the material while it is moved. For this purpose the tool has to be synchronised to the moving material web. After the machining process the tool has to be positioned to the starting position for the next machining process. Cross cutters operate in a rotary manner, flying saws based on linear movements.

The motion sequences of cross cutters and flying saws can also be used for machining processes which are not continuous, like for example welding, embossing and a synchronised infeed of material from a stack into a continuous production process. The drives in this category basically can also be referred to as "intermittent drives", as they synchronise a movement to the production cycle, carry out the machining process and then have to approach the next machining position for the following production cycle.

4.8.1 Application of cross cutters and flying saws

Cross cutters. Cross cutters are used on continuous production lines to separate the material. The cutter is placed on a roll. During the cutting procedure the roll has to move synchronously to the material web. The remaining part of the revolution is controlled in a way that the next cut can be executed according to the format set.

If there aren't any varying formats, a cross cutter can be designed in a way that the circumference of the roll corresponds to the cutting length. This roll then has to be moved synchronously to the material speed via a synchronous drive.

Many applications, however, require varying formats, and thus variations of the cutting length without changing the roll. For this purpose a motion sequence is used, enabling the roll with the cutter to move in a way that the desired cutting intervals are executed.

Cross cutters can be used for all materials which can be separated by a rotary knife. These are:

- paper webs, e. g. after reel fed printing, for the production of paper sheets, for the production of packaging material like cardboard or corrugated cardboard,
- metal webs,
- film webs.

The accuracy required varies; however, normally it is in the range of 0.1 to 1 mm. The cycle rate can be very high and can attain up to 600 cycles per minute. This corresponds to an interval of 0.1 s between the cuts. By this, material speeds of 400 m/min with a cutting length below one meter can be achieved. Table 4-15 provides an overview of the fields of application of cross cutters.

Table 4-15. Fields of application of cross cutters

Fields of application for cross cutters	Industry sector	Speed (m/min)	Accuracy (mm)	Cycles (per min)
Production of corrugated cardboards	Paper	100–400	0.1–0.5	60–300
Production of steel wires	Metal	40–80	0.2–0.3	60–300
Production of packaging material	Packaging	100–200	0.1	300–600
Cutting, embossing of cardboard, sheets, films	Paper, plastics	100–200	0.1	300–600

Flying saws. Flying saws carry out a similar task as cross cutters. They also separate moving material originating from a continuous production process. Due to the material thickness, a saw is required; the entire cutting process therefore takes longer than that of the cross cutter.

Fig. 4-88. Flying saw

In the case of the flying saw, the tool is positioned in the cutting position and is then moved together with the work piece (Fig. 4-88). After the machining process has been completed, the tool is placed in the next cutting position.

Flying saws are used for the following materials:

Table 4-16. Fields of application of flying saws

Fields of application for flying saws	Industry sector	Speed (m/min)	Accuracy (mm)	Cycles (per min)
Wood	Wood working	5–40	0.5–1.0	5–10
Insulating materials	Construction materials	2–25	0.5–1.0	10–20
Metal webs	Metal	5–50	0.5–1.0	4–10
Plastic profiles	Plastic	4–60	0.5–1.0	5–20

- wood,
- insulating materials,
- metal webs like thick sheet plates,
- extruded plastic profiles.

The accuracies are similar to that of cross cutters. The cycle rates, however, only attain values of 20 cycles per minute, so that a machining process takes more than 3 s. Thus the material speeds can also only be 40 m/min.

4.8.2   Structure of cross cutters and flying saws

Mechanical structure of cross cutters. A cross cutter consists of two rolls with one cutter, respectively. Between the two rolls there is the material web. The two rolls are moved so that during the cutting process the two cutters are positioned above each other and separate the material (Fig. 4-89). Alternatively there can be a cutter on one side and an impression cylinder on the other side.

Depending on whether the speed of the cutters during the cutting process corresponds to the speed of the material web, asynchronous and synchronous cross cutters are distinguished. If the cutting distance is longer than the circumference of the knife drum, the cross cutter has to be operated in a subsynchronous mode. If the cutting distance is smaller than the circumference, the cross cutter works in an oversynchronous mode.

Asynchronous cross cutters. In the case of an asynchronous cross cutter, the cutter moves asynchronously to the material speed during the cutting phase (Fig. 4-90). Thereby a subsynchronous speed is not usual for cutting, as the material to be cut would be compressed during the cutting process, and therefore upstream machining processes could be affected.

With an oversynchronous speed, the material is accelerated for cutting during the cutting process, and it may tear before it has been completely cut. With an over-synchronous cutting, the acceleration causes regular gaps between the cut pieces.

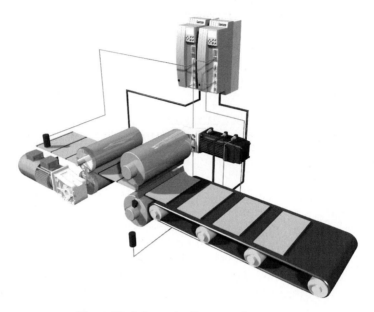

Fig. 4-89. Schematic diagram of a cross cutter

Due to the different speed of the cutter and the material web, the cutting quality is not as high as in the case of a synchronous cross cutter. However, some materials to be machined are so tolerant that the cut can be executed asynchronously. Usually asynchronous cross cutters are used, whenever the cutting quality or accuracy does not need to be very high.

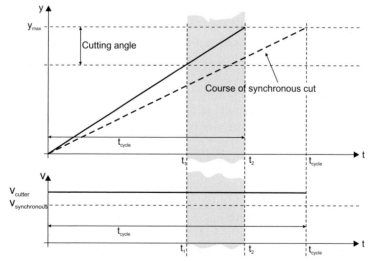

Fig. 4-90. Oversynchronous cutting length for the asynchronous cross cutter

The advantage of the asynchronous cross cutter is that different cutting lengths can be attained by setting the speed ratio between the material speed and the circumferential speed of the cross cutter differently. Thus a synchronous drive can be used, which is technologically simpler than a cross cutter drive. Also no high accelerations and braking processes are required, which reduces the energy required and the drive power to be installed.

For this reason, asynchronous cross cutters are often used for very short cutting lengths and high cycle rates, if this is tolerated by the material to be machined. The adjustment range of asynchronous cross cutters is very small.

Synchronous cross cutters. In the case of a synchronous cross cutter, the cutter and the material web move synchronously during the cutting process. By this, a very high cutting quality is achieved. During the cutting process the material is neither compressed nor torn.

If the circumference of the cutter corresponds to the cutting length, the knife drum moves synchronously to the material web (Fig. 4-91). In this case a synchronous drive is required, which, like in the case of an asynchronous cross cutter, does not have to carry out acceleration and braking processes. This construction is always used if the format is not changed. This is for instance the case for cutting papers after rotary printing. However, often it is required to cut different formats. For this the more complex motion sequences described below are required.

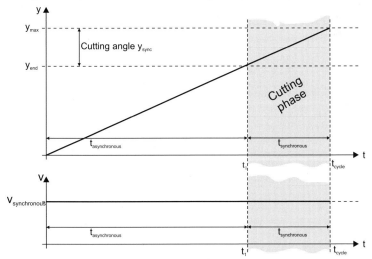

Fig. 4-91. Time flow for a synchronous cutting length

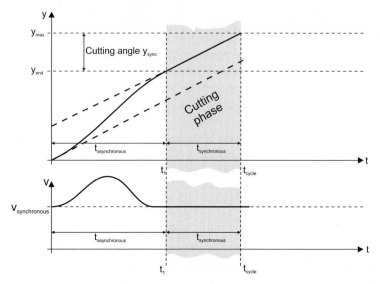

Fig. 4-92. Over-synchronous cutting length for the synchronous cross cutter

If the cutting length is smaller than the circumference of the cutter, the roll between the cutting processes first has to be accelerated and afterwards has to be braked to the material speed (Fig. 4-92).

If the cutting length is longer than the fly circle circumference, the roll first has to be braked and then accelerated to the material speed (Fig. 4-93). As soon as the cutting length is more than twice the fly circle circumference, the roll decelerates to standstill.

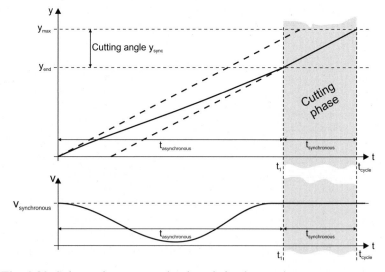

Fig. 4-93. Sub-synchronous cutting length for the synchronous cross cutter

From the motion sequences of the synchronous cross cutter it can be seen that the drive has to be very dynamic. For cycle rates of 400 per minute and therefore time intervals of 150 ms between the cutting processes, the drive has to be able to accelerate to its rated speed and decelerate from it within approx. 50 ms.

In order to achieve this dynamic performance, apart from a highly dynamic drive also a low mass inertia of the mechanics is required. At the same time it has to be possible to absorb these high acceleration forces. Therefore sometimes stiff and extremely light carbon fibre materials are used for the rolls of cross cutters.

Table 4-17 shows a comparison between the characteristics of asynchronous and synchronous cross cutters.

Table 4-17. Comparison of asynchronous and synchronous cross cutters

	Asynchronous cross cutter	**Synchronous cross cutter**
Cutting quality	Limited	Good
Cutting length	Smaller than the circumference of the cutter	Longer, equal, and smaller than the circumference of the cutter
Number of cycles	High	Medium to high
Drive power	Low	High
Drive function	Synchronous drive	Cross cutter drive

Flying saws. In the case of a flying saw the tool is moved synchronously with the moving material web during the machining process (Fig. 4-94). For this purpose, the tool first has to be moved parallel to the material web on a slide. In order to submit the feed of the tool during the machining process, a second drive moves it transversely to the material web. Here a positioning drive is used, which has to ensure a constant feeding speed during the cut and at the end has to execute a repositioning to the initial position. A third drive moves the tool (tool drive). For lifting the saw for the repositioning, different techniques are used.

The motion sequence of the drive of the flying saw comprises the following phases (Fig. 4-95):

- *Synchronising:* First the tool has to be accelerated from the initial position to material speed. In the moment where it reaches the material speed it has to be in the cutting position. For this purpose the acceleration distance has to be determined from the motion profile data and the acceleration process has to be started at the right moment.
- *Synchronous run:* Now the tool plunges into the material and then moves synchronously to the material speed until the machining process including the lifting of the tool has been completed.

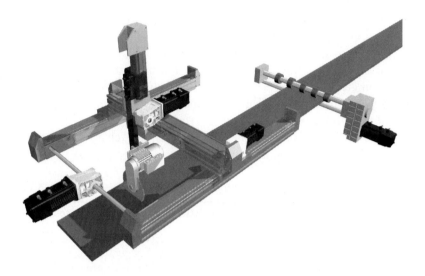

Fig. 4-94. Schematic diagram of a flying saw

- *Approaching the initial position:* Finally the tool is repositioned to the initial position. Normally here a highly dynamic performance is required, so that the minimum distance between the cuts is as small as possible allowing a greater range of cut formats.

The minimum cutting length depends on the ratio of the material speed and the duration of a cutting cycle. The duration of a cutting cycle is composed of the following:

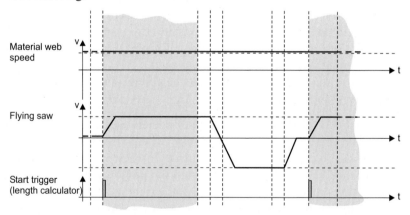

Fig. 4-95. Motion sequence of a cutting process of a flying saw

- the time for synchronising the tool to the material speed,
- the width of the material web and the feeding speed of the tool, which define the cutting time,
- the time for removing the tool from the material web,
- the time for repositioning the tool to the initial position for the next machining process.

In order to minimise the last two periods of time, a high dynamic performance of the drives is required.

On the basis of the speeds and accuracies required, toothed belt drives are a suitable alternative for the linear movements of a flying saw.

Sensors for cross cutters and flying saws. In order to be able to control the process, the following quantities have to measured or known:

- The speed of the material, and
- the position of the cutting positions on the material web.

The material speed can be transmitted electronically by the according drive using master signals or a communication system. If this, e.g. due to material expansions, is too inaccurate, a measuring wheel which runs on the material web can be used.

With measuring wheels, it has to be observed that the resolution of the incremental encoder is at least tenfold with regard to the accuracy of the cutting position required. Thus the measuring wheel should have a correspondingly small diameter resulting in the measuring resolution required.

If the exact cutting positions on the material web are not crucial and only the cutting length is important, the timing for the cuts can be calculated from the material speed and the cutting length. However, if the cutting positions have to be placed exactly in defined positions on the material web, a sensor has to detect corresponding marks. This for instance is required for cutting printed materials. The mark sensor signal can be detected very exactly via a touch probe input of the drive (chapter 3.4.3) and can be linked to the current position of the drive.

4.8.3   Drive systems for cross cutters and flying saws

Requirements. The specific requirements with regard to the drives of cross cutters and flying saws result from the motion sequence and their highly dynamic performance.

Dimensioning. The dimensioning of the drive components is effected according to the dynamic requirements for the motion sequence. For this, the speed-torque characteristic of the motions with the highest dynamic performance has to be determined.

In the case of cross cutters, the dynamic performance required is the higher the further the cutting length diverges from synchronous operation where it corresponds to the circumference. Particularly long cutting distances can only be achieved by braking to standstill before accelerating to material speed again. The minimum cutting length is determined by both a high maximum speed and a high acceleration. The longest or smallest cutting length determines the drive power.

In the case of the flying saw the repositioning requires the highest dynamic performance, and thus the drive power. Here as well the smallest cutting length which can be achieved is limited by the amount of installed drive power.

Gearboxes. The selection of the gearbox is determined by the high dynamic performance and the stability under alternating load. In cross cutters with very high cycle rates gearboxes can be dispensed with. Otherwise planetary gearboxes are advantageous due to their low backlash. For the linear movements of the flying saw normally toothed belt drives are used. In some applications they can be driven directly by the servo motor. In other cases a gearbox, normally again a planetary gearbox, is applied between the motor and the toothed belt sprocket.

Motor. Due to their requirements with regard to the dynamic performance and accuracy, synchronous cross cutters and flying saws can only be operated with servo motors. For smaller powers normally synchronous servo motors are used, achieving a very high dynamic performance by their low mass inertia and the high overcurrent capability.

At high cycle rates with wide webs, cross cutters can require very high drive powers exceeding 100 kW. In these cases asynchronous servo motors which are internally ventilated are used. Due to their compact design and low mass inertias, high acceleration capability can be achieved.

Inverters. The servo motors are combined with servo inverters and a servo control.

Cross cutters offer a special feature in the electric design. As a large amount of energy has to be exchanged between the DC bus of the inverter and the kinetic energy of the rotary knife for quickly accelerating and braking, it is suitable to dimension the DC-bus capacity so that it can absorb the entire kinetic energy. Thus, the conversion to heat via a brake resistor or the power recovery to the mains is omitted.

Motion control. The motion sequence of a cross cutter and a flying saw depends on the following parameters:

- the material speed,
- the cutting length,
- the position of the cuts,

- the accelerations and decelerations of the drive, which can be achieved.

In order to avoid excitation of the mechanics, which in turn can deteriorate the machining result, jerk-free motion profiles should be used.

The motion control can either be integrated within the drive, or in the higher-level control. The implementation by the servo inverter is a suitable alternative, as it can detect the signals to be considered, like for example the material speed and the mark positions, and the entire motion sequence only requires few parameters. Therefore a servo inverter can autonomously implement the "cross cutter" or "flying saw" machine function using the corresponding software function.

In the software of the drive the motion profile can be implemented via positioning profiles which are linked to each other. Thereby the material speed is included in the speed of the positioning process like an "override factor". Another possibility is the implementation via an electronic cam which, depending on the position of the web, defines the motion sequence required. Electronic cam drives are described in chapter 4.9. The provision of a specific function block for the motion sequence of cross cutters and flying saws, which then only requires few parameters for the setting, also is reasonable.

Interaction of drive and control. If the motion control is implemented within the drive – following a decentralised concept – the control interface is reduced to commands for the sequence control of the function (start, stop) and to parameters specified for a production lot (cutting length). A batch counter can either be implemented in the control or in the drive. Cross cutters and flying saws are predestined for a decentralised implementation of their motion function in the drive, so that an autonomous machine module is created.

Safety technology. If an operator has to intervene in adjustment or fault situations, it is required to implement safety functions. Also for drives of cross cutters and flying saws it is reasonable to integrate them within the drives. For a safe shutdown, the "Safe Torque Off" (STO) function is sufficient. If slow speeds have to be allowed for purposes of adjustment, the "Safely Limited Speed" (SLS) function can be used.

4.9 Drives for electronic cams

Martin Harms, Torsten Heß

Cams are gears with a non-uniform transmission. They can control complex motion tasks required for recurring intermittent operations in production and handling machines. In contrast to the time-controlled positioning technology (chapter 4.4) for cams a position reference according to the machine cycle is established by the motion control. This motion therefore can also be described as a function of a path-controlled profile generator.

4.9.1 Applications with cam drives

Drives with cams are used in fast running processes of production and handling technology. The closed and cycled motion sequence allows the execution of highly dynamic positioning tasks within the machine cycle during which material is locked into or out of a process, or a tool is positioned to the work piece for machining.

The production processes using cam drives include:

- cutting,
- embossing,
- gluing,
- welding,
- bending,
- forming.

For this, continuously rotating cylindrical tool holders move the tool to the work piece and keep it in synchronous position for the machining process. After machining they are accelerated or decelerated, in order to achieve an adaptation to process or format parameters. In material handling the cam technology allows the design of optimised cyclic motion sequences. Thus a product can be taken up from the running process and reintegrated in the process at synchronous speed in another position. Processes using this are:

- stacking,
- repositioning,
- rotating,
- reversing,
- sorting,
- locking into and out of a production process.

Often the cam drive is combined with mechanical components like roller tappets or linkage systems.

For the design of production machines running in fast cycles, mechanical cams have been a used for a long time. They provide for a high-speed and precise implementation of the processes described above. Their low flexibility is a disadvantage. If a format change is effected, either the cam has to be replaced, or its operation has to be adapted using other mechanical mechanisms. These adjustment procedures require manual operations and demand skilled machine operators.

Electronic cam drives represent the operating principle of a mechanical cam on a servo drive. As the curve profile can now be selected electronically and therefore at the push of a button, the flexibility of a machine for the adaptation to formats and processes increases considerably. It is also possible to reduce error sources due to adjustments that have been carried out inadequately.

Drives for electronic cams are often used in packaging machines, weaving machines, printing presses and book binding machines, automatic assembly machines, and in machines for cleaning, filling, and handling bottles for beverages.

4.9.2   Operating principle of cam mechanisms

In the following, the principle of cam mechanisms will be described, as this is the precondition for the operation of electronic cam drives.

Mechanical cam mechanisms. In contrast to gearboxes with a uniform transmission, which mainly operate as speed or torque variators, cam drives convert a linear motion (normally provided by the angle of rotation of the main shaft of a machine which is driven at constant speed) into a non-linear motion.

The most important physical quantity for the transmission of cam drives is the traverse path or lift of an output element as a mathematical function of the angle of rotation of the drive end. This relation describes the motion task of a cam drive.

Thereby the motion is isochronous to the drive end and thus synchronised – independent of the machine speed set. The upper dead centre is always reached in the same position of the machine cycle.

The curve characteristic is composed of sections with defined motion rules and values for the speed and acceleration at the transition of one section to the other. These transition points can be described as follows (table 4-18).

Table 4-18. Motion sequences of cams (VDI 2143)

Motion	Speed	Acceleration	Abbreviation
Standstill (latch-phase)	$v = 0$	$a = 0$	R
Constant speed	$v \neq 0$	$a = 0$	G
Reversal	$v = 0$	$a \neq 0$	U
Motion	$v \neq 0$	$a \neq 0$	B

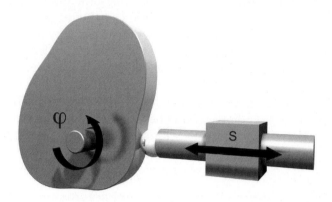

Fig. 4-96. Cam and tappet

For the motion sequence between these transition points there are motion rules resulting in the curve characteristic y = f(x). The motion rules are described further below.

The example in fig. 4-96 shows the function of a simple cam mechanism. The motion of the tappet and therefore the contour of the cam are composed of one reversal point (transition I to II) and two latch phases III and V (Fig. 4-97). The tappet reaches its upper dead centre at the reversal point and remains in one position during the latch phases.

The speeds and accelerations of this motion can be determined by differential calculus (Fig. 4-98).

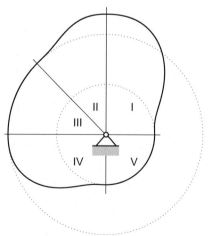

Fig. 4-97. Position-angle-diagram

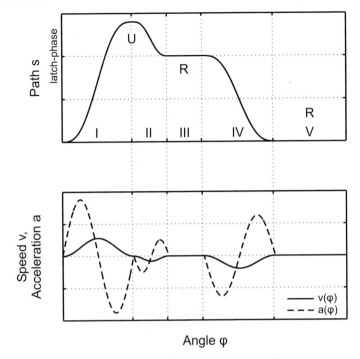

Fig. 4-98. Position, speed and acceleration diagrams

The analysis of the speed and acceleration characteristics shows the forces required and the way in which the machine is excited by the motion [No02]. Due to the linear acceleration characteristic during continuous operation, the present function will only generate a low excitation of oscillations in the machine.

Optimised motion profiles. The development of optimised motion profiles for production machines is a discipline of gearbox engineering. The experience achieved by ongoing development of the motion rules has been compiled in the VDI 2143 guideline which provides a comparison of these motion rules on the basis of characteristic values. The objective is to increase the performance of the machines by reducing vibrations of the moving parts and thus the noise and wear within the system, so that faster machine cycles can be executed [No98].

Due to the non-uniform motion of the machine elements and tools, often inevitably high accelerations with corresponding load alternations occur. Nevertheless, ever-increasing production cycles are expected. The design and dimensioning of the components has to meet these requirements.

However, there is a contradiction here. The demand for a higher production speed and thus a high production cycle requires an optimum design of the machine parts and the connecting elements for the transmission of the motions. At the same time, however, the design has to allow for a quick and easy retooling, in order to be able to flexibly react to the demand for ever smaller production lots with an increasing product variety.

A result from these requirements are low-jerk, or better still, jerk-free motion sequences which spare the mechanical structure of the machine and the product to be machined and thus provide for a high number of cycles. Optimised solutions for the design provide additional advantages here. Sliding block guides and forcibly guided tappets on cams allow for a high number of cycles.

These solutions, however, often make the planning and design of series machines difficult, as it is only possible to adjust the design to a limited number of different products.

Further restrictions are the result of ancillary conditions arising from the design principle of gearboxes with a non-uniform transmission. For example by the use of linkage systems it is difficult to achieve a linear movement that is required, although in turn they are very suitable for protruding movements in the plane and in the space, and linkages generate less friction than a guide.

For electronic cam drives, a lot of the restrictions resulting from the mechanics [LoOsSp06] no longer apply. By the software of the servo inverter it is much easier to optimise and change curve characteristics. It is possible to determine and use an optimised cam for each product without having to implement mechanical retooling or adjustment measures [Jo00].

Transition to electronic cam drives. The transition of a mechanic to an electronic cam solution is explained using an example.

In the case of the mechanic design of the stamping machine in fig. 4-99, two movements are derived from the rotation of the line shaft:

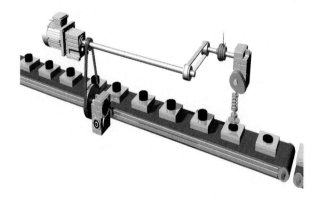

Fig. 4-99. Stamping machine with mechanic cam

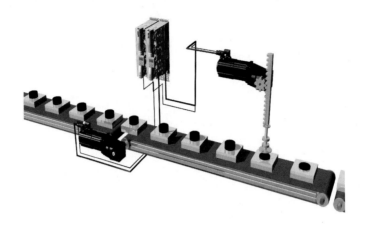

Fig. 4-100. Stamping machine with electronic cam

- The uniform motion of the conveying belt on which the work pieces are placed in regular intervals. One rotation of the line shaft moves the work pieces on by one cycle, i. e. by their distance to each other.
- The non-uniform linear motion of the stamp, which is generated via a cam. Thereby on the connection to the line shaft there is a clutch by means of which the stamping motion can be engaged and disengaged.

In the case of the electronic solution there no longer is a mechanical line shaft. The two movements of the conveying belt and the stamp are generated via two separate drives and their motion sequence has to be synchronised in order to control the production process (Fig. 4-100). The function of the mechanical clutch is taken over by the inverter [No99]. The design of the machine is simpler, the drive solution, however, is more complex.

4.9.3 Drive systems for electronic cams

The major requirements with regard to a drive system for electronic cams concern the generation of the motion sequence. Accordingly, we will be concentrating on this in the following [PaJo02].

Use of servo drives. Servo drives are used for electronic cams, as they alone meet the specifications with regard to the motion sequence and the precision required. The selection of a gearbox and motor depends on the dynamic performance and precision required and follows the notes given in chapter 3.7.1. For the dimensioning, the motion sequences are to be evaluated with regard to their requirement concerning the torque and speed, and components suitable for this purpose are to be selected.

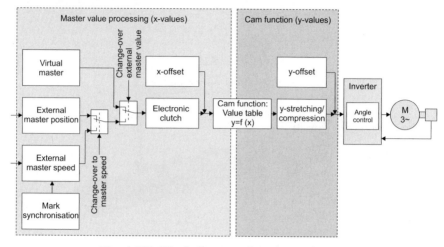

Fig. 4-101. Block diagram of an electronic cam

Block diagram of an electronic cam. Fig. 4-101 shows the general signal chart for the electronic cam function. All blocks have the task to calculate the angle setpoint for the control of the servo drive. From the motion sequence also further quantities for the feedforward control of the drive control (speed and torque characteristics) can be determined, by means of which the load on the drive control is reduced and contouring errors are minimised.

The functions of the electronic cam include the following groups:

- the generation of the master value which the cam function refers to,
- the implementation of the mathematical curve function,
- the modification of the output value by phase or axis trimming or factors for stretching or compressing the curve characteristic.

Calculation of the x position. The x position corresponds to the angle of rotation of the mechanical line shaft. The machine cycle represented by a 360° revolution of the mechanical line shaft is now mapped on values in a defined range 0 to x_{max}. At a constant machine speed, this variable then follows a saw-tooth characteristic (Fig. 4-102).

Virtual master. In machines for which the production cycle is not defined externally, this master position signal has to be generated within the machine control itself. This can either happen in the higher-level control or in the inverter. The software carrying out this function apart from the continuous generation of the master cycle can also take over further functions:

- altering the master speed v_x via adjustable acceleration and deceleration ramps,
- stopping in a defined stop position, e. g. the end of the cycle at x_{max},

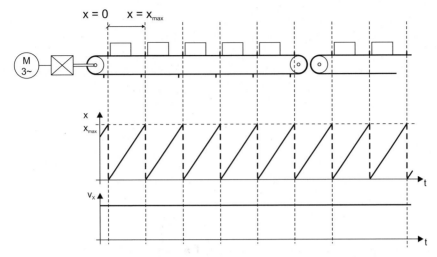

Fig. 4-102. Master cycle

- single cycle operation with passing through the value range 0 to x_{max} once,
- forward and backward inching of the master movement for the commissioning of the machine,
- simulation of the handwheel function for the line shaft,
- continuous change-over between the virtual and an external master position.

The position values at the output of the virtual master (virtual line shaft) serve all drives which are involved in the production process and which are synchronised as master value source. For this purpose separate digital frequency signals or the cyclic transmission via a communication system can be used.

Fig. 4-103 shows the generation of the master position for passing through six cycles. The acceleration with the acceleration ramp set is effected in the first cycle. In cycle 6, the speed is reduced to zero via the deceleration ramp. Exactly at the end of the cycle the standstill is reached.

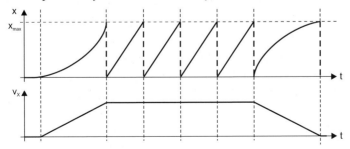

Fig. 4-103. Master cycle with acceleration ramps

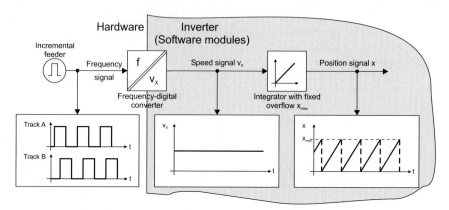

Fig. 4-104. Generation of the master position from encoder signals

Selection of the x position by means of an external position sensor. If the speed of the machine is detected by an encoder which for example moves along on the material web, its signals only contain incremental, but no absolute position values. They have to be generated by means of corresponding software functions (Fig. 4-104). For this, first the speed is determined from the A/B signals of the encoder, which then generates the x position via an integrator. By this the saw-tooth characteristic explained above is generated again.

The position reference of the integrator to the machine cycle, which generates the master position, is established by means of the following procedure:

- setting the integrator unit to a defined x starting position before starting the machine,
- cyclic alignment to a sensor signal, e.g. a mark sensor, during machine operation.

The evaluation of digital frequency signals from an upstream drive, which in their form resemble the encoder signals, is also effected as described here.

If an absolute encoder is used, its absolute angle information can be directly used as master positions.

Mark synchronisation. If the position of the work piece shifts within the machine and if an exact positioning is required, then the exact position of the work piece to the machine cycle has to be detected. For this purpose, sensors are used, detecting marks on the work piece. The detection can thereby be effected via touch probe inputs within the drive, which detect and store the angle position of the sensor signal with high precision.

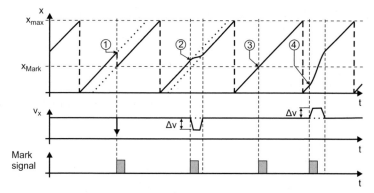

Fig. 4-105. Synchronisation to a mark

If a difference between the detected position of a mark and the setpoint is detected, it has to be brought to zero. For this, the speed controlling the master position integrator accordingly has to be increased or decreased. Fig. 4-105 shows the corresponding process. At point (1) a forced synchronisation to the mark position is effected, which normally is only carried out when the machine is started. At point (2) a correction in the reverse direction has to be effected, at (4) in forward direction. For this, the speed is correspondingly altered for a short time, so that the correction is carried out continuously. At (3) no correction is required.

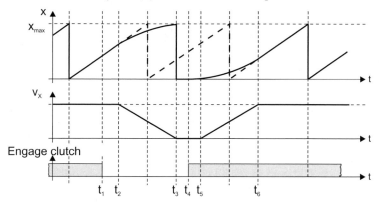

Fig. 4-106. Sequences for decoupling and coupling

Electronic clutch. Like a mechanical clutch, the electronic clutch allows for the decoupling of individual cam axes.

In this way machine units that are not required can be shut down or waiting cycles can be skipped.

The electronic clutch is a software function in the inverter. It does not cause wear and simplifies the machine structure. Via digital control signals the decoupling process from the master value (e. g. with standstill at the end of the cycle) and the coupling process to the current master value can be effected precisely.

The sequences in fig. 4-106 show a decoupling and a coupling process. The decoupling is triggered at point t_1. On the basis of the selected deceleration ramp and the x stop position, the braking process is started at point t_2, so that the position of the output end of the electronic clutch at standstill reaches the stop position at point t_3.

The coupling process is activated at point t_4. The electronic clutch calculates the starting time t_5 from the acceleration ramp and the continuously running x input position. At point t_6 it synchronises to the x input position.

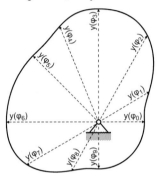

Fig. 4-107. Cam and resulting curve function

Curve function y = f(x). The calculation of the curve $y = f(x)$ is the core function of the electronic cam. This software function uses the x input position in the range 0 to x_{max} for calculating the output position (Fig. 4-107). The function $y = f(x)$ then normally is not an analytical formula, but is provided as a grid point table with two columns (x/y column).

From the tooth-saw characteristic of the x position, the cyclic characteristic of the y-output results from the access to the grid point table (Fig. 4-108).

Between the grid points of the table an interpolation of the curve according to different procedures is effected (e. g. linear interpolation, Spline interpolation). For this, it is reasonable to map curve sections with a strong curve above closely positioned grid points, whereas sections with a uniform motion or standstill can be represented by few grid points. Like this a curve section for a latch phase is clearly defined by two grid points. The interpolation procedures have to work in a way that the acceleration process is constant, so that there is no jerk, which in turn would induce vibrations within the mechanics. For this purpose there are Spline functions consisting of interconnected third-order polynomials which ensure the consistency of the speed and the acceleration at the grid points.

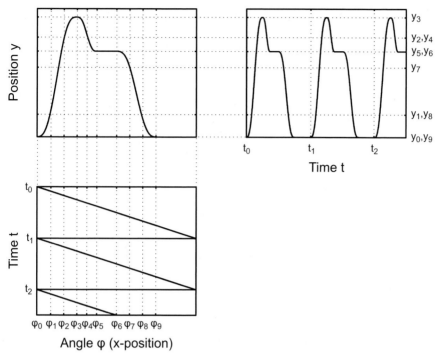

Fig. 4-108. Cyclic generation of the y-position

Angular trimming (x-offset). An angular trimming (phase shift) of machine units to the master position can be effected via an x-offset, and therefore the drive movement can be shifted in the master cycle. An x-offset can be visualised by the horizontal shifting of the curve (Fig. 4-109).

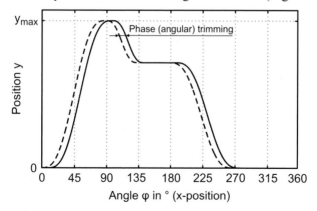

Fig. 4-109. Effect of the x-offset

During operation the x-offset may not be changed abruptly, as then an abrupt change of the y-position may result. An upstream profile generator provides for a smooth adjustment to the desired offset value.

The x-offset is used to adjust the exact machining position on the work piece.

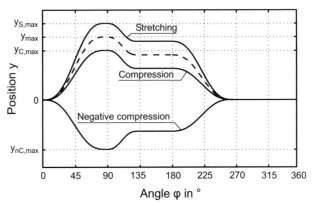

Fig. 4-110. Effect by y-stretching and compression

y-stretching or compression. The curve function can be stretched or compressed by the selection of a factor (Fig. 4-110). Thus, different motion sequences can be implemented by using one grid point table. Examples of stretching or compressing the curve are:

- a stretching/compression factor $K_1 < 100\%$ results in a compression of the curve (in the extreme case $K_1 = 0\%$ results in a standstill of the drive axis),
- a stretching/compression factor $K_2 > 100\%$ results in a stretching of the curve,
- the direction of movement can also be reversed by a stretching/compression factor $K_3 < 0\%$.

The stretching and compression can be used to adjust the tappet lift for a tool.

Shifting the axis position (y-offset). Apart from the angular trimming (x-trimming) there is a trimming which is considered as a y-offset in the signal chart (Fig. 4-111). The y-offset allows for shifting the operating range of the output axis.

Like the x-offset, the y-offset also may only be changed constantly during operation, as otherwise an abrupt change of the y-position will result.

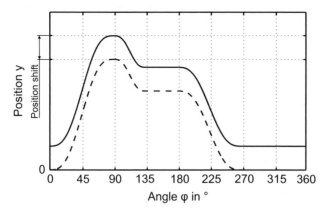

Fig. 4-111. Effect of the y-offset

Selection and creation of motion profiles. The selection of the motion profile is crucial for the quality and performance of the cam drive motion. Basically it has to meet the following criteria:

- The required function of the motion has to be fulfilled.
- The energy input, i. e. the accelerations required for the motion sequence, has to be as low as possible.
- The movement is not to induce vibrations into the mechanics. Thus the frequency spectrum of the motion sequence may not have any portions in the high frequency range.

There are several procedures which are partly documented in guidelines, by means of which such motion sequences can be created.

Standardised motion rules in accordance with the VDI 2143 guideline. By the use of the VDI guideline, motion profiles can be created very easily and, in particular, in a practical manner. The following motion rules which are based on different mathematical equations are included:

- power law, in particular high-order polynomials,
- trigonometric laws,
- combinations of polynomials and trigonometric laws.

On the basis of the good manageability and calculation of simple polynomials, motion profiles can be created relatively quickly. They often serve as an approximation for the creation of a cam mechanism or an electronic cam drive.

With high-order polynomials, further reaching requirements can be met, as by the higher number of degrees of freedom more complex conditions can be considered. The use of power laws in particular is suitable for intermittent motion sequences with temporary standstill (latch phase) of the drive.

The forces to be generated by the drive are lower compared to trigonometric laws. The fifth-order polynomial is to be emphasised here. It is easy to calculate, generates a jerk-free movement and in combination with linear sections it can be used in many applications with transitions in latch phases, or phases with constant speed.

Trigonometric laws can also be easily used. In contrast to the polynomials they are more suitable for designs which are susceptible to vibrations, but the drive forces required are a bit higher. Another advantage is the possibility of shifting the inflection point. Thus the characteristic of the movement becomes asymmetrical, and the maximum values for the speed and acceleration can be adjusted to specific requirements within the machine cycle.

Combinations of polynomials and trigonometric laws provide for an appropriate segmentation of the machine cycle into individual motion sections. According to the requirements and with regard to limiting conditions for the movement, which for example may include specifications for the maximum speed and acceleration, the standardised motion rules can then be combined in a reasonable manner.

Harmonic Synthesis (HS) profiles. In HS profiles the motion is described by means of a Fourier series. The term HS profile therefore is based on the procedure for the harmonic synthesis (HS) of the movement searched. In practice, the application cannot be as easily implemented as the use of standardised motion rules. Normally calculation programs are used for the creation and optimisation.

In contrast to the combined standardised motion rules with polynomials and trigonometric functions, the motion sequence for the HS profile is generated across the entire machine cycle and not only across individual sections.

The advantage of HS profiles with regard to the polynomials in particular is that they can always be constantly differentiated again. Relating to the higher derivatives, therefore there aren't any discontinuities at the transition points of the motion sections, which for instance will occur in the case of a profile of combined standardised motion rules.

If the resonant frequencies within a machine are known there is furthermore the possibility of suppressing harmonics by optimisation runs of the dimensioning software or of generally limiting the number of harmonics.

The optimisation of the motion profile is essentially based on the feasibility, not to pass *exactly* through angle or path positions or standstill phases (latch phases), but with a tolerance to be specified. If required, the tolerances are to be adapted to other drive elements operated within the system. Within the tolerances there is room for optimising the speed and acceleration. As a result, in particular the acceleration curve of the motions can be optimised.

Especially in flexible mechanical designs this has a beneficial effect on the tendency towards oscillation of the machine parts and thus provides for a higher number of machine cycles.

Online profile modification. In the case of an electronic cam drive the motion sequence can be altered. For adaptations during operation preferably the functions stretching, compression, and phase shift (x offset) relating to the calculated and optimised motion rule can be used. By means of these functions the parameters of the motion profile of the output element with regard to the master value can be altered during operation without changing the motion rule.

If extensive optimisations of the motion sequence are required for a format change, the machine control should contain the software required for this, and the corresponding optimised motion sequence should then be loaded into the cam drive. A change-over between two curves during operation requires that the curve characteristics match at the point of transition and that they have the same position, speed and acceleration.

4.10   Drives for forming processes

Karl-Heinz Weber

Many work pieces are produced by forming processes. Examples are the pressing of metal sheets and the extrusion of plastics. Drives which apply the energy required to form materials into different shapes are described in this chapter. In contrast, the following chapter 4.11 describes drives which carry out cutting processes.

4.10.1 Forming processes

Forming drives serve to produce, form, deform, compress, and press materials. This includes:

- the extrusion of plastics, rubber, caoutchouc products,
- extrusion of food and feeding stuff mixtures,
- extrusion of ceramic masses and clay,
- extrusion of aluminium,
- die casting or blowing of hollow pieces,
- edge bending and bending of materials,

Table 4-19. Forming processes with typical parameters

Application	Industry sector	Cycles (rpm)	Required power (kW)
Extrusion	Plastics processing	Continuously	1–600
Blowing	Plastics processing	10–20	2–15
Deep drawing of plastic parts	Plastics processing	< 60	1–75
Deep drawing of sheet metal parts	Automotive industry	< 20	5–75
Pressing	Automotive industry	< 20	10–650
Tablet presses	Pharmaceutics, food, luxury articles	Continuously	4–22
Double belt presses	Plastics processing	Continuously	5–20
Vibration	Cast stone industry	10–20	5–30

- pressing of powdery, solid, and paste-like materials,
- deep drawing of plastics and metal parts,
- compression by vibration and centrifuging.

As this is a very broad technological field of application containing both continuous and cyclic processes, and being used for very different materials, not all of the technological processes can be described here. The description focuses on processes which are carried out with controlled electrical drives. Items overlapping with hydraulic drives and other chapters of this book where for instance synchronous drives or positioning drives are described cannot be avoided. This chapter is to sketch the applications as a basis for the corresponding requirements with regard to drive technology.

Extrusion process of plastics, caoutchouc and rubber. Extruders consist of a worm shaft which is pivot-mounted in a massive, heated steel cylinder, and which is driven by an electric motor via special gearboxes or also directly (torque motor). At a high temperature (approx. 60 °C to 300 °C) and a very high pressure (approx. 1 to 30/70 MPa) they generate continuous profiles or mass flows from granulates or preplasticised materials. (Fig. 4-112 and 4-113)

The granulate is fed to the extruder screw via pipe systems and feed hoppers. Within the extruder the material is melted and homogenised, which then is pressed out of the forming opening (nozzle, tool) (similar to the principle of a mincing machine). A temperature-controlled heater encloses the extruder cylinder and provides the process heat required.

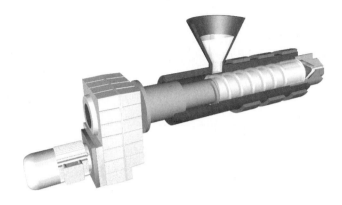

Fig. 4-112. Schematic diagram of an extruder

The extruder output is specified in mass per time (kg / h). Often different granulates are mixed and extruded. The extruder can consist of one or two screws. Extruders are also designed as multi-stage machines.

The material leaves the extruder nozzle (extruder tool) in a plastic state, thereby formed by this nozzle and then it is cooled down in the air or in a water bath before further processing.

Fig. 4-113. Two-stage extruder with the 2. stage opened

Extrudable materials. The following materials can be processed by extruders:

- plastic and mixed products, e. g. thermoplastic materials and wood fibres, granulates, rubber, and caoutchouc,
- silicone materials for industrial and medical purposes,
- aluminium,
- clays and ceramic masses like raw materials for the production of bricks,
- foodstuffs, feeding stuff mixtures,
- pharmaceutical products.

Extrusion products. The following products are produced by the use of extruders:

- foils, discs, tubes, pipes, bars,
- profiles for windows and doors,
- seals, cable ducts, fibres,
- tire treads,
- bicycle rims,
- V-belts and toothed belt with tear-proof inserts,
- yarns, tapes, transport and packaging tapes,
- shields of wires, cables, ropes and pipes,
- coatings of rolls,
- foamed continuous materials,
- confectionery (e. g. chewing gum, pasta, caramel),
- pet food.

Extruders are also used for the production of clay materials (e. g. bricks) to press the clay pulp into threads in a highly-compressed manner, from which then the bricks are cut to length.

Injection moulding. Injection moulding is a process where granulate is plasticised by means of a screw before – at high pressure – being injected via feeding channels into the cavities of a mould which can be opened (injection moulding tool). After it has cooled down, the mould is opened and the piece is removed. The forming of the material during the injection moulding process is similar to that during extrusion. Whereas extrusion is normally a continuous process, injection moulding is a cyclic process.

Injection moulding is a procedure to process or produce a very broad spectrum of materials and products in high quantities and in a cost-effective manner. The material spectrum comprises different plastics, metal powder, silicone materials, and ceramic materials. The product spectrum ranges from small parts of daily use of below 1 g to big parts of up to 100 kg for the following applications (Fig. 4-114):

Fig. 4-114. Injection moulded part

- consumer goods,
- automotive parts,
- household supply and toys,
- semi-finished industrial parts,
- hollow parts as containers, bottles, canisters, buckets,
- chairs,
- housing parts.

Thermoforming, deep drawing of plastics. Thermoforming is a process for forming thermoplastic materials (films, sheets) into moulded parts. The film is intermittently pulled from a roll, is preheated and in a soft state is pressed into the finished contour by a thermoforming tool. The final moulding is then achieved using high or low pressure. On the cooled contour of the forming tool the film quickly hardens in the desired shape, is removed and is punched out during the next operation cycle. Plastic sheets are used to produce moulded parts for packages, refrigerator walls, bath tubs, dashboards, lamps, and lots more.

Blow moulding. By means of the blow moulding technique, hollow parts from a hot extrusion tube are blown into a mould that can be opened. The work pieces produced like this range from ampoules, tubes, beverage bottles and cosmetics cases to canisters, barrels and fuel tanks.

Glass forming. Hollow glass (bottles for beverages and other materials) is produced in glass machines. For this the semifluid glass flows continuously away from the molten mass which has a temperature of approx. 1,400 °C and is separated into individual portions, one for each bottle, using a feeder cutting system. This semifluid lump falls into a closed blank mould, which consists of two half shells, via a series of channels and tubes. Now an oxygen lance is pushed into the liquid glass forming the hollow space inside the bottle and the bottle top. The semi-fluid bottle is transferred to a final mould (two half shells) where it is blown into its final shape using compressed air. The bottle, still red hot, is released from the mould and conveyed into a leer where it is cooled from approx. 900 °C down to room temperature. Finally, the bottles are checked for quality, packed onto pallets and transferred to temporary storage. The production process is highly automated and is controlled by servo drives (opening and closing of the mould shells, pivoting of the blank mould to the final mould, retraction of the blowing device, etc.). For this purpose pneumatic drives were used in the past.

Pressing of powdery substances. Powdery material is pressed into tablets in metal mouldings under high pressure. The tablets are used for a variety of purposes: foodstuffs and luxury food, pharmaceuticals, cleaning agents, and also sintered materials like sintered blank magnets.

Compression of precast concrete parts. Material compressions can be carried out by vibration, pressing, or centrifuging. During the production of precast concrete parts, vibratory tables are used for compressing cobblestones and curbstones. For round parts, like for example high voltage towers, telecommunication towers, and floodlight poles, round centrifugal moulds are used. In the centrifugal moulds the compression is achieved by centrifugal forces which are generated at the outer walls of the rotating mould.

Clinching. Clinching is a procedure for connecting metal sheets without the use of an additional material (Fig. 4-115). It is an alternative for welding. The connection is achieved by forming the material.

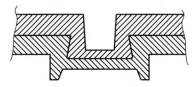

Fig. 4-115. Connection by clinching

An advantage of clinching is that no pre-piercing process is required. A clinching tool is composed of a mould and a die. The metal sheets to be connected are pressed into the mould by the die while being deformed plastically, similar to the process of deep drawing. By a specific design of the mould a shape similar to that of a button is produced, connecting the metal sheets to each other by form-fit and friction.

Clinching is used in automotive production to connect body components to each other. Further fields of application are the white goods industry and the field of mechanical engineering in general. The procedure partly replaces spot welding.

Pressing, edge bending and bending of metals. Other types of forming include edge bending and the pressing or bending of steel plates or profiles. Hereby the work piece, e. g. a metal sheet in a folding press is pressed into the target format by means of a moulding tool. Bending machines clamp one edge of a metal sheet then, using a pivot bar underneath the free part of the sheet, and bend the sheet to the desired angle.

Pressing with a continuous cycle. In order to press several layers of different materials together under high pressure in a continuous process (composite materials, laminate, coatings of chipboards and fibre boards, plates for the furniture industry, copper laminates for flexible printed circuit boards, floor covering, conveyor belts) and connect them under thermal impact (heating, cooling), double belt presses are used. Depending on the material to be connected, presses with a constant pressure or a constant material thickness are used.

4.10.2 Machine types for forming processes

Flat film extruders. If plates or flat films are to be produced, the extruder tool which forms the plates or films is designed as a flat die which can be adjusted in a slot. For the production of multilayer laminated films or plates, several mass flows are combined within the nozzle. In the following machines, calibration, stretching, cooling, edge trimming, rewinding, or separating plates into defined lengths is achieved.

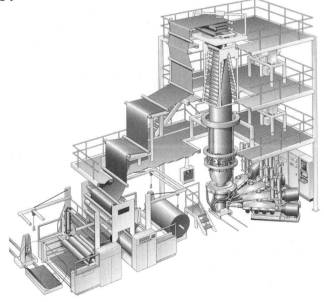

Fig. 4-116. Blown film production

Blown film extruder machines. Blown film extruder machines produce film tubes (Fig. 4-116). For this a die is placed at the extruder outlet, which when it is supplied with air develops a vertical film tube, that is then calibrated, cooled and rewound. By setting up several extruders and combining the individual melting flows, laminated films can be produced like this. The tube can also be sliced lengthwise to be then wound up as a film.

Extruders with downstream spinning pumps. For these units, a spinning pump is installed downstream to the extruder, producing fibres or threads, which are cooled down in water baths and then under the impact of temperature are extended and secured via godet drives. At the end of such a production line the threads are cut to defined lengths or wound up with winders for further processing.

Extruders for pipe or profile production. These extruder lines have tools such as outlet nozzles and produce pipes, tubes, or profiles of plastics (Fig. 4-117). After leaving the extruder tool, the continuous profile that is still in a plastic state is calibrated in order to maintain the form; then it is cooled with water and transported further by caterpillar or belt haul-offs for separation to defined lengths by a flying saw. Pipes with a diameter of up to two metres are produced.

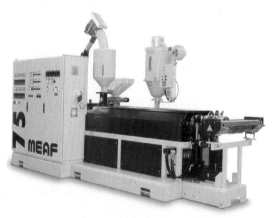

Fig. 4-117. Extruder production line

Calenders, chill roll casting lines. Calenders (Fig. 4-118) and chill roll casting lines are a parallel roll arrangement which absorbs the plastic melt from extruders or downstream kneaders and form films, deep drawing films, or plates from it.

By the chill roll procedure cast film is produced by casting a melt with up to nine layers on to a chill roll and cooling it down abruptly, from which the outstanding optical and mechanical film characteristics result.

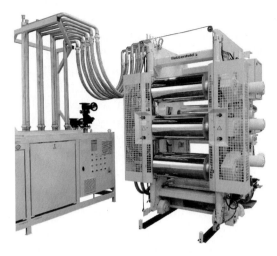

Fig. 4-118. Calender roll rack

The distances of the rolls to each other (roller clearance) are adjusted by motors or hydraulically. A continuous thickness measuring acts on the production process and provides for a constant material thickness. After the forming process the plastic material runs over heat and cooling rolls, is shrunk or extended and is wound up in a solidified state as film or, in the case of plates, cut to defined lengths for further processing. Depending on the use of the films or plates, the drives have to run very smoothly and have to feature a high speed stability, so that no chatter marks occur on the product surface. For this purpose high resolution encoders and high quality servo controls with low torque fluctuations are used.

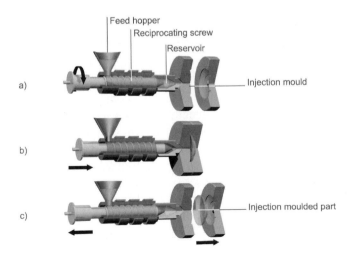

Fig. 4-119. Process steps of an injection moulding machine

Injection moulding machines. In injection moulding machines the plasticising screws, the injection, and the handling of moulded parts and end products are driven. The screw takes up the granulate from a feed hopper, plasticises the material and transports it into a cavity (Fig. 4-119 a). Via an axial piston movement of the screw (Fig. 4-119 b) the material is injected into the negative mould. A non-return valve prevents the material from flowing back into the screw. After solidifying (Fig. 4-119 c) the mould is opened for the removal of the moulded parts. As well as hydraulically operating systems, increasingly more common are energy-efficient electrical drive systems, sometimes with water cooled direct drive technology (torque motor, inverter).

Blow moulding machines. With the blow moulding technology hollow parts are blown from a hot extrusion tube, by applying compressed air, into a mould that can be opened. As well as the drives for the main and secondary extruders, servo drives are also used in the forming process and in the handling equipment for unloading the machine.

Tablet presses. The pressing of tablets is a continuous process. The powdery material or granulate is fed to the continuously rotating press via pipe or metering systems, is pressed by dies at high pressure and is then ejected from the pressing moulds again.

The machine movements for the pressing process are achieved through mechanically coupled cams. The main drive is an inverter-driven, speed-controlled three-phase drive with an angled hollow-shaft gearbox and a belt transmission. Geared motors are increasingly replaced by gearless, synchronous direct drives.

Concrete vibrators. For producing and compressing concrete parts, among other things vibratory tables are used. The vibrations required for this are generated by four servo drives driving adjustable eccentric masses according to a specific motion pattern, so that vertical and horizontal accelerations with a variable amplitude and frequency are generated.

Compactors for spun concrete. These units compress freshly-mixed concrete by the centrifugal forces generated by the quick rotation of a steel mould into which the concrete flows. Thereby very high rotatory energies occur (just like in the case of a centrifuge). The rotary motion is transferred to the cylindrical steel mould by friction wheels. During very long acceleration and deceleration phases (sometimes several minutes) the drive has to bring the mould into rotation. For reasons of energy efficiency, the braking energy should be fed back to the mains.

Fig. 4-120. Clinching tool for robots for vehicle body production

Machines for clinching. Clinching tools (Fig. 4-120) in principle are pliers where a moving die enters the metal sheets to be connected, which are placed above one another, into an oppositely placed mould. Drives for clinching can be hydraulic, pneumatic or electromechanical (Fig. 4-121). For electromechanical drive systems, highly-dynamic synchronous servo drives with planetary gearboxes, toothed belt transmission and a leadscrew are suitable.

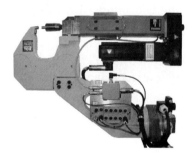

Fig. 4-121. Machine with servo drive for clinching

Eccentric presses. Eccentric presses are used for cutting-punching operations and embossing-bending operations with small slide strokes. The pressing procedure is achieved using an electric motor driving a flywheel via a belt drive, from which the energy is transferred to the eccentric shaft via a combined clutch-brake system. A push rod on an eccentric bush converts the rotary motion of the flywheel into the linear motion of the tappet. The motor is operated directly from the mains or via an inverter with variable speed.

Screw presses. As an alternative to hydraulic presses, presses are also fitted with synchronised spindle drives which bring the two pressing tools together like a bench vice pressing out moulded parts or stamping out parts. High pressing forces of up to several hundred tons and high numbers of cycles of up to 100 per minute can be achieved. The material in the form of steel plates or plates of non-ferrous metal is fed discontinuously.

Sheet metal folders. The bending of metal sheets is also referred to as folding; this procedure is carried out by means of sheet metal folders. One part of the metal sheet is held in a clamp while the free part is folded to a specific angle over a straight edge using a bending flange. Typical bending products are control cabinets, housings and metal doors.

The drive for the upper flange can be a hydraulically or electrically driven spindle drive. The drive for the bending flange is usually a variable three-phase drive which acts on the flange via a geared motor.

Folding presses. In folding presses metal sheets are pressed into a target mould by means of a moulding tool. A loading robot or manipulating device grips the metal sheet to be processed and places it in the open folding press. The press closes and moulds the metal sheet. For large sheets the manipulating device during the pressing procedure supports the parts of the sheet protruding from the press. After the pressing procedure the press opens and the manipulating device removes the moulded metal sheet from the press. The press drive can be a hydraulic drive, a spindle drive as gearless direct drive, or also a cable winch system with a high reeving factor.

Double belt presses for continuous pressing processes. These presses consist of an upper and a lower belt which are pressed together with a constant and evenly distributed pressure across the entire contact surface. Between these belts the unrolled layers to be connected are pressed under thermal impact. The end product is then wound up or separated.

4.10.3 Drive systems for forming processes

Continuously operating forming drives. These drives predominantly run at a constant speed which is specified by the process. The drive is operated in a speed-controlled manner, with a speed sensor, or, if the requirements to the speed accuracy are low, without a sensor. Within extruders the load does not change very quickly. For low speeds and for start-up, a high torque is required. However, for presses fast alternations of the load have to be compensated quickly.

Discontinuously operating forming drives. For these processes highly-dynamic drives with high overload capability are required. They have to provide for a medium to a very high dynamic performance and have to control high traverse speeds for opening or closing moulds or tools.

Specific notes for the drive dimensioning. For extruder drives the determination of the torques and speeds required results from the design of the extruder and the output per time. This is the basis for the decision whether the screw speed can be implemented via a geared motor with a special gearbox or via a direct drive.

In the case of drives for rolls the drive torque required is determined from empirical data, tensile forces, pinch pressures, material combinations, and the energy needed for the deforming process. In the case of discontinuously cyclic forming drives the drive dimensioning is based on the kinematic motion profile of the masses to be moved and the energy needed for the deforming process.

Gearboxes. For extruders special gearboxes are used as helical or helical-bevel gearboxes which can absorb the immense axial forces of the extruder screw and can deliver high output torques (e. g. 200 kNm). Furthermore a low-noise operation and a long service life are required.

For the other forming drives axial and right-angle gearboxes, or planetary gearboxes are used; sometimes using low backlash versions.

Compact geared motors. Also compact geared motors in hollow shaft versions with liquid cooling are used, where in one housing for example four motors are fitted in a planetary arrangement which act on a joint driven gear. Also single-stage and two-stage designs are used.

Motors. For extruders asynchronous motors with special gearboxes and increasingly also direct drives in solid shaft and hollow shaft versions are used. Water-cooled direct drives are especially compact and have low noise emissions, as there is no fan. Direct drives have a low power consumption and run smoothly. For the other continuously operated forming drives standard three-phase AC motors and asynchronous servo motors are usually used. In discontinuously operating forming drives, however, synchronous servo motors are usually used.

Electrical energy flow. In the case of continuously operating forming drives like extruders or calendars only motor operation occurs. Discontinuously working forming drives, however, operate in all four operation quadrants, so that regenerated energy has to be converted to heat by means of a brake resistor or has to be fed back to the mains via a regenerative supply unit. For multi-axis applications connecting the DC buses of the drives together allows for an energy exchange between the axes.

Drive control. For continuously operating forming drives vector control without and with speed feedback is used for drive control. If the requirements are high, servo control with a position sensor is used. For high quality extrusion processes high speed accuracies, long-term stability and very good characteristics with regard to smooth running are especially important. For breaking free the extruder screw in all situations the drive has to be provided with a sufficient overload capability over several minutes. Repeatable and precise torque limitations are important for the protection of the extruder against overload.

Discontinuously operating forming drives are predominantly operated with a servo control.

Motion control of the drives. Extruders are operated in a speed-controlled or pressure-controlled manner. In the case of pressure control, the actual value is detected via a pressure sensor in the extruder head. The pressure control utilises an external process controller, or one that is integrated within the inverter, controlling the speed. A lower speed limitation prevents impermissible operating states.

Extruder downstream units like calenders, chill-roll casting lines are operated as electrical shafts in parallel and cascaded structure. For double-belt presses electrical shafts, torque ratio controls in master-slave structures and symmetrical torque distribution with anti-slip control are used. Further notes for this can be found in the description of the synchronised drives in chapter 4.6.

Cyclic forming drives operate in a position-controlled manner and are often equipped with cam functions. This was described in chapter 4.9.

Diagnostic functions. Extruder drives are operated in continuous operation, 24 hours per day and nearly 365 days per year. The reliability and availability has to be very high. An interruption, even if it is very short, or a standstill due to the failure of a drive component is associated with high costs and therefore has to be avoided. Thus, prognostic concepts for diagnostics, service, and remote maintenance are of particular importance for these drives.

The monitoring of important process factors like speed, torque, current, and pressure is a basic required function. Also monitoring functions on the drive states like the motor and inverter temperature, the form and amount of the mains voltage and asymmetries of the motor current are required for a trouble-free operation.

4.11   Main drives and tool drives

Martin Ehlich

A main drive is the central drive of a machine or system. With its speed, the main drive defines the process speed. Thereby it moves the masses to be processed and thus provides the basic power for the process.

A tool drive defines the speed of a tool and therefore provides the machining power. With regard to machining, there are operations where material is separated, or where material is removed.

In the case of machines where the main drive itself carries out a machining process, there is a smooth transition between both term definitions.

4.11.1 Applications for main drives and tool drives

In nearly all machines and systems there are main drives. Therefore there are hardly any restrictions with regard to the materials to be moved or machined. The typical materials are:

- metal,
- wood,
- stone,
- glass,
- paper or cardboard,
- plastics of all specifications.

The tool drive as well cannot be assigned to any industry, but is to be found in all machining processes. Typical examples are driven tools in machine tools, which have been used for over 100 years in all kinds of series parts production. But also in production machines tool drives are required for separation of materials and machining processes. The processes listed in the following are executed by tool drives:

- drilling,
- milling,
- screwing,
- grinding,
- polishing,
- breaking,
- sawing,
- deburring.

Table 4-20. Process requirements of main drives and tool drives

Process parameter	Main drive	Tool drive
Power requirement in kW	10–400	0.5–5.5
Maximum speed in rpm	1,000–8,000	6,000–30,000
Torque requirement in Nm	100–4,000	1.5–20
Load matching, J_M / J_A	1 : 10–1 : 100	1 : 1–1 : 4
Acceleration times in s	1–50	0.02–0.50
Mounting of the motor	Outside of the workspace	Within the workspace

Table 4-20 compares typical ranges for the process parameters of main drives and tool drives.

4.11.2 Machines with main drives and tool drives

Machines with main drives. The term main drive is used in many industries. For instance it is typical for

- printing machines with central drives,
- presses,
- test benches,
- machine tools,
- wood working machines,
- stirring machines (Fig. 4-122) and milling machines,
- centrifuges.

Basically main drives are used as direct drives or as a drive unit consisting of a motor, transmission element and output mechanics. Due to their complex design, direct drives as yet have only been used in printing machines, machine tools, and wood working machines with a high number of pieces.

The typical main drive generates a rotation at the output end. Therefore linear motors are irrelevant here.

There are designs where the speed of the output shaft is not transmitted with positive connection. An example of this is vehicle test benches where the vehicle tyres are coupled to an electrically driven roll via static friction.

Fig. 4-122. Main drive in a stirring machine

Fig. 4-123. Tool drive in a grinding machine

Machines with tool drives. The most well-known and frequently used machine types equipped with tool drives are the following machine tools:

- machining centres,
- milling machines,
- lathes and turning machines,
- drilling machines,
- grinding machines (Fig. 4-123),
- polishing machines,
- sawing machines.

These machines are not necessarily to be found as a self-contained unit in the form of a machine tool, but can be found as machining stations in various production processes. Tool drives, however, are also found in construction site equipment for roadworks, mining and tunnel construction.

In order to establish a close relationship between the tool and the drive motor, the direct drive is a suitable alternative for tool drives. On the output end always a speed is required. The rotational movement does not have to be converted. The desired high process speeds can often only be provided by special motors. For most of the machine functions mentioned hollow shafts are required to place the clamping unit for the tool and ducts for the coolant. Furthermore a tool drive can be produced as a completely preassembled unit and, by the use of few adaptations, can be mounted in completely different machines and systems, so that the implementation of an integrated motor spindle also turns out to be profitable from an economic point of view.

For tool drives, the tool is ideally directly coupled to the drive motor, so that no transmission elements affected by wear and losses are required. A minimum required space and a low weight are important criteria, as the tool drive is directly placed in the workspace and often is mounted on a slide which has to be traversed very quickly to keep the non-productive time of the machine low. When tool changing, traversing long distances to the tool storage magazine is often required.

4.11.3 Drive systems for main drives and tool drives

Because of their tasks, main drives and tool drives often are the most powerful drives of a machine. Thus they have a significant impact on the manufacturing costs, the required space, and the availability of the machine. It therefore is important for every design engineer to deal with the drive solutions for main drives and tool drives in detail.

Dimensioning of main drives. Many processes require a continuous basic power over a longer period of time. Some productions are never shut down after initial commissioning. In this respect the acceleration time only plays a minor role for the main drive. Drives with preferably low losses in the operating point, a sufficient continuous power including a reserve for voltage and load fluctuations, and a high load stiffness are required.

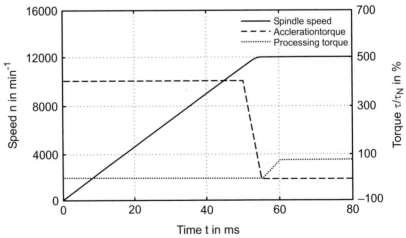

Fig. 4-124. Acceleration and torque requirement of a tool spindle

Dimensioning of tool drives. The external mass to be driven – consisting of the clamping unit, tool holder and tool – for tool drives is rather small compared to other drive solutions. Furthermore the machining power is typically generated by high speed, so a drive with a high acceleration capability is required here. A typical acceleration procedure with the corresponding time and torque requirements is shown in fig. 4-124. The dimensioning in many processes is effected according to the acceleration time. The motor and inverter thereby are operated within an overload range which they can only provide for a short time. The prerequisite is that the machining power thermally can be provided continuously by the drive.

Gearboxes. For main drives standard gearboxes like helical and right-angle gearboxes are used if a reduction of the motor speed is required. This depends on the machining process. The selection of the gearbox depends on the mounting situation, whereby a helical gearbox in most cases is the solution preferred.

In some applications for main drives a direct drive is used instead of the gearbox. The main reason for this often is the energy efficiency.

Due to the high speeds, tool drives normally do not contain gearboxes, but are designed as direct drives.

Motors for main drives. Basically all motors which are appropriate with regard to the power and speed range are suitable. The most important criterion for comparison is the power loss of the motor, as it is permanently used. In most applications a standard three-phase AC motor with high energy efficiency is selected. An expensive synchronous motor is only worthwhile if the installation space for the motor is particularly small.

A high internal inertia is advantageous so that the load-matching factor between the motor moment of inertia and the load moment of inertia (converted to the motor side) remains in the range 1 : 10 to a maximum of 1 : 50. The motor should be operated as far as possible in the field weakening range to optimally use the inverter power.

The cooling of the motor is to be designed as simple as possible: in the ideal case naturally ventilated, otherwise surface-ventilated or with internal ventilation by fans. A water or oil cooling is used in special applications.

Motors for tool drives. The main requirements of tool drives, like the implementation of high speeds, a good dynamic behaviour, and an integration of clamping units and the coolant duct determine the selection of the motor. Basically all motors which are appropriate with regard to the power and speed range are suitable. Motor spindles preferentially are designed as direct drives in asynchronous or synchronous technology [GrGe06].

For high-speed tools there are specific asynchronous motors with rated speeds of 10,000 to 20,000 rpm, which have a slim-line design.

The synchronous motor provides the advantage that there is no heat dissipation within the rotor, and therefore no heat flow to the tool is developed. The motors are operated in the top field weakening range, as the machining power is mainly provided by the speed and less by the torque. The motors should feature a low internal inertia, as the external inertia is rather low. The motor cooling is achieved by the use of water or oil if simpler solutions like air cooling would significantly increase the installation space or the weight.

Brakes. Normally no brakes are used for main drives and tool drives, but the drive is decelerated by the inverter. Only if it is required to lock the drive in standstill, holding brakes can be used.

Inverters. As main drives and tool drives put high demands on the smooth running performance and speed stability, a vector control is required compensating the torque fluctuations faster than a V/f control. Servo control with a speed feedback is required for applications with particularly high demands with regard to speed stability. Normally frequency inverters are used which, if required, can also evaluate a speed sensor.

In some applications for tool drives the high motor speeds, and thus the high output frequencies of the inverter are to be observed. Output frequencies in the range of 500 to 1,000 Hz are often used here. The implementation of these output frequencies in the frequency inverter often requires specific procedures for the calculation of the switching patterns and the actuation of the power switches in the inverter. For this reason, sometimes special designs are necessary for this application.

In some machines with further drives apart from the main drive it can be reasonable to connect the inverters via the DC bus. In this case the main drive also supplies the other drives, an energy exchange between the drives is ensured, and components like the brake transistor and the brake resistor only have to be installed once.

Drive control of main drives. Main drives operate in a speed-setting mode. Besides start and stop signals main drives are not controlled within the higher-level control. The (speed) control loop is closed in the inverter. In the simplest case there is no mechanical or electrical coupling to other drives of the machine or machining units. The task is the provision of the process speed at the right time and irrespectively of load fluctuations resulting from the process.

In highly-automated processes there are an increasing number of interrelations between main drives and other movements. Examples of this are the C-axis drive of main drives (position control with the objective of positioning a work piece, e. g. with lathes) and the angular synchronism for the transfer of parts to a downstream machining station.

Drive control of tool drives. Like main drives, tool drives operate in a speed-setting mode. Normally there are no interrelations with regard to other drives. The control requests the tool speed required at a specific point of time. The tool drive has to ensure that the speed is provided as quickly as possible and that it is kept at a constant rate during the machining process.

In few cases a positioning is required for the tool drive. These are integrated machining centres for metal or wood working carrying out an orientation process of the tools for milling and edge working.

An important additional function for the tool drive is the monitoring of the torque requirement during the machining process to discover possible wear or breaking of the tool.

Integration into the control. As main drives and tool drives do not carry out motion sequences with a highly dynamic performance and do not feature any interrelations with regard to other drives, the connection to the higher-level control is relatively simple. Via the speed setpoint it gives the starting command for the acceleration and for the stopping of the drive for both the normal operation and for the case of a fault, where usually shorter deceleration ramps are used. Normally a PLC is sufficient for controlling main drives and tool drives. The control signals are in some applications still connected by means of parallel wiring, however, normally via an electronic communication system. Cycle times in the range of 1 to 10 ms are sufficient.

4.12   Drives for pumps and fans

Markus Kiele-Dunsche

In production facilities there are many applications where gases or fluids are conveyed, gases are compressed, or the pressure level of fluids has to be adapted. For this purpose pumps, fans and compressors are used. The drives for these units are described in this chapter.

4.12.1 Conveying and compressing fluids and gases

Fluids and gases. Pumps, fans, and compressors have the task to move a medium with a specific density against a stationary pressure (Fig. 4-125). The power P which for this purpose has to be input hydraulically or pneumatically can be calculated by means of the differential pressure Δp and the flow rate Q or the volume flow dV/dt.

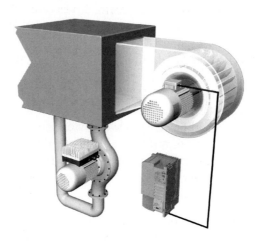

Fig. 4-125. Schematic diagram of pumps and fans

$$\text{translational: } P = F \cdot v \tag{4.11}$$
$$\text{hydraulic: } P = \Delta p \cdot Q = H \cdot g \cdot \rho \cdot Q \tag{4.12}$$
$$\text{pneumatic: } P = \Delta p \cdot \frac{dV}{dt} \tag{4.13}$$

Hereby ρ is the specific weight of the medium. As there are many applications conveying water, for fluid pumps the pressure generated by a pump is expressed in the height of the water column H_{H_2O}. Pipe friction losses are also described as a loss in the height of the water column. Thus, during the projecting process it can be directly evaluated which differences of level can be overcome when water is pumped, or which pump is required.

In the case of pneumatic processes, a differentiation between the conveying and compressing of gaseous media is made. For conveying it is assumed that the density of the medium upstream and downstream of the processing machine does not change substantially. A pressure ratio of above $p_A/p_B > 1.3$ between the inlet and outlet is then referred to as compression.

During compression a gaseous medium is restricted to a narrower space, which results in an increase in density and therefore in the relative weight. During compression apart from the density also the energy state of the medium changes. This is a thermodynamical process. If a closed space contains a gas, a specific amount of energy can be stored within the gas, depending on the pressure and temperature. If a gas is compressed, the pressure and temperature increase simultaneously. The change in thermal energy thereby corresponds to the mechanical power required to make the space smaller. If the gas is expanded, it performs work, causing the pressure and temperature to drop. This process for instance is used for refrigerators. By these Carnot processes it is possible to transport heat energy between different temperature levels by the use of mechanical energy.

During compression, the volume flow is different before and behind the processing machine. The mass flow, however, remains constant. The power required does not only depend on the density of the medium, but also on the gas characteristics and the compression or expansion process. Like this, for instance an uncooled compression at ideal gas behaviour at average pressure conditions is described as isentropic compression, and at higher pressure conditions is referred to as polytropic compression. Thereby correspondingly different powers result [BoEl04]. This complex subject is not to be described in detail here, as it is not important for the drive behaviour.

Controlled process factors. For the process factor which is controlled by the conveying and compression by means of pumps, fans and compressors, there are basically two different situations:

- A process factor of the hydraulic or pneumatic system is controlled directly. This can be the pressure or the volume flow.
- A parameter of a process, which is influenced by the pneumatic or hydraulic system, is controlled by the conveying of the volume flow. This can be a temperature.

This is to be explained by some application examples:

- If a fluid is to be conveyed, or a gas is to be compressed, the capacity is defined by the amounts required and by the pressure to be overcome. Here the pressure serves as a control variable.
- In another process, for instance, exhaust gases are to be extracted by suction, or fresh air is to be supplied. The variables required thereby depend on regulations and the pollution that is produced. The flow rate is used as control variable.
- In other processes heat is to be extracted or supplied. As pumping medium air or water is used. The demand depends on the amount of heat or air to be pumped and the usable temperature difference. The pressure to be overcome is defined by the gas or fluid friction within the pipe system and within the heat exchanger. Using the pump or the fan the temperature is controlled.
- If the pump of a circulating system is to convey heat to different places where the amount of heat is controlled individually, like in the case of a heating system, the pressure difference across the pump is kept constant. By means of a choke, the individual consumers can remove as much heat as required. The control variable is the difference pressure across the pump, which can be directly measured at the pump and adapted via the pump.

- During a process of drying, a specific amount of water is extracted from a space or a product. The amount of the water and the humidity of the air provided define the volume flow required. The air humidity is the control variable.

The examples show that for an optimum solution it is necessary to consider the whole process.

Closed and open systems. The behaviour of hydraulic and pneumatic systems can be divided into two main groups: closed circuits and open circuits.

In closed systems only frictional losses build up a back pressure. The medium is conveyed within a circuit where no geodetic pressures arise. Typical closed circuits are circulating systems or ventilation systems. Ventilation systems are closed systems, because the medium to be pumped is removed from one environment and is conducted to another environment with a similar pressure level.

If a medium is to be conveyed against a pressure, for example the pumping of water into a higher placed basin, the system is referred to as open. These systems are characterised by the fact that the conveying process already at low volume flows has to be achieved against a pressure. All compression processes or pumping stations for filling elevated tanks are open systems.

Fig. 4-126 clarifies this differentiation of closed and open circuits of hydraulic systems.

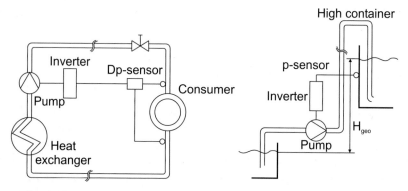

Fig. 4-126. Closed (on the left) and open (on the right) hydraulic circuit

Requirements. The requirements with regard to the processes described in this chapter can include the following items:

- a high process accuracy,
- ensuring that a minimum amount of the medium is conveyed within a time defined,

- minimising the energy input.

A dynamic speed control can be used for systems requiring a high process accuracy. Any intermediate buffers used to take up pressure fluctuations can be reduced or eliminated as the pump can directly compensate for fluctuations instead.

Energy efficiency. As the speed of asynchronous motors is defined by the mains frequency, it is very difficult to influence their speed without electronic measures. For the power control of pumping and ventilation systems therefore the simple principle of throttling is used, by which a choke (narrow point) is installed in the piping system, which increases the flow resistance, and by means of which the desired volume flow can be set. Chokes are actuators with losses, which deteriorate the efficiency of the pumping or ventilation system.

Pumps and fans have to be dimensioned for full load, even if this operation is rarely needed. An operation at partial load with a high efficiency requires an adjustment of the speed of the drive.

The minimisation of the energy input for fans and pumps can be implemented by means of speed adjustment. This property in particular is used in the field of building equipment like in the case of heating circulation systems or ventilation systems. For a low to medium volume flow or flow rate requirement with regard to the peak demand within a closed system, potential savings can be particularly high. This is due to the frictional losses which within a closed pneumatic or hydraulic system are predominantly responsible for the pressure build-up and which increase significantly with a higher volume flow. In the case of turbomachines not only the system losses (within the piping system and caused by consumers) increase disproportionately, but also the losses within the turbomachine itself. In the case of displacement systems the frictional forces within the machine are rather constant, i. e. irrespective of the flow rate. For centrifugal pumps in closed circuits, for instance, the potential savings, depending on the demand characteristic for the volume flow, is up to 60 % by the use of a speed control of the pump [EmKo02].

System and machine characteristics. In order to be able to determine the working points in a process, system characteristics are used. The system characteristics show the pressure which builds up at a specific flow. Fig. 4-127 shows the system characteristics of an open and a closed circuit.

As pump and fan applications have a non-linear behaviour, a process control does not act equally well for all situations. It is important to know which situation is the most crucial one to be controlled, so that the controller can be adapted to it.

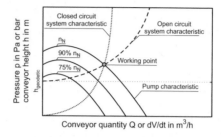

Fig. 4-127. System characteristics of a closed and an open circuit

Application examples. In production facilities pumps and fans are either used as a part of the production process (often for process engineering), or they provide an infrastructure for the production process in auxiliary processes (compressed air for pneumatic systems, heating and air conditioning, conveying of exhaust air, oil pressure for hydraulics). Below some fields of application are listed:

- refrigerating machines,
- chemicals and food industry (conveying, dosing, filling),
- increase of pressure (technical gases, air),
- water supply,
- building equipment (air conditioning, heating, ventilation) (Fig. 4-128),
- vacuum generation,
- industrial processes (drying, cooling, heating, lubricating, suction),
- waste water treatment.

Pumps, fans, and compressors provide the pressure and flow in a preferably autonomous and reliable manner. In order to achieve a simple and energy-efficient controllability of the pumping or ventilation system on the one hand and a high availability on the other hand, pumps and fans are often operated in parallel. This parallel operation one the one hand provides for the control of the flow rate by switching on or off individual units, and on the other hand the parallel operation ensures that in the cases of failure or maintenance of one unit the system is at least available with decreased performance.

4.12.2 Mechanical structure of pumps and fans

Power ranges. The pump is one of the oldest work machines. Today pumps can be found in all power ranges and in the most different applications:

- pumped-storage power stations: 250 MW,

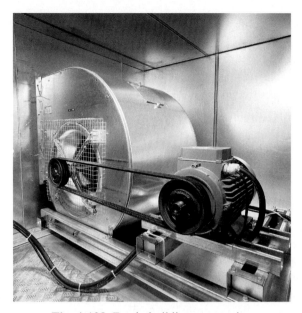

Fig. 4-128. Fan in building automation

- boiler feed pumps in thermal power stations: 10 MW,
- dewatering pumps in mines: 4 MW,
- pumps in production facilities: 100 W to 100 kW,
- circulation pumps in heating systems: 100 W to 10 kW,
- insulin pumps in medical devices: several mW.

Fans also cover wide power ranges:

- exhaust ventilation in power stations: 4 MW,
- fans for air conditioning systems: 200 W to 10 kW,
- radial ventilators for electronic devices: 0.5 to 20 W.

Pressure ranges. Compressors or vacuum pumps have the task to alter the pressure, or to work against a pressure. For instance, a compressor fills a tank with air generating a multiple of the atmospheric ambient pressure up to 10 or 15 bars.

In some applications in process engineering, pressures of up to 3,000 bars are required, which are generated by appropriate compressors. In other applications a low pressure below the atmospheric pressure, and even resulting in a vacuum, has to be generated. A gasless space where the pressure corresponds to a value of zero is referred to as a vacuum.

Principles of operation. With regard to pumps and fans, the following principles can be differentiated [Bo04]:

- principle of volume displacement, e. g. for piston, gear, or diaphragm pumps,
- axial principle for turbomachines, e. g. for axial fans,
- radial principle (centrifugal principle) for turbomachines, e. g. for centrifugal pumps or centrifugal fans.

According to the direction of the energy conversion, all turbomachines are divided into turbines and processing machines. Turbines convert a pneumatic or hydraulic power into a mechanical, rotary power. In the case of the processing machines, the power flow is the other way round. In this respect, turbomachines are processing machines like centrifugal pumps and fans.

Principle of volume displacement. The principle of volume displacement is one of the oldest drive principles. It is for instance used for a bicycle pump or handle pumps. By alternately expanding and reducing a space and only controlling the inflow and outflow by valves, there is a discontinuous material flow. This principle works for fluids and gases, whereby very high pressures can be generated. As gases, in contrast to fluids, can be compressed, the gas pressure to be implemented depends on the ratio of the big to the small volume. In general, this ratio is also referred to as compression ratio.

As the linear motion of a piston has to be converted from the rotary motion of the motor, positive displacement pumps have been developed which generate material flow from a rotary motion. Fig. 4-129 to 4-131 show two gear pumps and one rotary lobe pump. The material flow of such a pump is still discontinuous, but not as much as in the case of a piston pump.

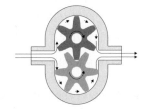

Fig. 4-129. Gear pump

Fig. 4-130. Gear pump

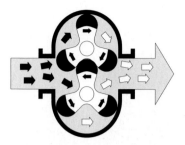

Fig. 4-131. Rotary lobe pump

Positive displacement pumps have the advantage that the interrelations between the flow rate and speed and the pressure and torque are proportional and thus very simple. Furthermore the pressure is maintained if the pump is in standstill. In the case of turbomachines, however, the medium would possibly flow back.

Positive displacement pumps have the disadvantage of the higher wear and in general cannot be designed as sturdily as turbomachines. The flow rate is discontinuous. If a constant pressure is required, a hydraulic or pneumatic damper has to be integrated in the system.

A further aspect is the locking due to a blockage in hydraulic systems. If in such a case the pump continues to work, a high pressure would build up, possibly resulting in damage of the system. Here the pressure can be limited by a torque limitation, so that the machine in the case of such a fault reduces the speed to zero.

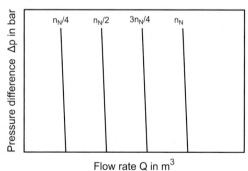

Fig. 4-132. Idealised characteristics of positive displacement machines

Fig. 4-132 shows the proportional interrelation between the speed and flow rate and between the torque and pressure for positive displacement machines.

Axial principle and radial principle for turbomachines. The operating behaviour of axial turbomachines and radial turbomachines is very similar. Therefore the principles of both machines are described in one.

Fig. 4-133. Centrifugal pump

As a rule, turbomachines feature a constant flow behaviour. The medium flows freely through the machine, whereby by means of a skilful arrangement of blades, a pneumatic or hydraulic mass flow and pressure are generated from a rotary movement. For both the axial and the radial principle, the mass inertia of the medium is used.

In axial turbomachines the medium tries to avoid the rotating blades which, due to energy reasons, inertia and blade configuration, is easier to achieve axially rather than radially in the direction of the blade movement. Sometimes behind the rotating blades stationary guide blades are mounted, preventing the medium from being set into rotation.

In the case of the centrifugal principle, e. g. for a centrifugal pump, the mass is set into rotation, so that a pressure is built up by the centrifugal force that is generated. Fig. 4-133 shows the profile of a centrifugal pump. A multi-stage centrifugal pump operates just like a single-stage one, however, in a relatively compact design and with higher pressures [Wa04a].

These interrelations can be gathered from fig. 4-127. From the pump characteristic for maximum the speed n_N the achievable flow rate can be determined. From the system characteristic for the open circuit it is possible to see the speed range within which the pump operates when the flow is specified. In the example in fig. 4-127 the pump requires a speed of a bit less than $70\% \cdot n_N$, to bring up the pressure without conveying the medium, and at n_N it has reached the flow required.

As a rule, the pressure within a turbomachine increases proportionally to the square of the speed. As can be gathered from fig. 4-127, the pressure that builds up is not solely defined by the speed, but also by the system characteristics, and thus by the behaviour of the counter-pressure. If a closed, hydraulic circuit is operated by a pump, and if the hydraulic resistance is so that the pump generates its rated pressure at the rated speed, then the following applies:

- The volume flow is proportional to the speed ($Q \sim n$).
- The differential pressure is proportional to the square of the speed ($\Delta p \sim n^2$).
- The power absorbed is proportional to the third power of the speed ($P_{mech} \sim n^3$).

For an open circuit the interrelations are not as simple. It is advisable to analyse the operating behaviour of the system by means of the system and machine characteristics.

Normally machine characteristics are provided by the manufacturers. They contain the Q / H or $(dV / dt) / \Delta p$ characteristics and in addition also the power consumption for selected operating points.

Parallel operation of turbomachines. In most cases only one pump or one fan is required to operate within a desired power range. The parallel operation of pumps or that of fans offers the following advantages (Fig. 4-134).

The power range of a system can be implemented more efficiently by installing two small pumps or fans instead of one big pump or one big fan. Such applications provide a high availability of the system, as if a pump or a fan fails, a minimum operation of the system can be ensured.

Fig. 4-134. Parallel operation of pumps

A pump exchange management is necessary for this operation, which also reports the fault, so that the system can then be completely recommissioned again.

4.12.3 Drive systems for pumps and fans

The main requirements to be met by a drive system for pumps and fans are:

- reliable service over a long period of time, exceeding ten years,
- low cost of components when they are purchased,
- high energy efficiency during operation for the purpose of lowering operating costs.

In this respect, the use of an inverter plays a key role in lowering energy consumption. An inverter is a good solution for an interface between the automation system and the process as it can execute decentralised process control and monitoring functions [BePr06]. In Fig. 4-135 these tasks are shown by a motor inverter.

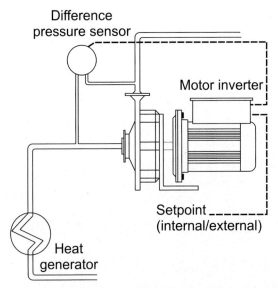

Fig. 4-135. Decentralised differential-pressure control

Basic principles for design and dimensioning. When dimensioning drive components, the assumption is made, as described in chapter 3.7.2, that a representative cyclical profile for the movement can be established. In the case of pumps, ventilators and compressors, such a profile does not usually exist, as a process controller specifies the rotational speed according to external factors. Consequently, the assumption is warranted that, over a relatively long period of time, the process requires a maximum operating point. For that reason, the components are designed for continuous operation.

Dimensioning of drive components. The power requirement for the maximum operating point can be ascertained from the machines' data sheets and characteristics. The components consisting of geared motors, motors and inverters are dimensioned to ensure that the continuous rating is lower than the components' rated quantities.

Without having specific knowledge of the process, it is possible to calculate an approximated rating for the drive power on the basis of the quantities involved (volume flow and differential pressure) and the typical efficiency of the pump η_{pump}.

$$P_{mech} = \frac{P_{fluid}}{\eta_{pump}} \tag{4.14}$$

$$M_{mech} = \frac{P_{mech}}{\omega_{mech}} = \frac{P_{mech}}{n_{mech} \cdot \frac{2\pi}{60}} \tag{4.15}$$

Start-up and braking are special operations. Generally speaking, the demands made on the start-up are not very high. In some cases, an increased torque may be needed as the work machine is blocked after a relatively long standstill. This should be taken into account when an inverter's parameter settings are made: the motor should be driven with an increased voltage at low speed. No specific demands are made on braking. If pumps are run in parallel or in the case of booster stations, check valves have to be fitted in the pipeline system to ensure that the pump can be shut down.

Motors and gears. In the case of pumps, the electromechanical drive solution consists, most frequently, only of a motor, as the speed range of pumps matches the rated speed of a standard three-phase AC motor - 1,500 rpm or 3,000 rpm (system frequency: 50 Hz). Consequently, it is not usually necessary to use a gearbox. The motor is directly connected to the pump.

Fans need gear mechanisms more often to adjust the speed of the motor to that of the fan. But there are often special-purpose turbomachines, which are directly connected to the motor. Hence, external-rotor motors, for example, are used for ventilators; in this case, the blades are attached directly to the rotor. This is a very compact type of design in which the motor is cooled directly by the current of air.

In pumps, wet-rotor motors are often used, as in this case expensive shaft sealing can be dispensed with. That makes sense particularly if, e.g., the medium being delivered constitutes an explosion hazard or if the pump works under water, e.g. a submersible pump.

A special type of leak-proof pump for chemical applications is the canned motor pump. In this pump, the stator is hermetically sealed off from the rotor of an asynchronous motor by a tube in the motor's air gap, so that the squirrel-cage rotor is in the medium being delivered while the stator remains dry.

In these special types of machines, a number of limiting conditions have to be complied with which can be implemented in the monitoring functions of an inverter. Thus, for example, a pump can be damaged if it runs dry due to its special bearing, so that it is necessary for the pump to be shut down if the load falls below a certain limit. It may also cause damage to the bearing if the speed drops below a certain value, as in that case lubrication is no longer assured.

If no three-phase supply system is available, capacitor motors are often used in the lower-end power range. In the lowest power range, other types of motors are to be found as well.

As a rule, pump and fan motors need no mechanical brakes as no torque has to be maintained when they are stationary. Nor are external fans required since, as described above, due to the machines' characteristics the torque delivered is considerably reduced at low speeds, so that the built-in fan can provide enough cooling capacity.

Actuation of the motor. In many present-day applications, pumps and fans are still operated at constant speeds. This involves connecting the asynchronous motor directly to the mains or starting it up via a star-delta connection or soft-start devices.

Compared to speed control, the use of chokes and by-passes to adjust the motor to the process requirements is disadvantageous for reasons of energy efficiency [Wa04]. That is why inverters are being used to an increasing extent to adjust the speed. Speed adjustment may either be necessary on account of the process requirements or the power savings it brings make the system considerably more economical. These are the advantages of speed adjustment:

- energy savings in partial-load operation, as power input increases disproportionately high to the rotational speed,
- noise reduction in partial-load operation,
- less wear,
- hydraulic dampers can be reduced in size or dispensed with,
- pressure surges in the system can be avoided or reduced.

In Germany alone, the potential savings for pump systems are around 15 billion kWh per year, which, assuming an electricity price of

8 cents / kWh, corresponds to an annual cost reduction of EUR 1.2 billion for operators.

In Europe, the total power saving potential for pump systems in all industries is about 40%. This potential is attributable to the following possibilities for optimisation:

- speed control of pump systems: 20%
- improved system design: 10%
- choice of the right size of pump: 4%
- decision in favour of pumps and motors with higher energy efficiencies: 3%
- optimised installation and maintenance: 3%.

The use of inverters for speed control. The use of an inverter to achieve speed control is the most elegant method to adjust the mechanical output of pumps, fans and compressors to the process. At the same time, no high dynamic demands are usually made on the drive system. In these applications, neither position nor speed sensors are needed as no high-precision speed or position control has to be deployed.

Generator operation. As the rotating parts of pumps have a relatively small weight compared to the output they produce, a pump slows itself down in an adequately short period of time. Even if the process does involve a great deal of kinetic energy, no active braking operation is generally necessary.

Fans and ventilators, on the other hand, have relatively high moments of inertia compared to the output they produce. The result is that, in some cases, fans would decelerate for several minutes if the drive were to be disconnected from the power supply when running at full speed. That is why the use of inverters to execute an active braking operation is a good way of shortening the rotor's deceleration period. At the design stage, a check must be made to see whether or not the kinetic energy of the medium in the system also acts as a generator during the braking operation, thereby delaying the braking action.

An initial rough estimate has to determine the time required for the kinetic energy present in the system to reach zero at a constant rate of power dissipation. The power dissipation can be assumed to be the power requirement existing at the operating point when the system is in an active state; from this, the time lapse until standstill is reached can be estimated. If this period is more than double the time available for braking, additional energy and power have to be allowed for when dimensioning the brake components.

In the case of fans and ventilators, kinetic energy can be calculated by adding up the energy values by means of equation (4.16).

$$E = \sum_i \frac{1}{2} \cdot m_i \cdot v_i^2 = \sum_i \frac{1}{2} \cdot \delta_i \cdot l_i \cdot \frac{\dot{V_i}}{A_i}^2 \qquad (4.16)$$

Using the media, masses m_i can be determined by means of the volume and the densities δ_i. Velocities v_i being calculated from the quotient of flow rates dV_i / dt and the line cross-section A_i. l_i denotes the length of the lines.

Drive control. To enable fans, pumps and compressors with their specific load characteristics to be operated as economically as possible with the help of inverters, there are special V/f types of characteristics. Given a closed-circuit characteristic of the type shown in Fig. 4-127, the power requirement will vary in proportion to the cube of the speed; the torque required being the square of the speed.

$$P \sim n^3 \qquad (4.17)$$
$$M \sim n^2 \qquad (4.18)$$

The V/f characteristic normally being used for a constant torque application (V/f ~ V/n = constant) would result in a low efficiency in the partial-load range, when applied to turbomachines. Therefore a square-law V/f characteristic curve (V/f² = constant) is often used for voltage control.

Other inverter functions for pumps and ventilators. In addition to setting the speed, inverters can also execute other functions for pumps and fans.

- integrated process control (PI controller in the inverter),
- flying restart circuit for the purpose of determining the speed of a motor coasting down after a restart,
- integration into the higher-level automation system,
- remote maintenance possible,
- use of the inverter for motor overload protection,
- additional functions such as providing protection for the machine or reducing machine wear (e.g. dry-running protection, V-belt monitoring),
- functions that control parallel operation of pumps or fans (e.g. sequence control circuits).

With the description of drives for pumps and fans, the presentation of the twelve drive solutions is now concluded. It is now to be demonstrated how these solutions can be applied to typical systems belonging to production and logistics plants. The aim is also to show that the various types of applications can be classified according to these patterns and furthermore to show the distribution of the drive solutions in terms of the required quantity.

4.13 Examples of applications in production and logistics plants

Hans-Joachim Wendt

In the preceding sections, twelve applications were presented, leading in each case to different drive solutions. Observations in production facilities and logistics centres show that eight to ten of these drive solutions are frequently required for the processes of manufacturing and packaging goods, and for in-house logistics. However, there are great differences as regards the quantities of their use.

To give some examples, five plants involving different production processes and speeds are to be examined with regard to the drive solutions employed in them:

- newspaper production,
- manufacture of laminate flooring,
- beverage bottling,
- the logistics centre of a supermarket chain,
- car production.

4.13.1 Newspaper production

Process description. Printing plants for newspaper production belong to the category of highly-automated production facilities, running at a production cycle of between five and ten newspapers per second.

The offset printing production process starts with the unwinding of the rolls of paper and the feeding of the paper webs into the printing machine positioned above. The typographic image is applied to the paper using aluminium plates mounted on cylinders which apply the inks to the paper by means of water. Each printing cylinder can accommodate a total of eight printing plates (four juxtaposed and two over the circumference of the cylinder). That means that 16 pages are printed simultaneously on the front and reverse side of the paper, these then being laid individually in the folding unit and folded together to form two newspaper sections of eight pages each.

A 32-page newspaper is produced on two printing towers with a joint folding unit and then drawn onto a chain conveyor, with which the newspapers are conveyed to the automatic insertion machine for advertising supplements and then finally to the packaging systems. A compensating stacker with caterpillar haul-offs produces piles of 20-50 newspapers, which are subsequently sealed in weatherproof plastic film. As the finished newspapers are usually despatched immediately after being manufactured, a high-bay storage facility can be dispensed with.

The aluminium printing plates are exposed to a violet laser beam directly in the print house, with several hundred printing plates being produced every night depending on the page count and the number of different local editions.

Fig. 4-136. Conveyor system in a newspaper printing plant

Drive solutions. The newspaper production plant described above is fitted with about 150 drives, 80% of which are controlled by frequency inverters or servo drives. Nine different drive solutions are used.

The unwinders on the double turret winder work as centre winders, which have to be able to control high mass moments of inertia. To bring the rolls of paper weighing up to 1,000 kg to a standstill if an emergency stop occurs, drives with a torque of up to 350 Nm and a rated power output of 75 kW are used. A very disadvantageous ratio between the inertia of the paper rolls and that of the motor ($k_J > 20,000$) necessitates optimally-coordinated control loops and mechanisms without backlash.

In the printing units a total of 16 angular - synchronous drives implemented as direct drives without a gear ratio, drive each printing cylinder separately (Fig. 4-137). In addition to the speed and position control, a superimposed register control system compensates for the stretching of the paper, which arises during the printing process due to the addition of the inks and water. The axes are synchronised by means of a real-time communication system.

Drives with a square-law V/f control matching the load characteristic of pumps convey the inks and the water.

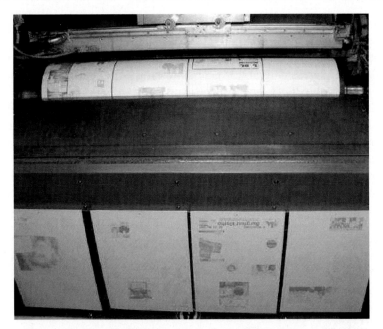

Fig. 4-137. Finished printing plates for use in newspaper production

Four synchronised servo drives in the folding unit run in angular-synchronous mode with the drives of the printing units. To separate the pages of newsprint, two servo drives operate as cross-cutters.

At the chain conveyor, inverters with sensorless closed-loop vector control drive the transport chain (Fig. 4-136) and provide the set-values for the synchronised-speed drives in the insertion machine. At the packaging station, unwinders and cross-cutters are used to feed in the plastic film. Conveyor drives move the newspaper stacks on roller conveyors to the waiting pickup trucks, which take the newspapers to the distribution points.

For the production of the printing plates, 15 servo position drives with synchronous servo motors in a power range below 100 W are used to place the printing plates in exactly the correct position in relation to the laser.

4.13.2 Production of laminate flooring

Process description. Short-cycle press lines for the production of laminate flooring are manufacturing systems with very high productivity and production rates of up to 50 million square metres per year. A production line can thus deliver more than 10 % of today's European demand for this product.

The raw material consists of 3 to 40 mm thick wood fibreboards. In successive stations, these boards are stacked together with vapour sealing film underneath and printed paper and cover film for surface protection above and fed to the press (Fig. 4-138). The pressing is done in a process lasting 24 s at more than 200°C and a specific pressure of 300 to 600 N/cm². More than 225 cycles per hour and thus more than 1 million pressings per year can be achieved. When the pressed material has cooled down, it is sawn to the base board size by cut-to-size saws.

The milling of the grooves and edge profiles along the four sides of the boards is done with throughput speeds of up to 200 m/min. Finally, the boards are stacked in retail packing units, provided with labels and packed with heat shrink wrapping.

Fig. 4-138. Short-cycle press line for the production of laminate flooring

Drive solutions. The press line described above is equipped with approx. 200 drives, of which 80 % are controlled by frequency inverters or servo drives. Eight different drive solutions are used.

The source materials are stacked using servo position drives and geared asynchronous servo motors with resolvers in order to achieve the required accuracy of approx. 1 mm. The point-to-point positioning is carried out by the servo drives, i.e. without additional control hardware being required. Altogether, approx. 80 position drives are used. The cycled feeding of the stacks via roller and belt conveyors is executed with vector-controlled frequency inverters. All inverters used in the press line are equipped with the drive-based safety function "safe torque off".

The press is loaded via a slide weighing several tons, which has to cover a distance of more than 10 m within approx. 3 s. For this purpose, a position drive with two 90 kW helical-geared motors connected in parallel on a single shaft is used. The force is transmitted via toothed belts.

The press for the forming process is operated hydraulically.

The pressing section is followed by a cooling section, where the boards are put upright and then moved slowly through a cooling zone. After this, they are sawn to the final board dimensions. A board stacker stacks the boards on a pallet, which is then stored in a high-bay warehouse for conditioning via a completely automated driverless transport system and a storage and retrieval unit. Here the boards remain for 3–4 days.

Depending on the chosen profile, the milling of the long and short board sides requires six to eight tool drives for each side. These tool drives are operated at speeds from several thousand to several tens of thousand revolutions per minute. Here special high-speed asynchronous three-phase motors are used, which are operated in the field weakening range with the help of frequency inverters. The milling tools are attached directly to the motor shaft.

In the packaging station, winding drives are installed for unwinding the packing material and the heat shrink film, and a cam drive controls the welding bar.

Travelling and hoist drives are used for the in-house transport with industrial trucks and storage and retrieval units.

4.13.3 Fruit juice production and filling

Process description. The filling of beverages into plastic bottles, e.g. PET bottles takes place in cycles of approx. 5 to 14 units per second. The production rate of a plant amounts to several million units per day and may comprise not only plastic bottles but also carton containers and glass bottles.

The fruit juices or concentrates are delivered by trucks and stored in tanks. Depending on the product, the fruit juice or nectar is produced by mixing the concentrate with water and, if required, sugar under sterile conditions. All fluids are transported through pipes.

The PET bottles are delivered as moulding blanks already provided with moulded screw threads for the caps. The bodies of the bottles, however, are formed on site in blow moulding machines. The finished PET bottles are transported to the filling station by means of compressed air with the bottles hung up by the thread and guided in a rail.

The filling of the bottles is a continuous process comprising several steps: the sterilisation, the filling, the capping and the labelling (Fig. 4-139).

Fig. 4-139. Bottle filling

Then the bottles are transported with chain conveyors and grouped in units of six or eight with the help of guiding plates and belt conveyors. These units, corresponding to the ones to be found on supermarket shelves, are put on cardboard trays and wrapped in heat shrink film, which is shrunk in a continuous furnace.

Finally, the trading units are palletised and stored in a high-bay or flow-through warehouse until they are loaded on HGVs for delivery.

Drive solutions. The production line described above is equipped with more than 1,000 drives, of which 60 % are controlled by frequency inverters or servo drives. Ten different drive solutions are used.

The fluids are transported by means of pump drives with an asynchronous motor acting directly (i.e. without gearbox) on the pump. The power range extends up to 50 kW. The pumps are controlled by frequency inverters operating with square-law V/f characteristic adapted to the load.

The drive solution used most frequently is the conveyor drive. Conveyor drives transport the individual bottles from one processing station to another as well as the trading units and the pallets. Due to the vast extension of the system, distributed drive technology with motor inverters and starters is selected in order to save on control cabinet segments and long shielded motor cables. The power range extends to approx. 4 kW. The geared standard asynchronous motors are mainly fitted with bevel or shaft-mounted helical gearboxes with the hollow shaft being connected to the input shaft of the conveyor section. The closed-loop controlled conveyor drives are accelerated via s-shaped acceleration ramps to prevent the pallets from tilting. All drives are controlled from a central control room via fieldbus systems.

The trading units are palletised by handling robots (Cartesian robots or articulated robots) equipped with coordinated servo drives.

Winding drives and cross cutters are used for the feeding of the packing film in the packaging stations. Hoists and travelling drives can be found on the storage and retrieval units.

4.13.4 Logistics centre of a supermarket chain

Modern logistics centres of supermarket chains supply a catchment area of several hundred supermarkets for a population of up to ten million people. A logistics centre not only has to store goods temporarily, but also has to provide the trading units on rolling containers, according to the daily orders received from the supermarkets. The logistics centre described uses a state-of-the-art automatic order-picking system and achieves a picking rate of up to 240,000 units per day.

A high availability is one of the most important requirements since many conveyor sections and hoist units are not backed up by redundant units.

Fig. 4-140. Roller conveyors used in a logistics centre

Process description. The goods coming in on euro-pallets are fed into the system via an input station and then transported to the high-bay warehouse via roller conveyors, accumulating roller conveyors, rotary tables, traversing carriages and hoists. Storage and retrieval units load and unload the storage cells used in the picking process.

Automatic depalletising stations precede the picking process. Here the packing units for the supermarkets trays are reloaded from the euro-pallets to individual trays and brought to a second storage via various conveying facilities, especially roller and belt conveyors, ejectors and continuous vertical conveyors (Fig. 4-140). The typical conveying speed is 1.2 m/s.

The automatic picking process consists of the order-related picking of the trays by storage and retrieval units, the subsequent sorting for optimising the stacking sequence and finally the automatic stacking on roller containers for the transport to the supermarket.

Drive solutions. The system comprises more than 5,000 drives, of which approx. 50 % are operated with inverters. The drive solutions required are, for the biggest part, conveyor drives, but positioning drives, hoist drives and travelling drives can also be found.

The conveying facilities for the pallets and the trays are almost completely implemented with conveyor and hoist drive solutions. Roller conveyors with centre drive equipped with helical geared motors as well as conveyors with head drive equipped with bevel gearboxes in hollow shaft version are used. The standard three-phase AC motors used are mainly in the power range below 1 kW. Additional mass inertias are provided on the motor shaft in the form of cast metal fans to limit the acceleration of the conveying sections in order to prevent loads with a high centre of gravity from tipping over. In this way a soft start can be implemented without using frequency inverters.

Due to the high requirements regarding the availability of the system, the brakes and sealing elements of the gearboxes have an industry-specific design. In the event of a failure, complete modules can be replaced and the MTTR can be kept low.

The frequency inverter controlled drives used for transporting containers are run in sensorless vector control mode which optimally adjusts the excitation of the asynchronous motors to the different load conditions of the trays.

For depalletising with special gantry robots and for the automatic stacking of the roller containers, servo drives with integrated point-to-point positioning are used. The position control is equipped with resolvers for the feedback signal. Synchronous servo motors are used to enable an optimum load matching via the gearbox ratio, in the power range below 1 kW for the x and y axes and 20 kW for the z-axis.

DC bus coupling allows energy to be exchanged between the axes. Braking units with resistors convert excessive braking energy into heat during the dynamic deceleration while the target position is being approached.

4.13.5 Automotive production

An automotive production can be split up into the sections pressing, body-in-white, paint shop and final assembly. The number of controlled drives alone significantly exceeds 10,000. In this chapter we are going to have a closer look at the drive solutions used in the sections body-in-white and final assembly.

Description of body-in-white production. The sheet metal parts for the car bodies are assembled by articulated robots in enclosed cells. For welding, the robot arms are equipped with special welding guns (Fig. 4-141). The sheet metal parts are fed with the help of power & free conveyors attached to the factory ceiling or via loading magazines loaded by means of industrial trucks. For loading the magazines, there are roller shutters in the wire mesh guard of the cells. Several hundred robots set the – depending on the model – approx. 7,000 welding spots of a complete car body. Here cycle times of approx. one minute are reached. A robot transfers the completely welded car body parts to the conveyor system attached to the factory ceiling which transports them to the next cell. In addition to spot welding, sometimes laser welding is used as well.

Fig. 4-141. Welding robot used in automotive production

Description of final assembly process. The final assembly comprises the line assembly of the interior components. During this process the cars are continuously moved by carriages with a speed of 3 to 5 m/min. At the different processing stations, preassembled components like e.g. the complete dashboard are fitted manually. Often special handling gear is used to handle the components, e.g. for fitting the windscreens.

The material is mainly transported by forklifts or by circular conveyor systems. Hoists lift the finished cars to a higher level where they are transported by Skid conveyor systems.

Depending on the model, a manufacturing cycle takes between 50 seconds and 2 minutes.

Drive solutions. The robots used in body-in-white processing are equipped with special multi-axis servo inverters. The servo inverters are connected to the DC bus so that the regenerated energy during braking is available for the other axes. Due to this and to the fact that the load is almost always divided on different axes, the DC bus capacity, the brake modules and even the heatsinks of the multi-axis servo inverters can be optimised for common operation.

To meet the high requirements for availability, the inverters and all electrical cables have a pluggable design. This results in a MTTR of less than ten minutes.

The motion control executes a coordinated motion in 3D space (trajectory control) calculated by the robot control and transferred to the servo inverters as torque setpoints (which therefore work as torque-controlled actuators).

The motors used are synchronous servo motors with special cycloid gearboxes. The resolvers integrated into the motors are used for both the servo and the trajectory control. In this way accuracies of down to 0.1 mm can be achieved.

The welding guns are driven by synchronous servo spindle motors. The motor shafts of these motors are provided with a spindle for a defined traverse path [Hi02]. Compared with pneumatic drives, the guns can be opened and closed faster, and the quality of the welding spots increases due to the exact torque control by the servo inverter.

The roller shutter doors used in the body-in-white section to reload the loading magazines are equipped with geared standard three-phase motors. These motors are controlled by decentralised motor controls which can include frequency inverters or motor starters. All connections (mains cables, motor cables, control cables and position sensors) have a pluggable design in order to meet the MTTR requirements. The majority of these drive solutions has drive powers of up to 4 kW.

4.13.6 Distribution of the drive solutions

The analysis of the production and logistics facilities shows that the twelve drive solutions described above are appropriate for structuring the multitude of individual drive tasks and the corresponding implementations selected.

Many of the analysed production facilities including their internal logistics require eight to ten different solutions. A single machine, however, often only requires three to five drive solutions so that the mechanical engineering companies do not have to hold the entire know-how.

The numerical distribution of the drive solutions varies widely. In controlled drives, complex motion controls can be found less often than the drive solutions which are not so sophisticated. Fig. 4-142 shows the distribution of all drive solutions for controlled drives, which can be estimated with an annual requirement of more than five million axes for the European factory automation at the moment.

Table 4-21. Overview of the drive applications used in the examples

	Newspaper production	Production of laminate flooring	Beverage filling	Logistics centre	Automotive production
Conveyor drives	++	++	+++	+++	++
Travelling drives	+	+	+	++	+
Hoist drives	+	+	+	++	+
Positioning drives	+	++	++	++	+
Coordinated drives for robots			+		+++
Synchronised drives	+++		+		
Winding drives	++	+	+		
Intermittent drives for cross cutters and flying saws	+	+	+		
Drives for electronic cams		+	+		
Drives for forming processes					
Main drives and tool drives	+	++			
Drives for pumps and fans	++	+	++		

The distribution also shows that the proportion of conveyor drives is dominant, since in highly-automated production and logistics facilities work pieces have to be transported within the production machines and from one processing station to another and finished products are conveyed to and from high-bay warehouses.

For the transport of continuous materials like paper, textile or film, synchronised drives and winding drives take over the task of the conveyor drives. Winding drives can, however, also often be found at the end of discrete manufacturing systems, where they are used in packaging stations.

The high proportion of the positioning technology can be explained by the requirement for an exact positioning of work pieces and tools in automatic production machines.

Complex motion controls (e.g. in cam drives, flying saws or cross cutters) or coordinated drives are used less often.

Numerous applications – also in mechanical engineering for factory automation – require drive solutions for pumps and fans, e.g. for the supply of coolant.

Compared to this, the number of drives for the machining of work pieces with tool drives or forming drives is relatively small.

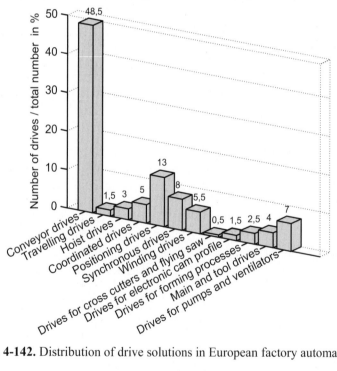

Fig. 4-142. Distribution of drive solutions in European factory automation

4.14   Development trends

Dr. Edwin Kiel

At the end of this chapter we will have a look at some development trends. We will start with the so-called mega-trends, which also affect the use of drive technology in production and logistics [ZV06b].

Globalisation. Globalisation means that the global borders restricting the exchange of goods have been abolished. Productions can be distributed all over the world, separated markets no longer exist. Therefore every production has to compete with competitors on a worldwide scale.

Taken as a whole, globalisation leads to a larger market, but potentially also to more competitors. Up to now, globalisation has brought more advantages than disadvantages for the European manufacturers of drive technology and machinery. The market has increased and the competitors have not yet gathered the knowledge and the technological level for being a real threat. It has been possible to shift simple production processes to countries with lower labour costs and thus to reduce the manufacturing costs for the products.

Regarding the workforce, the situation is more differentiated. New automotive factories tend to be established in countries with lower labour costs which at the same time are large growth markets. In the traditional production countries, which have saturated markets, it is usually not possible to keep the employment level in the production sector stable. Regarding the drive technology, the result of globalisation is that the factories in which drives are used are spread – stronger than before – all over the world.

The increasing goods traffic, which grows faster than the global gross national product, also leads to high investments in the logistics sector. The numerous container terminals, which usually are highly automated, are only one example for this process.

To what globalisation may lead, take the sector of the textile industry [Ri06]. This market has already been globalised for 200 years, and as a result there have been strong shifts in production locations and trade flows. Today, the production is linked on a global scale. The main part of the machinery is still produced in Europe, but textile production can be found here only to a small extent. The productivity increase of the textile machinery, however, to which drive technology has contributed, has been so high during the last decades that the efficiency increase of the machines has significantly exceeded the demand. As a result, the textile machinery manufacturing sector has clearly contracted. The production of sewing machines has already been shifted to those places where the machines are mainly used – the countries of East Asia and Eastern Europe.

At the moment Europe still has a globally leading role in most of the mechanical engineering sectors. Machines are exported worldwide so that everywhere the same concepts are used. This means that globalisation primarily is a growth opportunity for most mechanical engineering sectors. This advantage has to be continuously defended by means of innovation. Here the drive technology has to make an important contribution.

Automation. The automation level of production and logistics centres increases continuously. It seems as if there is no counteracting force to oppose this trend. While the costs for manual labour increase (also in countries which today still have a low wage level), the costs for automation decrease due to the continuous productivity increase in the manufacturing of electronic products including the automation and drive systems.

An increase in automation leads to an increasing number of drives being required. This results in the number of inverters sold growing by approx. 15 % per year as a long-term average. A saturation is not foreseeable. The more the knowledge grows on how production and logistics tasks can automatically be carried out by software programs, the more drives are required to implement the commands of these programs.

Energy efficiency. Electrical energy is still comparably cheap. Only in a few cases the energy costs become significant compared to other costs. Due to the climate discussion and to the increasing raw materials and energy costs, the awareness for efficient dealing with energy grows. For the discussion about the reduction of CO_2 emissions, the reduction of the energy consumption plays a role as important as the generation of energy from renewable sources. This is supported by the European framework legislation.

Electric drives are a significant consumer of electricity. Energy saving concepts utilised in electric drive concepts make an important contribution to overall energy consumption reduction.

Smaller production lots. In many product areas, the product variety increases. For cars, for instance, the number of designs and types as well as the equipment variety has grown. The number of variants of some types is so high that the same variant is rarely produced twice. Here the trend is towards customised mass production.

The consequence for the machinery manufacturers is that they have to design their machines in such a way that product changes can be carried out fast and reliably. High costs and efforts for manual retooling and running-in in the event of a product change are no longer competitive. Here the automation, too, plays a role in supporting and enabling the development. Again this results in an increased number of controlled drives being required.

These four mega-trends form the basis for the future technical development of drive systems [Ki05].

More controlled drives. Today, in Germany approx. 30 % of the newly installed three-phase drives are operated in speed-controlled mode. This proportion has tripled during the last eight years. The driving factor for this is, of course, the continuous cost reduction of inverters, which is based on the progresses made in microelectronics ("Moore's Law") affecting the costs of power electronic components. It is foreseeable that the proportion of closed-loop controlled drives will increase further because this offers many advantages:

- The speed of the drive can be adjusted and the drive can be optimised to the operating point of the process.
- The torque can be controlled at any time so that accelerating and braking does not depend on process parameters, but on the values specified by the control. This enables defined motion sequences to be generated.
- Positions can be approached very precisely.

The increasing automation and the increase in energy efficiency result in an increasing use of inverters. Eventually, one can assume that in 10 to 15 years practically all newly installed motors will not be started directly across the line, but will be controlled electronically.

Decentralised concepts. With the growth of automation particularly in extended systems, decentralised concepts are more and more being used. Long conveyor sections and pumps in pipe systems are ideal for the use of decentralised inverters which are mounted on the motor or close to it. The advantages provided by this concept are mainly based on the simplification of the installation.

Direct drives. The proportion of linear and rotary direct drives is still low at the moment, but it will rise continuously. The disadvantages of mechanical transmission elements including gearboxes can be eliminated. The use of direct drives is hampered by the higher costs and by the difficulty to define standardised components which are suitable for many applications.

Integrated drive systems. For some applications with high numbers of work pieces being processed, the obvious solution is to integrate all components – including mechanical transmission elements – into the drive so that they form a physical unit. Alternatively, the drive can be integrated directly into mechanical parts of the machine. Examples of such solutions already being implemented today are decentralised servo drives with integrated gearbox and electronics, motors integrated into the rollers of roller belts [MaVo06] and the integration of gearbox, motor and inverter into a single housing. All these mechatronic units offer the advantage of direct use.

Power recovery. Regeneration of energy into the mains today only constitutes a small proportion. However, the trend towards energy-efficient concepts will lead to an increase of this proportion.

Drive control. The major developments in the closed-loop control of electric three-phase AC drives were made in the 1980ies. Besides the understanding of the theoretical interrelations (field-oriented control), it was the availability of powerful and affordable digital signal processing systems which made the implementation into products possible. Today we can say that the digital control of three-phase AC motors is very mature. There are only few fields left for improvement. These are mainly the control methods without speed sensors and the handling of oscillatory loads. The performance of synchronised drives, too, can be improved by appropriate methods. The increasing computing power based on the developments made in microelectronics will lead to the use of more complex methods and thus to the further optimisation of the drive control.

Motion control. In addition to the drive control, the motion control has a major influence on the performance of a drive in an application. For a long time, the limited computing power has restricted the quality of the motion control. Unlike in drive control, in motion control there is still much room for improvement. The desire to increase the productivity of machines leads to lighter-weight and more flexible machines in order to reduce the masses to be accelerated. At the same time these masses have to be accelerated faster. Therefore the importance of methods for controlling the handling of oscillatory loads grows. These methods, however, have to be designed in such a way that they can be used easily. For this purpose it would be useful to have identifiers for the system behaviour in order to limit the efforts required for the process adjustments [KnSch06].

Diagnostics. Controlled drives have to be equipped with sensors which provide them with information about the process to be controlled. This fact can be used to establish preventive diagnostics and maintenance concepts [De06]. Today such concepts are only used rarely because the diagnostics processes do not yet have the robustness required to make their reactions reliable. Usually a premature reaction is as harmful as a reaction coming too late when the machine already has been damaged. This opens a wide field for the use of computing power, intelligent methods and algorithms to receive reliable information for process monitoring [Wa02, Wa04b].

Communication. Today, almost every second inverter is controlled by a communication system. This proportion has grown significantly during the last few years. This trend is driven by the advantages during the installation, by the technical options for the interaction of control system and drives and by the decreasing costs for microelectronic components. It can be assumed that in 10 to 15 years the majority of the drives will not only be controlled electronically, but will also be connected to communication systems.

Communication systems are especially suitable when methods for fast product changes and process diagnostics are to be used. Basically, communication systems provide the possibility to use the drive not only as an actuator, but as an active element of the entire automation system.

Safety engineering. Safety engineering is a sector which has reached the drive technology rather late. This is even more astonishing as drives are the parts of the machines causing the greatest hazard. The use of drive-based safety is just getting started. Due to the high development efforts, today software-based safety functions are still rather expensive. At the moment, not all of the technical options provided are used to the full extent. With increasing use, however, the costs will decrease. It can be assumed that all drives generating motion which could be dangerous to people will be provided with a drive-based safety system closely interconnected with the machine automation.

Reliability. Initially, the poor reliability of inverters was an obstacle to their wide-spread use. However, the knowledge about the reliability of their design has grown enormously and today inverters are significantly more reliable. Nevertheless, there is still room for improvement.

Electronic systems consist of more components and thus have more failure mechanisms than mechanical systems. Furthermore, the knowledge about these mechanisms determining the reliability is not as mature as it is in mechanics. Since inverters are used in mechanical systems and their reliability must be comparable with that of electromechanical systems, the benchmark is set very high. The period of operation, too, is longer than for other products (e.g. in consumer electronics), and since the majority of electronic components are used in consumer electronics, this sector plays a decisive role in their development.

Drive manufacturers spend a high proportion of their development and production costs on guaranteeing high product reliability. This is essential for the increased use of controlled drives.

Simple usability. Controlled drives can carry out numerous functions. At the same time, the number of applications is very high (as shown in the description of the twelve drive solution in this chapter). The need to design inverters which are as simple as possible to use is ever more important. Simple usability of drives and their improved reliability are the keys to accelerating their use. The manageability of a drive system is evaluated according to the efforts and costs it causes during its life cycle.

The engineering of drives, i.e. the work required during their operation and the tools simplifying this work, as well as the determination and optimisation of the costs across the life cycle are the topics dealt with in the final chapter 5 which will show ways to simplify the use of drive systems and to reduce their costs.

5 Engineering and life cycle costs of drives

In the previous chapters we looked at where drives are used, how a drive system is structured and how drive solutions are developed in production and logistics applications. This next chapter is devoted to engineering and the life cycle of drive systems.

The term engineering covers all the activities involved in selecting and commissioning the right drive solution for an application so that it executes its function when the machine is running. Engineering is therefore a process that is undertaken by people with various roles and tasks and is supported by tools, generally in the form of computer programs. In chapter 5.1 we look at how this process can be made efficient and reliable.

A second important aspect relating to the use of drive systems is their complete life cycle, from selection, through commissioning and use, to disposal. In evaluating a drive system, an all-round consideration of its life cycle costs is becoming increasingly important as a means of identifying efficient and economic solutions. Chapter 5.2 looks in depth at how life cycle costs can be calculated, which components have to be taken into account and how costs can be reduced over the life cycle as a whole.

5.1 Engineering of drive systems

The engineering of drives is a process that can be broken down into a series of phases (Fig. 5-1):

- Selection of the drive system based on the task executed in the machine
- Dimensioning of the drive components based on the performance data for the application
- Configuration of the inverters for defining the function of the software
- Commissioning of the drives in the machine. This process is often subdivided into commissioning by the machine manufacturer and commissioning at the installation site by the machine operator, during the course of which functional tests and acceptance measurements are performed, and

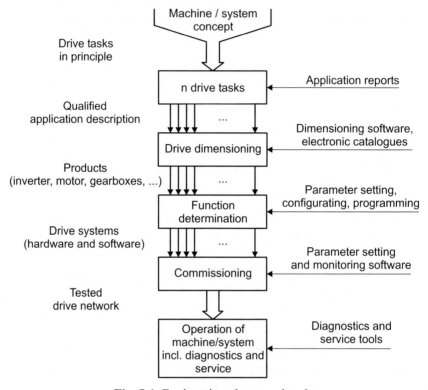

Fig. 5-1. Engineering phases and tools

- Diagnostics and servicing of the system during use by the machine operator.

5.1.1 Selection and dimensioning of drives

Olaf Götz

The first phase of the engineering process for the drive system involves selecting and dimensioning the drive components. It ends with the purchase from the drive manufacturer of those components, which after delivery can be installed in the machine and then commissioned.

During this phase many decisions are taken that are of fundamental importance for the function of the machines. Therefore we will start by looking at the machine manufacturer's development and design process in order then to illustrate how the selection of the drive system corresponds to that process.

Development process in engineering. The numerous models available for the development and design process in engineering are broadly similar in scope and describe the following principal phases (VDI 2221/2222) [Li07]:

1. Defining and clarifying the task
2. Determining functions and their structures
3. Identifying solution principles and their structures
4. Breaking them down into feasible modules
5. Designing the key modules
6. Designing the complete product, and
7. Preparing the design and usage specifications

Taken as a whole, the development process progresses from broad outlines to fine details and from defining requirements to creating specific solutions.

As we saw in chapter 2.1.3, machine manufacturers generally have only low to medium quantities. Annual production figures of 10 to 100 machines are usual. The machines commonly have to be customised to the machine operators' specific requirements, since they naturally also want to secure advantages over their competitors. However, at the same time, a machine manufacturer cannot completely redesign a plant each time. It has to use its designs repeatedly in order to remain efficient and cost-effective and to make optimum use of its capabilities.

The consequence is that in building an actual machine the manufacturers can generally use existing modules and designs for around 80% of the machine, while the remaining 20% is tailored to the specific order. This ratio often applies both to the mechanical construction and to the control programs which largely shape the overall function of the application.

Machine modules that are used repeatedly generally undergo a complete redesign by their manufacturer at intervals of around ten years. In between these times, only minor modifications are made to the modules in accordance with specific application cases. If the function of the machine module is linked very closely to the functionality of the drive system, then the machine manufacturer is naturally also reliant upon the system being delivered from the supplier during this time. This point strongly influences the machine manufacturer's expectations regarding the availability of the drive products.

Overall there are therefore two approaches to the general development and design process by the machine manufacturer.

Execution of a machine order. When an order is placed for a machine, it is manufactured as much as possible from existing modules. The task is defined together with the machine operator according to his use of the plant. The determining factors are the products being produced, their variation, the performance data for the production plant (production rate and availability of the plant) and the resulting costs of the plant. On the basis of the machine operator's requirements, the machine manufacturer has to be able to calculate an accurate costing and quote for the plant. A residual uncertainty will of course remain in the case of any specific requirements.

Determining the functions and the structure involves assembling modules to form a solution. Individual modules often have to be linked together. Materials handling equipment, which itself has to be assembled from modules in accordance with the specific application, plays a key role here.

Designing the key modules and determining the design specifications involves design modifications, which often result in a plant-specific dimensioning of the drives.

When designing an actual machine, the choice of drive system for the machine modules is usually predetermined. The actual number of drives and their function depend on the overall structure of the plant. Substantial variations can be introduced by the use of interconnecting materials handling equipment, by the automation of materials handling processes, and by drives that automate the adjustment of the machine to a product. By contrast, the structure of the drives for the main production processes generally does not change.

The main attention of the machine manufacturer when executing an order for a machine is focused on being able to calculate the implementation costs. There is generally no time or budget available in this phase for experimenting with new functional principles. At the same time, a machine manufacturer may have to bring in innovations, and as we explained earlier, this is achieved through the fundamental revision of machine modules and new machine concepts.

Development of a machine module or new machine concept. Unlike a specific order, where the emphasis is on meeting the requirements specified for the plant, new machine concepts and machine modules are developed with the aim of being able to use them for diverse applications. Here the machine manufacturer has to be able to recognise similarities between different application cases and to translate them into machine structures. As a rule, these similarities centre around the process and the physical mechanisms of action. The core expertise of most machine manufacturers generally relates to these processes. The manufacturers are able to translate their understanding of the process into machines which then mechanise and automate its execution.

In fundamental revision and new development, the early phases of the design process are vital. Defining and clarifying the task now encompasses the full extent of the application to be covered. The manufacturer has to identify and define not only the commonalities between applications but also their variations.

When it comes to determining functions and their structure and identifying solution principles, there is much more room for freedom and creativity than there is when executing a specific order. Engineers have devised methods of developing and evaluating alternative solutions for tasks. One example is the morphological box, which helps with structuring partial solutions and combining them to form complete solutions.

In this phase the choice of the drive system plays a critical role. Drives are central to many machine functions, since many functional principles are based on motion. The classification of the twelve drive solutions in chapter 4 showed how machine functions are implemented by drives. In particular, manual setting and adjustment operations can be replaced by positioning drives. Also possible is the release of mechanical couplings in synchronised single drives with synchronous or cam functions. This provides a good deal of freedom when it comes to determining structure and solution principles, which the machine manufacturer can use to its advantage. Handling a wide variety of products on a machine, with minimal downtime allowed for switching between them, is normally only possible with a large number of controlled drives that are operated by the machine control.

Tasks involved in selecting drive components. The process of selecting the components to be used for constructing a drive system for an application involves the following steps:

- Definition of the drive system
- Basic choice of drives
- Selection and dimensioning of the drive components
- Application, drive and process simulation, and
- Determination of design data (CAD data for the engineering design, electrical connection plan).

Definition of the drive system. In defining the drive system for a machine module, it can be helpful to follow the steps outlined below:

- Defining the *functions* that are to be performed by *individual drives*. Here it is decided how many drives are used in the machine module.
- Describing the *task of the individual drives* in the module. It is also decided how the tasks of multiple drives interrelate, since this has an im-

pact on the function of the drive and on the automation concept. From the description of the task it is usually possible to establish to which of the twelve classes of drive solutions outlined in chapter 4 this drive belongs. The description of the drive solution can then be used as a guide for defining the drive system.

- Based on the task of the drive, its *basic performance data* is then described. Aspects such as dynamic performance, accuracy, setting ranges for speed and torque and the basic motion sequence have to be considered. In addition, the functions of the drive in operating situations outside of automated production (set-up, fault) have to be established. These points then influence the basic *choice of drive components*, as covered in chapter 3.7.1. It is generally decided at this stage which motor type and gearbox is used and whether a frequency inverter or servo drive is necessary.

Since the machine manufacturer generally has an expert knowledge of the machine processes but not of the options offered by drive technology for their implementation, close cooperation is necessary at this stage between the machine manufacturer and the drive supplier. Detailed information about system requirements is needed in order to be able to select a drive that is appropriate for the application.

Defining the drive system is also a mechatronic task. Controlled electric drives can execute very precise motion sequences defined by a software program. The implementation of machine functions thus becomes a combined effort arising from the interaction between mechanical systems, electronic systems and software. To this end the corresponding disciplines (mechanical engineering, electrical engineering, computer science) have to work together. At an operational level this is generally achieved by establishing a team drawn from mechanical engineers, electrical engineers and software developers, working in close cooperation with the drive supplier, who will usually also have an electrical engineering background.

Choosing the drive system for a machine module normally involves making some important decisions that have a critical impact on its function.

Basic choice of drives. In analysing the task and developing the mechatronic drive concept, there is a fundamental need for information. What is the performance expected of the drive components? How much space is available for the components? Roughly how much will the drive cost?

This need for information can be satisfied in the first instance by roughly calculating (e.g. with a pencil and paper) the physics of the application. Product catalogues (in paper or electronic form) can be used to select potential products and to find out costs and sizes. Technical documentation helps to establish specific product features.

The outcome of the basic planning stage is thus an outline solution which in the following stages can be filled out, tested for feasibility and then finally also recorded in the form of a specification.

Selection and dimensioning of the drive components. The drive components have to be selected and accurately dimensioned. The objective is to check whether the chosen components can fulfil the physical demands of the application (power, torque, speed) and the required machine task, without being overdimensioned. This last point is aimed at the desired motion function and qualitative requirements such as positioning accuracy and quality of drive control. The steps involved in the drive dimensioning process are covered in chapter 3.7.2.

The drive dimensioning process often involves a number of loops in order to arrive at the optimal solution. Since there are usually several possible solutions for an application, it is advisable to establish alternatives comprising different solution concepts, drive technologies and products. The various solutions can then be compared with one another in terms of technical advantages and disadvantages and costs, in order to establish the best solution from both a technical and an economic perspective.

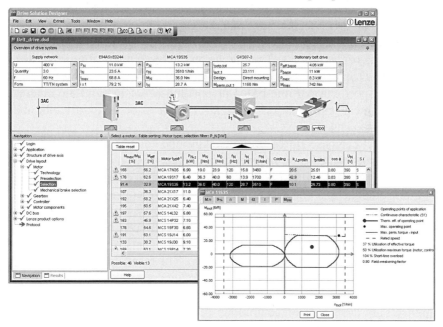

Fig. 5-2. Drive dimensioning software

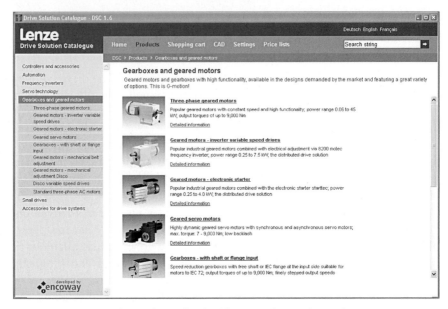

Fig. 5-3. Product selection from an electronic catalogue

A number of drive technology manufacturers have developed software tools for drive dimensioning (Fig. 5-2) in order to speed up and improve the accuracy of the process. They focus on the basic applications that were outlined in chapter 4. The mechanical variables (masses, friction) and the individual speed curve (time-based or position-based motion) are entered. The mechanical and electrical drive structure can be tailored to the machine requirements.

A product database stored in the background enables the physical drive dimensioning results to be mapped against the product range in order for the appropriate products to be selected. The software tool includes expert help on applications and electromechanical drive components, which the user can consult interactively [Ar+01].

Using these tools, high-quality drive solutions can be generated very quickly which both meet the physical requirements and are achievable from a manufacturing and sales perspective. The results are presented in an informative technical report and are easy to understand.

It is not always necessary to dimension all the components. In simple applications designed for continuous operation, for example, it is often sufficient to select the products from a catalogue on the basis of key process data.

An electronic product catalogue represents a useful advance over paper catalogues in engineering (Fig. 5-3). The entire product range is available for selection and for product configuration. The search results in the shopping cart can be sent to the supplier in the form of a request.

Both software tools, the drive dimensioning program and the electronic catalogue, are based on a product configurator which allows users to select from the full range of products with all its variants. The product configurator uses the data from the product kit, from which the individual drive systems can then be put together by combining the various blocks. As we saw in chapter 3, a mechatronic product construction kit contains a large number of alternatives, creating a huge range of possible variants. Table 5-1 provides an overview of the range of possible variants [Wi04, LoKi05, Ki06].

Table 5-1. Range of possible variants of a mechatronic drive kit

Block	Variants
Gearbox series	6
Gearbox sizes	5-7
Gearbox ratios	30
Driven shafts	1-4
Gearbox options	2
Motor types	2-3
Associated motor frame sizes	2-3
Brake options	2-3
Encoder systems	5
Inverter series	3-6
Associated inverter power	1
Filter options	3
Safety modules	2-3
Communication modules	4-7
Basic software configurations	1-10

In total, a kit of this type offers 10^8 to 10^9 variants. Several thousand different modules are used to implement these variants.

Application, drive and process simulation. Building on the results of the drive dimensioning process, in highly dynamic motion sequences it is necessary to examine the control aspects of the drive system. Simulation programs that can forecast control quality, vibration response and precision (speed, position) are used for this purpose.

However, complex applications such as non-linear mechanical systems can also be simulated in advance in order to obtain input data (torque and speed characteristics) for drive dimensioning.

Simulating process sequences is a useful way of checking or visualising sequences in machines with many subfunctions or in situations where a number of machines are working together. Inefficient motion sequences or collisions within the process, for example, can be detected in this way.

Simulation tools have the advantage that the design model and the control structure can be created and adjusted flexibly and in accordance with individual requirements. Complex technical interrelations can be assessed and quickly optimised before a prototype machine is built. The risk of the machine not being suitable for the specified task is minimised.

Creating the simulation model takes expert knowledge of the machine, the drive components and their product features in order that simulations are not restricted to a limited user circle. Therefore they are only used in tasks which cannot be analysed with adequate precision using a conventional design.

Dimensional drawings, CAD data and electrical connection plan. In the planning stage it is necessary to consider the geometry of the drive components in relation to the design of the machine. To this end dimensional drawings must be provided which can be integrated into CAD programs.

The wide variety of geared motor models available has an immense impact on the engineering and on the resulting dimensions. An additional benefit of a product configurator is therefore that CAD data can be generated in various formats and views, corresponding to the selected products. This functionality is available in electronic catalogues.

The electrical connection plan also has to be considered at this stage in order to be able to plan the control cabinet.

Mechanical and electrical counter-checks can cause the product selection to be modified.

Tools for the product selection and procurement tasks. Table 5-2 provides an overview of the tasks involved in selecting and dimensioning the drive components, and of the tools that can be used for these tasks.

Table 5-2. Comparison of product selection tools

Tasks \ Tool	Paper catalogue	Electronic catalogue	Technical documentation	Application database	Dimensioning software	Software for cam profile motions	Product configuration	Simulation software	CAD-data software	ERP-system (e.g., SAP)
Task analysis	●	●	○	○	○				○	
Drive concept			○	○	○				○	
Product selection	●	●	○	○	●		●			○
Drive dimensioning			○	○	●	○	○	○		
Product configuration		●			○		●			
Simulation								●	○	
Scale drawings, CAD, connection diagramme	○	●	○				○		●	
Offer		●								●
Order	●	●								○

(Rows grouped under: Mechatronic system consideration)

● required and/or useful
○ sometimes required and/or useful

5.1.2 Configuration

Sebastian Lülsdorf, Mathias Stöwer

Once the hardware components of a drive have been selected, the function of the inverter software has to be defined for the task in hand. As we saw in the individual sections of chapter 4, inverters today can take on substantial motion control and drive control tasks in order to solve the corresponding drive task. Since they receive extensive information about the process via their sensors, they can also perform diagnostic and monitoring functions.

Depending on the complexity of the required application, there are various ways of defining the operating mode of the software. First of all, the drive manufacturer supports this process by offering predefined solutions and configurations for standard applications. In many cases, however, it will be necessary to make adjustments and changes to the function of the software on the basis of this predefined solution in order to adapt it to the task defined by the mechanical engineer. There are three basic processes involved in this stage:

- Parameter setting,
- Configuration, and
- Programming.

Parameter setting. Parameters are variables which can be changed by the user in order to set the function of the software. Each parameter includes the following data:

- The actual value
- A unique parameter number, often in the form index.subindex
- The data type and hence the resolution of the value
- A descriptive name
- Limits for the minimum and maximum value
- Information relating to how the value is displayed (unit, decimal places), and
- The parameter type (when can it be modified, is it a display value?)

Typical parameters for an inverter are as follows:

- The motor data
- The maximum speed
- Acceleration and deceleration times
- Choice of operating mode for the drive control, and
- Parameters for setting the individual controllers (P gain, integrator time constant).

Parameter setting is always sufficient as long as no fundamental changes have to be made to the signal flow structure in the inverter. In some cases the signal flow too can be selected from a set of predefined configurations by means of a parameter. The advantage of parameter setting lies in its simplicity. Its limits are reached when the number of parameters becomes too large. This can be overcome to some extent by organising the parameters into function-oriented groups (e.g. for motor data, for drive control, for the setpoint channel) so that the parameters can be accessed quickly.

The parameters can be set by an operating unit which is either built into the inverter or connected to it. Simple operating units merely comprise a numerical display with a few special characters. The parameter is selected by its number and can then be altered by means of keys (normally one key to reduce and one key to increase the value). This type of operating unit is inexpensive and ideal for simple devices with a limited functional range.

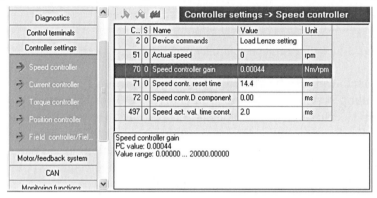

Fig. 5-4. Setting speed controller parameters via a table

More user-friendly operating units include a text display. With this type of operating unit parameter groups can normally be selected too. Once a text display is included, however, the question of language arises. Inverters are generally used worldwide. The skill level of the operating personnel in the factories has to be taken into consideration. It is therefore important to offer operating concepts that can be used intuitively anywhere in the world.

PC programs can provide an easier way of setting parameters. The individual parameters are then either listed in tables (Fig. 5-4) or in graphical parameter dialog boxes (Fig. 5-5). Interfaces that are tailored to the drive function in the application and to the pictographic language of the user mean that the process of defining the software function can be made very intuitive and efficient. Compromises always have to be found between offering a wide range of functions that can cover many different applications and ensuring that the system is easy and reliable to use.

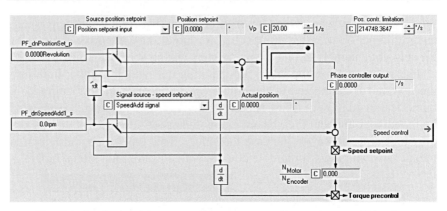

Fig. 5-5. Setting position controller parameters in a parameter dialog box

Parameters can also be altered using communication systems. Most communication systems, as described in chapter 2.3.5, include services which can be used to alter the parameters. This allows the machine control to adjust the function to the current situation, such as the amount of material produced. Another application is the reading and writing of all parameters of an inverter to allow its application data to be stored centrally in the control system and reloaded if the device is replaced.

In principle it is possible to set all the drive solutions described in chapter 4 by means of parameter settings. The inverters must offer predefined configurations for the individual solutions, the parameters of which can then be set. Only if the inverter is required to execute tasks going beyond the basic functionality of the solution, as with additional process controllers for example, a solution for altering the signal flow has to be provided. Methods other than parameter setting have to be used in such a case.

Configuration with a function block editor. If there is a need to change the predefined signal flow of the setpoint or actual value path for motion control and drive control, a graphical programming language based on function blocks can be used. Function blocks are basically software blocks with the following elements:

- Signal inputs
- Signal outputs
- Parameters that influence the function of the block, and
- Internal statuses.

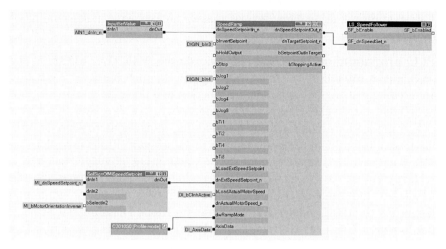

Fig. 5-6. Example of a function block interconnection

Function blocks encapsulate complex functions. By connecting signal outputs to signal inputs, and thereby interconnecting the blocks, the entire signal flow can be defined (Fig. 5-6). The project planner can choose from a set of blocks in order to expand the predefined solution. A typical set of function blocks includes blocks for process controllers, brake controllers, various ramp function generators and selectors for the signal flow.

For the function block configuration, a central question that has to be answered is that of the functional scope of the individual blocks. If it is very small and already includes logical and arithmetic basic functions, then signal flow structures can be made very flexible. However, they then become very large and bulky. The function definition process becomes very time-consuming and efficiency is reduced. This type of function definition is only really suitable for experts.

If the function blocks tend more towards larger function units, such as generating a positioning sequence, the function of an electronic gearbox which in addition to the gearbox factor also includes functions for clutching and declutching, then the entire function of a drive can usually be constructed from fewer than 20 blocks. The flexibility is then reduced, however, because most of the functionality is contained within the blocks rather than being obtained by the interconnection of many blocks.

In order to achieve high efficiency for the function definition process, a drive manufacturer has to define function blocks in such a way that they already include much of the functionality required by the various drive solutions. In this way the need to expand them is significantly reduced.

This requires the drive manufacturer to have an extensive understanding of the use of drives.

PLC programming languages in accordance with IEC 61131. If the possibilities offered by parameter setting and function block configuration are no longer sufficient to define the function of a drive for a task, the third option that can be chosen is programming.

For applications in automation engineering, it makes more sense to offer the programming languages that are used to program PLC controllers rather than one of the standard computer programming languages (such as C or Java). As we saw in chapter 2.3.1, in addition to manufacturer-specific languages, the standard IEC 61131-3 has defined graphical programming languages such as Ladder Diagram, Function Block Diagram and Sequential Function Chart, as well as textual languages such as Instruction List and Structured Text.

If a drive has the ability to execute control programs, then in addition to motion control and drive control it can also execute other control functions. For servo drives in particular, which carry out distributed machine functions via their motion sequences, this can be extremely useful. The use of a small PLC controller can often be avoided in this way [KiLo98].

Functions that modify and optimise the drive function by means of parameter changes and functions for analysing actual values that are used for process diagnosis can also be implemented in this way.

If additional functions are to be carried out by an inverter, it is always important to consider what the reasonable limits are. All functions that are closely related to the motion generated by the drive can reasonably be performed by an inverter. All functions that relate to the overall function of the machine and plant, on the other hand, should be implemented in a central machine control system. Central data storage and machine operation should also be implemented centrally.

Network configuration. In order to connect various automation components in a machine or plant, it makes sense to use a communication system, as described in chapter 2.3.5. For the function definition the devices have to be configured as network nodes in an appropriate manner. The detailed settings depend on the network type. As a general rule the project planner must plan the following elements and ensure that each device is set up correctly:

- If the baud rate is variable it must be defined, and a unique address assigned for each node

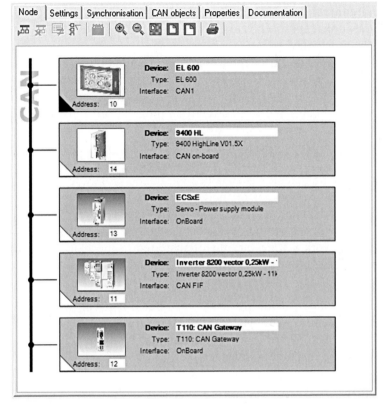

Fig. 5-7. Network configurator

- The user data from the applications to be transferred cyclically must be configured as process data
- Node synchronisation must be set up via the network if necessary, and
- The initialisation of the network must be ensured when the machine is switched on.

If devices from various manufacturers are to be connected via a network, the project planner may have to use different, manufacturer-specific tools for the individual devices. Planning the network then requires pencil and paper and is difficult to describe in simple terms.

It is much easier if the network configuration of all devices can be done using a network configuration tool (Fig. 5-7). The network configurator then plays an important part in configuring the overall automation of machines.

Central control concept. For machines with a central controller (PLC), the conventional approach is to use the programming tool for the controller to configure the network.

Field devices can be integrated into the programming tool by reading the device description files. These are available in various network-specific formats. For CANopen, for example, the EDS format is defined by the DS301 standard. Work is currently in progress on converting the EDS format to XML format, which is based on ISO 15745. This ought to provide more detail about the device functionality. Device description files are normally provided by the device manufacturer.

Each time the machine is switched on, the control system initialises all connected field devices via the network. The only requirement here is that the address and baud rate of the field devices are set correctly. This can be done using switches on the device or via the operating unit, for example. All other network settings are then transferred to the individual field devices on initialisation. In CANopen this is done by means of service data access.

Distributed control concept. In the distributed concept the overall function of the machine is distributed between several devices, which may also include inverters. There is no natural, central entity to initialise the network when the machine is switched on. In this scenario the commissioner transfers the network settings to each individual device and stores them there so that they are safe against mains failure. When the machine is switched on, each device uses the settings that it has stored.

In this case too, a central engineering software tool is ideally used to configure the network.

Integrated engineering of various components. When selecting the optimal components for an automation solution, technical, economic and strategic considerations often result that components from various manufacturers are used [ArKo01].

Each component manufacturer usually also provides commissioning software for its components. However, it is in the machine manufacturer's interests to use as few software products as possible, since each program is associated with licence costs and integration and administration time. Above all, however, it can save a lot of engineering time, if a machine consisting of multiple components can be configured using an integrated tool. This has particular advantages over individual tools in terms of, for example:

- Setting up a distributed application with network communication
- Comprehensive project data management, versioning and storage, and

- Documenting the complete project for the end customer.

Component integration methods. A software tool for engineering complete systems must offer standardised, open interfaces for integrating components supplied by various manufacturers. A number of different device integration methods are described below.

Device description files. The engineering software configures the device on the basis of a device description that it reads from a file. There are a number of different standardised formats for describing devices. In semantic terms they differ in the extent of the device features that can be described and the elements that can be configured. A device description file does not necessarily describe the complete functionality of a device, only those aspects that can be configured using the central tool [DiSiRi99].

The EDS format (Electronic Data Sheet) is used for example for integrating devices into network configuration tools in accordance with CANopen DS301. An EDS file allows configuration of the process data to be transferred and parameterisation by the engineering software using the service data. It describes:

- Variables from the application
- Process data objects provided by the interface
- The standard configuration for the process data configuration, and
- Other device parameters.

FDT-DTM. With this method a frame application can call individual manufacturer-specific parameterisation tools. The individual tools appear with their user interfaces in the frame of the complete tool and work on a database supplied by the complete tool. The abbreviations FDT and DTM have the following meanings:

- FDT = Field Device Tool, i.e. the complete tool.
- DTM = Device Type Manager, i.e. the device-specific individual tool.

Tool Calling Interface (TCI). TCI allows PROFIBUS and PROFINET device parameters to be set without leaving the engineering tool of the automation system. It is a simple, standardised calling interface for calling parameterisation tools from other device manufacturers from the central engineering system. Unlike FDT-DTM, TCI does not require centralised data storage.

5.1.3 Commissioning

The software function definition steps described so far generally take place before commissioning of the machine and its automation. Once this preparatory work has been completed, the function can be tested. Commissioning can take place in two different situations:

- Initial commissioning involves the testing of a new machine function
- Standard set-up involves the commissioning of functions that do not differ substantially from previously tested configurations

Procedure for initial commissioning. Initial commissioning takes place on the machine. The design engineer has prepared the application and, where possible, tested it on a model.

Commissioning begins by installing the drive solution onto the various inverters. This is done using a commissioning program, which is usually installed on a laptop.

The commissioner first commissions the inverters individually, taking care to disconnect the motor from the mechanical components of the machine to prevent any damage in the event of a malfunction. Initially the setpoints for the motion sequence are not supplied by the PLC but are defined manually. If there are limit switches in place for a limited traversing range, they are checked to ensure that they are working correctly.

This is followed by connection to the fieldbus and hence to the control system. A program that logs network traffic is a useful aid for error analysis here.

If the network connection is working and if the first functional test in conjunction with the control system is successful, then the machine is tested in the network. To this end the motors are connected to the mechanical components. Since the load has now changed significantly, the speed controller of the inverter must be configured to the machine. The commissioner does this by setting setpoint steps (within the permissible limits) and observing the transient response of the drive. The oscilloscope function offered by many inverters is a useful tool here. The transient response can be optimised by changing the controller parameters.

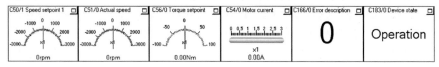

Fig. 5-8. Monitor windows in the commissioning program for continuous monitoring of the main status values of the inverter.

During the whole commissioning process, key variables such as motor current, torque and setpoints are continuously observed via monitor windows in the commissioning program (Fig. 5-8).

Finally the machine is run with material and the accuracy of the production process is checked. At this stage too the inverter parameters need to be fine-tuned until the desired result is achieved.

Standard set-up. The procedure for a standard set-up is usually different from that for initial commissioning. The aim here is to transfer the application together with standard settings to the drives as efficiently as possible. Since this may be done by less experienced personnel, it is important to use the simplest possible PC program. Drive manufacturers can supply special standard set-up tools for this purpose.

Another possibility is that all inverter parameters are stored in the machine's central controller and transferred to the drives via a network.

5.1.4 Diagnostics and maintenance

Drive replacement. If an inverter fails during the production process, every minute of downtime can lead to high financial losses. Therefore service and maintenance personnel must be able to replace a faulty drive with a new one as quickly as possible. The machine must then resume production from where it left off.

First of all the inverter model must be able to be identified quickly. This is normally done from the nameplate, which must be easily accessible. It must include all the information needed to obtain a replacement from the manufacturer, including product ID, model code and version.

When replacing a drive, the mounting and connection method for the inverter is important. There should be as few electrical connections as possible. Connections with non-interchangeable plugs that require no special tool save a lot of time here.

There are a number of different methods that can be used, either separately or in combination, to transfer the configuration data needed for operation to the new device quickly and easily.

Using a memory module. Drives that contain a memory module are particularly easy to replace. The operator simply removes the memory module from the faulty drive and inserts it into the new drive to transfer the complete data set with all parameter settings. In the case of simple inverters with no special memory module, plug-in operating units which store the parameters can be used.

Initialising the drive from the master control. Machines with a central controller normally test and initialise all connected drives after a restart. To do this they must be able to communicate with the drives via the network. The extent of this initialisation is typically limited to setting parameters. If the drives require further data, it can be uploaded by the controller. For machines with many field devices, an appropriate memory location must be allocated in the controller for storing all the device data. PC-based controllers can be advantageous in this regard.

Transferring data with an initialisation program. If the machine has no central controller and the drive does not have a removable memory module, the user must transfer the data to the new device from a PC. The service technician should be able to do this with a simple initialisation program that is easy and intuitive to use.

If this method is to be used, the data must be carefully archived at the end of the commissioning process so that it is immediately accessible if needed. It is standard practice to store it on a CD in the control cabinet along with the documentation.

If a replacement drive is supplied after an extended period, it may contain a more recent software version. Although new software versions have to be backward-compatible, there may be limits to this compatibility if several versions have been skipped. It is important for the initialisation program and the drive to be able to detect an incompatibility of this type and to display a warning message. The user then loads the correct software version for his application onto the replacement device.

Remote maintenance. With remote maintenance, error correction measures can be started more quickly in order to reduce downtimes and thus save costs. In addition, productivity can be increased by monitoring operation during the production process without an on-site presence by means of improved process control.

Remote maintenance requires a data connection to be established between a service centre operated by the machine or plant manufacturer and the machine or plant. This has traditionally been done via the telephone network, but nowadays the internet is more commonly used, which is generally available wherever the machines are used.

A second requirement is a data connection within the machine between the connection point to the telephone network / internet and the individual automation devices, including the inverters.

The existing communication system for the controller can be used for this purpose. Additional functions many need to be installed in order to pass on the service data requests for remote maintenance (Fig. 5-9).

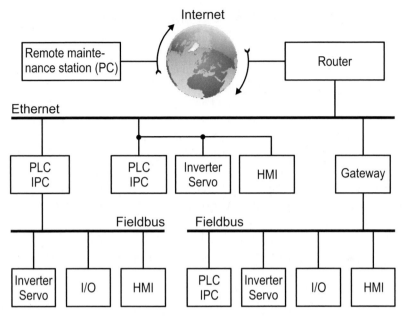

Fig. 5-9. Remote maintenance configurations

Additional devices may also be installed to connect the telephone modem or internet connection to the communication system in order to pass on these data services.

Finally, another possibility is to install a dedicated communication network for remote maintenance purposes. However, this will only be necessary if no communication system is used for automation or if it does not have sufficient services or capacity to transfer data for remote maintenance purposes.

On machines and plants that include a central PC for operating and diagnostic purposes, remote maintenance is often combined with the PC's functions and communication systems. On the whole, the information that is needed centrally and the information that is needed for remote maintenance purposes are largely identical. Specialist remote maintenance personnel are generally able to provide more accurate analyses than operating personnel on site.

5.2 Life cycle costs in drive technology

Volker Bockskopf

Drive systems are one of the most important functional elements in machines and plants, having a critical impact on their productivity and reliability. They generate costs not only at the procurement phase but throughout their entire life cycle. Nowadays purchase decisions are influenced not only by the functionality of the drive but very considerably by its procurement costs. However, these factors are no longer sufficient for choosing the best possible system. For instance, the use of distributed concepts cannot be economically justified on the grounds of procurement costs alone. The economic advantage of these concepts comes from the savings that are made during installation and commissioning. Buying a frequency inverter for a pump is expensive, but the cost is offset by the energy saving. These two examples show that a comprehensive consideration of costs over the entire life cycle is needed in order to be able to evaluate drive systems and concepts.

In this chapter we look at how life cycle costs (LCC) and the total cost of ownership (TCO) are calculated, which factors are involved, and how systems can then be compared with one another in order to find optimal solutions.

Manufacturers of large machines and plants already use these figures in their sales pitches. In return they expect similar information about the drive systems they use. The calculation of life cycle costs and total cost of ownership, and their use as a sales tool, is therefore growing in importance.

Life cycle costs are also important from another point of view. Drives are one of the largest consumers of electrical energy. Industry in Germany consumes approximately half of all the electricity generated, and around two-thirds of that is converted by electric drives. The rest of the electricity consumption is used for heat, lighting and information processing.

Energy-saving drive systems can help to reduce this energy consumption, and this is becoming increasingly urgent in the light of the debate on climate change and the drive to reduce CO_2 emissions. In addition to motors with improved efficiency, such systems include speed control through the use of frequency inverters, allowing the optimal adjustment of the operating point of a drive to the process.

Here too, however, the use of energy-saving drives increases procurement costs and only proves more economical once the entire life cycle costs are calculated. Thus life cycle costs are also the key to reducing energy consumption and hence to reducing CO_2 emissions.

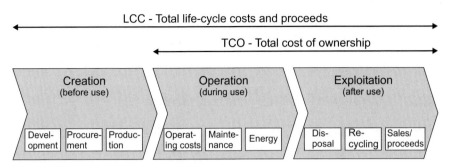

Fig. 5-10. Differences between TCO and LCC

5.2.1   Definition of LCC and TCO

Two abbreviations are used in the calculation of costs over the entire life cycle:

- LCC (Life Cycle Costing) and
- TCO (Total Cost of Ownership).

LCC is a method of calculating and optimising the entire life cycle costs of a product or plant. LCC comprises the three phases:

- *Before* use (due to development, production and procurement)
- *During* use (operation with operating costs, servicing and energy usage), and
- *After* use (recovery with disposal, recycling and sales revenue from the old plant).

LCC thus includes the manufacturing costs of a machine or plant and therefore provides a comprehensive picture "from cradle to grave".

The definition of TCO is largely the same as that of LCC, except that it only considers the costs on the operator side. It does not take into account the costs prior to use in the development phase, e.g. production and design costs.

Fig. 5-10 shows the scope of LCC and TCO with reference to current German and European regulations.

Machine operators are understandably more interested in TCO as a means of optimising their costs over the usage phase. The subject of M-TCO (maintenance TCO) is currently a source of debate in the automotive industry in particular, although this is limited purely to maintenance costs and plant availability.

Before we look at ways of reducing life cycle costs through the drive system, we need to consider the different models that are used to calculate these costs.

5.2.2   Overview of LCC forecasting models

DIN EN 60300-3-3:2005. DIN EN 60300-3-3 (2005) is a standard for reliability management. The guidelines highlight the costs associated with reliability. Life cycle costs are defined as the cumulative costs of a product from design to disposal. According to these guidelines the aim of LCC is to provide data on which to base a decision for or against a product.

The guidelines identify six main phases of LCC:

1. Conception and definition
2. Design and development
3. Production
4. Installation
5. Operation and servicing, and
6. Disposal.

The standard describes the use of cost breakdown structures to identify costs. The explanation of relevant costing methods provides a good overview of the various methods of cost determination so that a method appropriate to the individual application can be selected.

It describes the influence of data integrity on statements regarding availability, use of relevant information, optimism/pessimism and forecasts (e.g. exchange rate fluctuations). Sensitivity analysis is recommended as a means of reducing risk.

Extensive appendices provide detailed examples of individual topics (e.g. for typical cost-generating activities or a parameterised calculation of costs).

Overall these guidelines represent the most comprehensive description of the elements involved in life cycle costing. Having studied them, the user will be able both to define the subject of LCC and to understand the typical elements of goals, phases and more detailed aspects of conceptual design and modelling. He will also be able to set up and evaluate various models and costing structures.

DIN EN 60300-3-3 can therefore be regarded as the basic standard, an essential source of information on all aspects of LCC.

VDI 2884:2005. VDI Guideline 2884, issued December 2005 "Purchase, operating and maintenance of production equipment using Life Cycle

Costing (LCC)", describes the basic elements and gives practical tips. It considers the following two elements:

- Firstly it provides a methodological framework to support the owner in selecting production equipment. It also supports the manufacturer in choosing innovative configurations. In particular the guideline helps to highlight the influence of recurring costs and their consequences.
- Secondly it proposes a method of calculating life cycle costs.

It defines LCC as the "total" costs and revenues that a system generates throughout its service life from the perspective of the machine operator.

It also includes decision-making tools and indices for calculating LCC (e.g. high follow-on costs, long service life, anticipated or non-obvious follow-on costs in operation).

Consideration is also given to establishing a maintenance strategy in conjunction with the customer or subsequent operator of the plant, and to the need to define the service life and operating conditions.

Three tables corresponding to the three typical LCC phases are listing the usual costs to be taken into consideration in the phases before, during and after use. Gathering the necessary information for forecast costs and revenues is regarded as one of the most complex stages, and the three tables offer a pragmatic approach to this process.

The final section considers the subject of decision-making. Risks such as unreliable data and the prognostic nature of the LCC method are highlighted, which can be reduced substantially using VDI 2884. The guideline concludes by mentioning the advisability of carrying out a sensitivity analysis, which is also covered in DIN EN 60300.

To summarise, VDI 2884 is a good source of information about the basic elements of LCC. It offers plenty of food for thought (decision-making, sensitivity analysis) but fails to provide a specific example of a forecasting model. Instead it offers a lot of suggestions (e.g. unreliable data) to support the owner and manufacturer, without giving any specific guidance how to proceed.

VDMA 34160:2006. The forecasting model for the life cycle costs of machines and plants provided by VDMA 34160:2006-06 describes a standard, engineering-specific calculating tool for forecasting LCC, which addresses the interests of both machine manufacturers and machine owner.

Life cycle costs are defined in the standard sheet as the sum of all costs necessary for the proper use of a machine or plant, from procurement to disposal (Fig. 5-11).

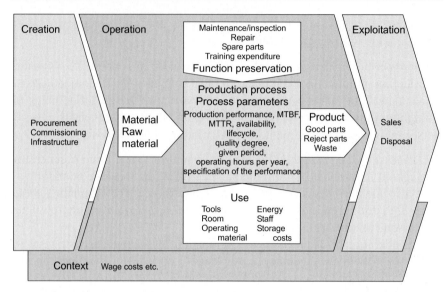

Fig. 5-11. Cost components of LCC in accordance with VDMA 34160

The forecast can be used by the manufacturer as a marketing tool and as the basis for quotations with a comprehensive estimation of costs and by the operator as the basis for a standardised, comparable request for proposals for capital investments.

Described over 14 pages, the forecasting model identifies three phases: development, operation and exploitation. The individual relevant types of costs for each phase are established and then the life cycle costs are calculated as the sum of the costs for these three phases. It is the first model to take into account any revenues obtained in the exploitation phase.

The model is structured in such a way that the types of costs can be systematically broken down into their components.

The scope of consideration is clearly from the perspective of the operator, so the model is aimed primarily at the user. Cost elements such as development costs, design costs and production costs, which would have to be included in a comprehensive consideration of LCC, are not included. The period under consideration is explicitly defined as being the section of the machine's life that begins with its procurement and ends after its service life. The model introduces an element of uncertainty here, since costs before or after the period under consideration are only to be included if they "have a cost influence on the service life". No details are given on how that influence might be identified or estimated.

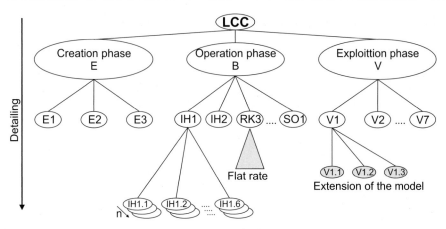

Fig. 5-12. VDMA standard sheet 34160

For each phase the model defines the types of costs and gives instructions for their calculation. In the development phase the focus is on procurement, commissioning and provision of the necessary infrastructure.

In the operation phase the model identifies four aspects: "Material", "Product", "Use" and "Maintenance of function". The underlying data is provided by customer-defined conditions such as average repair time (*MTTR*), availability and period under consideration.

In the exploitation phase the model considers both costs arising from disposal (i.e. recycling, destruction or dumping) and possible revenues obtained from the sale of the used machine on the basis of its residual value.

The types of costs are broken down and the standard sheet gives details of the relevant costs involved. The individual cost items are gradually added together to give the overall LCC.

The model is structured in such a way that it can be broken down into its component parts if necessary and the level of detail increased or reduced as required (Fig. 5-12).

The description also explains the *boundary conditions* (e.g. load spectrum types) and the necessary *basic data* (e.g. service life, operating hours per year). Detailed tables set out the individual cost elements for the three phases. These are defined by the operator for the operation of the machine. The model also includes illustrative examples and forecasting forms.

5.2.3   Using drives to optimise life cycle costs

The drive technology used in machines and plants has a big influence on life cycle costs. As such we need to look at the three phases of the life cycle in more detail. We will also look at how concepts and procedures can optimise costs.

Development phase. During the development of a machine or plant, drives contribute to the costs in the following ways:

- Through the selection, dimensioning and configuration of the drive system (part of *development costs*),
- Through the procurement costs for the drive systems (part of *material costs*) and
- Through the installation and assembly of the drive components and their commissioning (part of *manufacturing costs*).

The *development costs* of a manufacturer of production line machines average out at around 5% of its turnover, sometimes even higher if customised modifications have to be made. Activities associated with the drive systems in the mechanical engineering stage (selection and dimensioning of products, entry in design drawings and bills of material) and in the electrical engineering stage (selection of electronic products, planning of control cabinets and installation, entry in control programs) account for around 10 to 20% of the overall development costs.

The following concepts and procedures can help to reduce these costs:

- Reliable advice on the choice of drive concept
- Quick and reliable product choice
- Ready-to-use solutions that reduce the need for an individual function definition for each application [Ha06] and
- Powerful software tools for defining the function of the inverter software.

The *procurement or material costs* for the drive systems account for about 5% of the sales revenue for a machine or plant (this is an average figure, which can vary substantially in individual cases). Drives almost always number among the five most important categories of materials that machine manufacturers have to buy. Therefore the procurement costs of a drive system have a major economic significance.

The costs of a drive system can be influenced significantly by the right choice of technology and components. In choosing a technology (e.g. standard three-phase AC motor or servo motor, geared motor or direct drive, frequency inverter or servo drive) it is important to base the choice on the requirements of the application and not to setup restrictions. This does of course assume that the drive manufacturer has a correspondingly wide range of products.

Accurate dimensioning of an application that satisfies the physical requirements and does not introduce any overdimensioning can also help to reduce costs.

In addition to the procurement costs for the drive components, the costs for the control cabinets, cables and other power supply components must also be taken into consideration.

The *manufacturing costs* of a machine manufacturer are heavily dependent on its level of vertical integration. If we look at the entire manufacturing sequence for a machine or plant, including subcontractors' manufacturing costs, then manufacturing costs for the drives derive from installation of the electrical components, assembly of the mechanical components and commissioning.

Since modern machines and plants need a large number of drives, their share in the overall costs for electrical installation is around 30%. In order to reduce these costs, the electrical components need to be easy to install. One job which is often extremely expensive and can be simplified considerably by the use of easy-to-install concepts is the connection of the shielding of electrical cables, in particular that of the motor cable.

Assembly of the mechanical drive components, on the other hand, is not a major factor because they make up only a small proportion (< 1%) of the total number of mechanical components in a machine.

It is generally not possible to differentiate between the commissioning costs arising from the drives and those arising from other elements. Since drives have a very substantial influence on the function of a machine, much of the work involved in the commissioning process relates directly to the drives.

There is another aspect that has to be taken into account with regard to commissioning. Payment for a machine or plant is normally linked to acceptance procedures, in other words the function and agreed performance data have to be proved. However, the majority of the costs of manufacturing the plant have already been incurred by the start of the commissioning period. Therefore the commissioning time ties up capital and capital expenditure for the machine manufacturer. It has a fundamental interest in reducing the time for commissioning. On the drive technology side, all concepts that reduce the time and cost of function definition and that generally ensure reliable operation help to speed up this process. Diagnostic concepts for evaluating and optimising the mode of operation can also be of benefit.

Distributed concepts. At this point it is worth looking in more detail at a concept whose advantages only emerge when the costs of a plant are considered as a whole. Large-scale plants contain large numbers of drives that have to be controlled electronically. This can be done either by means of control cabinets with point-to-point wiring to the drives or by means of distributed inverters which are connected to the motor or directly to the mechanical components [Ro99, RoStWi00].

Distributed inverters will always cost more to buy than devices that are installed in the control cabinet because they require a higher degree of protection. On the other hand, however, there are many other costs that are reduced by this concept:

- Costs for the control cabinets are eliminated (as is the space they take up on the factory floor)
- Point-to-point wiring with expensive shielded motor cables can be replaced by a linear, unshielded network connection to the distributed inverters
- Whereas in a central concept the wiring between the inverters and the motors can usually only be carried out at the plant installation site, where extensive testing is required, in a distributed concept this testing can be carried out when the machine modules are assembled by the machine manufacturer. As a general rule all jobs carried out at the machine manufacturer's premises can be organised more logically and therefore more efficiently than they can at the installation site.

However, other aspects with reference to distributed concepts during the operation phase of the plant must also be borne in mind. For instance, central installation in a control cabinet may offer advantages when it comes to fast replacement of drives. In order to assess the relative merits of these concepts, careful consideration must be given to the machine owner's service and maintenance schedules.

To summarise, the drives and their associated costs are a factor for consideration throughout the entire development process of a machine and a plant. All cost components of the machine manufacturer are significantly influenced by the drive system and the operations relating to its installation.

Operation phase. The following cost aspects play an important part in the operation phase for the machine operator:

- The operating costs
- The servicing costs
- The energy costs

The **operating costs** of a plant comprise the following components:

- The costs for *depreciation* of the procurement costs. In addition to the procurement costs, consideration must also be given to how well the plant can be used. The downtimes for retooling in connection with product changes, which can be heavily influenced by the level of automation and the drive concept used, are a critical factor here. A more highly automated machine with more drives, which is more expensive at the procurement stage, can reduce the running costs purely by allowing a longer machine time.
- *Personnel costs*, which are very dependent on the level of automation of production. The use of drive technology is one way of automating manual tasks and reducing personnel allocation.
- *Material costs for wastage.* If production operations generate only a low added value relative to the value of the material, as is the case with many finishing, packaging and assembly processes (where material costs can represent over 80% of the overall production costs), then avoidance of wastage is critical. Manufacturing processes with drives whose function is reproducible, which do not have to be "run in" after a product change and which can be adjusted to process variations, have an important role to play here.

The following factors influence *servicing* costs:

How much *regular and preventive maintenance* and servicing is required? Drives contain wear components (see chapter 3.8), but their service life is normally planned so that this is of no relevance in the typical operating life of a plant. This assumes, however, that functional principles that lead to a high level of wear are replaced by other technologies. One example is the frequent switching of brakes and clutches, which can be replaced by starting and stopping drives with inverters. The load of a drive also affects its service life.

What costs arise due to faults and their elimination? The more drives there are in a plant, the more critical is the average failure rate at the bottom of the "bath-tub curve" (see chapter 3.8). Production plants or logistics centres with 5000 to 10,000 drives are no longer a rarity, as we saw in chapter 4.13. A statistical failure rate of 0.5% would then result in 25 to 50 failures per year. Such highly automated production plants have to be brought back into service very quickly. Shutting down parts of the plant is not usually an option, as that would require a corresponding redundancy in the production sequence or flow of goods, and that is not possible in every case (the aisle of a warehouse normally contains only one storage and retrieval unit, for example). If a fault occurs, the time taken to eliminate it is critical because it affects the loss of production.

Drives with a high degree of reliability (high MTBF) make the greatest contribution to reducing interruption costs. The concepts that allow this level of reliability to be achieved were covered in chapter 3.8. In addition to the engineering and production of components, operating conditions also play an important part for increasing the reliability:

- Elevated temperatures, as a result of either high loads or high ambient temperature, always reduce the reliability of drive components. Accurate drive dimensioning is vital here.
- A clean installation reduces influences arising from electromagnetic compatibility (EMC) effects, which can in turn lead to sporadic faults that are difficult to diagnose. Concepts for fast and reliable installation, as described above, can help in this regard.

If a fault occurs on a drive component, it is usually replaced. The time taken to identify the defective product, to acquire a replacement product, to install it and to bring it back into operation makes a major contribution to the cost of fault elimination (MTTR). The costs of maintaining adequately trained service personnel (usually for three shifts per day) and the capital tied up in stocks of spares held on site must also be taken into consideration. The costs of the spare parts themselves often make up only a small proportion of the overall costs.

In order for devices to be replaced quickly, appropriate systems must be put in place as described in chapter 5.1.4. These must also take account of the level of training of the service personnel. Complex sequences of operations using PC programs are probably not appropriate for fault elimination during the night shift. In the case of mechanical components it is worth checking the extent to which plug-in connections are necessary to allow the component to be replaced quickly and safely.

Another critical factor for the running costs are *energy costs*. As we have already seen, two-thirds of electricity consumption in industry is due to drives. Energy-saving concepts and products can make a significant contribution to cost reduction. Given the importance of these issues, we will return to them in more detail at a later point (chapter 5.2.4).

Recovery phase. For the recovery phase after use, it is important for drives to be constructed in such a way that they can easily be recovered or their materials reused (recycling). Avoidance of toxic substances during production is important here, as is the ability to dismantle the drives easily in order to be able to remove recyclable materials such as iron, aluminium and plastics. Of course some components necessarily contain a mix of materials (motor stators containing sheet steel and copper, electronic circuit boards and their components), which have to be processed accordingly.

Table 5-3. Costs and their optimisation in the life cycle phases

LCC phase	Controllable costs	Ways of optimising costs
Procurement	• Procurement (i.e. costs incurred by the machine manufacturer for development, equipment and production) • Installation • Commissioning	• Test different drive technologies • Easy product selection • Exact product dimensioning • Implement system solutions • Increase usability • Use ready-to-use solutions • Easy installation • Easy commissioning
Operation	• Maintenance, inspection • Training • Spare parts • Energy • Servicing	• Increase reliability • Raise MTBF • Reduce MTTR • Increase motor efficiency • Increase ease-of-maintenance • Diagnostic tools • Increase maintenance intervals • Reduce product range to reduce stocks of spare parts • Fast availability of replacement products worldwide • Easy product replacement in the event of service
Recovery	• Disposal • Dismantling • Recycling • Revenue	• Avoidance of pollutants • Easy to dismantle • Recycling-friendly • Acceptance of used product and removal

To summarise, drives contribute to the costs of a machine or plant throughout all phases of its life cycle. Table 5-3 provides an overview of ways in which costs can be optimised in the various phases of the life cycle.

5.2.4   Energy-saving drive concepts

We have already mentioned that two-thirds of the electricity consumed by industry is associated with drives. There are only a few processes that are not executed either directly or indirectly by drives. Even pumps and compressors for hydraulic and pneumatic systems are electrically driven machines.

The energy input can be influenced substantially by the drive systems. Two aspects are of critical importance here:

- How efficient is the drive system at the actual operating point?
- Is the mechanical energy output by the drive system used as efficiently as possible, or is the efficiency intentionally reduced (e.g. by throttling)? This also includes the energy removed from the mechanical process by braking.

Efficiency. We will start by considering the factors that influence the efficiency of a drive system, looking at the individual components of the drive train.

The efficiency of the *inverter* is dependent on the following factors:

- The components that are used for the conversion of power. Nowadays IGBTs are almost exclusively used for the power switches in the inverter. Their efficiency has increased steadily over the years. With regard to the rectifier it should be noted that diodes are more efficient than IGBTs that are used for a rectifier that is capable of energy recovery.
- The switching frequency. High switching frequencies above the limit that is audible to the human ear lead to high switching losses, which reduce efficiency. They should only be used if the application requires it. That is generally not the case in production environments.
- The motor current. Drive concepts that require a lower motor current (e.g. the use of synchronous motors instead of asynchronous motors) also lead to lower power losses.
- The internal consumption of the inverter. At low power levels this can be significant (around 10%), whereas at power levels above a few kW it is no longer relevant.

The efficiency of inverters is normally between 94% and 97%. It cannot easily be increased by means of engineering measures.

The efficiency of the *motor* is dependent on its principle, its structure and its operating point. Asynchronous motors and hence the widely used standard three-phase AC motors are available in various energy efficiency classes (see chapter 3.3.2). Lower losses can be achieved by means of higher material costs and alternative materials (copper instead of aluminium in the rotor). A motor with high energy efficiency is therefore significantly more expensive, but its use pays off in all applications that have a long running time and a high continuous power. Consideration is now being given to setting a minimum efficiency for three-phase AC motors in order to eliminate the use of motors with poor efficiency.

In asynchronous motors the operating point also has a substantial influence on the efficiency. The reactive current that has to be applied in order to obtain magnetisation leads to a power loss as a result of ohmic losses that is largely independent of the load. This increases its relative size in partial load operation. Selective voltage reduction and hence reduction of magnetisation in partial load operation can reduce losses here, although this leads to a reduction in dynamic response as the magnetisation has to be built up again first in order to build up the torque.

A synchronous motor is more efficient than an asynchronous motor because no current is needed to build up the magnetisation; instead it is provided by permanent magnets. It is therefore superior not only in dynamic applications but also in terms of energy efficiency, although it is more expensive to manufacture. In specific application cases it is worth checking whether the use of a synchronous motor might not be advisable in terms of energy savings. All modern designs of energy saving drives for high-volume applications such as fans and pumps and compressors in household appliances (fridges, air conditioners) are based on controlled, permanently excited synchronous motors (also known as EC motors).

The efficiency of a *gearbox* is dependent on the number of gear steps and on the type of gearbox. As we saw in chapter 3.5.4, worm gearboxes have poor efficiency and should therefore not be used in applications with a long running time. The higher costs of a bevel gearbox are quickly offset by the cost savings from energy consumption and also by the fact that it requires a smaller motor and inverter.

There are no significant differences in efficiency between spur gears and bevel gears. A power loss of about 2% should be allowed for each gear stage. It is important to use the correct type of lubrication and the correct amount of oil in the gearbox to prevent excessive churning losses.

Mechanical processes. The mechanical processes that are controlled by the drive have a greater influence on the overall energy audit of a drive system than the efficiency of the drive system itself. In many processes the energy requirement is dependent on the status at any given time. Uncontrolled drives can only influence the output power by varying the torque. In the case of pumps and fans, the volume flow remains constant unless mechanical influencing measures are provided. As we saw in chapter 4.12, the flow losses also remain constant. Fans generally supply too much air, while in pumps the pressure is reduced in the mechanical components. Both result in a poor energy balance.

For these reasons the speed control is the decisive factor for high energy efficiency. The two variables that determine the mechanical power (torque and speed) are then adjusted to the operating point of the process. In many applications this can result in an energy saving of up to 60%. The additional cost of the frequency inverter for setting the speed is usually offset after a short time.

Another aspect in connection with mechanical energy is the use of recovered drive energy. In machines with many drives it is always advisable to connect them to one another in a voltage DC link (see chapter 3.4.1). This results in an energy exchange between the drives, so that only process power and power losses are drawn from the mains.

If braking energy is recovered to any significant extent, then the use of active power recovery to the mains is worthwhile. Regenerative modules can be used for this purpose. This is particularly beneficial in hoist drives without a counterweight. In winding drives and loading drives on test benches, on the other hand, a power exchange with the drive motors is preferable.

(Europe)	Savings potential in million tons CO_2	Savings potential in %	
Increased use of energy saving motors	4.0	10%	Electro-technology
Electrical speed control instead of mechanical control	12.0	30%	Electro-technology
Mechanical system optimation	24.0	60%	Mechanics, process
Total	40	100%	

Fig. 5-13. CO_2 saving potential using electric drives

The evaluation of the energy saving potential using electric drives in Fig. 5-13 shows that the effect of speed control and the optimisation of mechanical processes, which often goes hand-in-hand with the introduction of speed control, have the greatest energy saving potential [ZV06a].

To summarise, there are many concepts and products that can be used to significantly reduce the use of electrical energy in drive systems. Their use is worthwhile from both an economic and an ecological perspective. An assessment of overall costs, using life cycle costing methods, can identify the most appropriate concepts and help with the decision-making process.

5.2.5   Comprehensive evaluation of drive systems

We have seen in this chapter that drive systems have to be analysed and evaluated on a comprehensive basis in order to identify and to use the most appropriate concepts. They have a tremendous influence on the overall productivity and reliability of a machine and a plant and on all the costs arising during their life cycle.

There are some significant models available for analysis and evaluation, and there are also many ways of reducing costs in all life cycle phases. Nevertheless, if they are to be implemented it is essential that decisions are taken on the basis of all these factors and not solely on the procurement costs for drive systems.

This consideration also forms the logical end for this book: that wants to explore the ways in which drives are used today in large numbers to achieve a high degree of automation of production and logistics processes, also how these drive systems are structured, how drive solutions can be configured for specific applications, and how their installation and use can be simplified by engineering tools.

Glossary

This glossary contains definitions of important terms used in this book. Rather than providing general definitions, it approaches the terms from the point of view of how they are applied to this specific context.

Actuator. Actuators are the final control element of a mechatronic system. They convert signals from the control system into forces to control the process. Simple actuators convert binary data. They include hydraulic or pneumatic valves for fluid drives or electromechanical contactors for switching electric motors or electromagnets. Electronically-controlled drives are also actuators but support additional functions for motion control.

Angle sensor. Angle and speed sensors are measuring systems which detect the rotor position of the motor both for motor control and as an actual value for motion control. A distinction can be drawn between magnetic (resolvers) and optical (incremental encoders, encoders, SinCos encoders) measuring procedures. Angle sensors which are able to detect one or a number of revolutions (single-turn, multi-turn) are another type.

Assembly line production. In assembly line production, stations for discrete production steps are set up and arranged in line. The work piece is conveyed or transported continuously and at constant speed from one station to the next, where each production step in the manufacturing process is executed. Assembly line production is used in mass production. The most well-known example is final assembly in automotive production.

Asynchronous motor. On asynchronous motors, the magnetic field in the motor is generated by a magnetising current from the electrical supply. By design, asynchronous motors work with a slip; in other words, a load-dependent speed deviation of the rotor speed in relation to the frequency of the motor voltage. In the form of standard three-phase AC motors, asynchronous motors are the motor type most frequently used in industrial applications. Asynchronous servo motors are also used as servo motors with increased accelerating capability and reduced mass inertia.

Cam profilers/Cam mechanisms. Mechanical cam mechanisms are gearboxes with non-uniform ratios. They usually convert a uniform input speed into a non-linear output motion. The term electronic cam profiler or drive is used to describe the motion control of an inverter that executes the non-uniform transmission as a function of the master motion and moves the connected motor accordingly. Electronic cam profilers offer advantages in respect of machine structure and flexibility.

Communication system. Communication systems are used for data exchange between electronic devices (e.g. controllers, inverters, command stations, sensors, actuators). Due to their application at field level, they are also known as fieldbuses. In the field of automation, several different technologies are established for various requirements, bandwidths and functions.

Continuous production. Continuous production produces endless material as part of a process which is also continuous. Continuous production is used for example to manufacture paper webs, in stranding machines, in the production of sheet glass and in rolling mills.

Controls. Controls execute sequence control and motion control in a machine or system and execute the communication and visualisation functions of an automation system. Different types of control include programmable logic controllers (PLCs), numerical controls (CNC), motion controls, special machine controls (embedded controls) and industrial PCs with soft PLC or soft CNC.

CNC control. A computerised numerical control (CNC) is an electronic device for motion control used in particular for co-ordinated drives (e.g. in machine tools). Motion is described in the form of NC blocks defining a machining step as a relative motion for several axes in space. For the calculation a high computational power is required. The generated setpoints for the drives have to be transferred to the drive controllers synchronously in time.

Conveyor drive. Conveyor drives are used to transport goods or persons over limited distances and primarily on predefined routes. They are used in production systems to transport work pieces from one machine to the next and also in logistics systems. Typical areas of application are roller and belt conveyors as well as escalator.

Cross cutter. Cross cutters are machines which make a shear cut through continuous webs (continuous material feed) crosswise to the feed motion. In drive engineering and automation, the term cross cutter is also used to describe the motion control for synchronising the positioning movements required for this machining process.

Drive control. Control of the power conversion of the inverter in accordance with the required setpoints for the speed, torque and/or angular position of the motor. The different methods used are "V/f control" (open-loop), "sensorless vector control" without speed feedback from the motor and "servo control" with measurement of the angular position and rotary speed of the motor. Drive control is almost always implemented within the inverter.

Drive dimensioning. Dimensioning of a drive solution. Sizing of all the components in a drive system with regard to their relevant load conditions (permanent load, dynamic load, etc.), ambient conditions and their cost-effectiveness.

Drive elements are mechanical transmission elements for transmitting mechanical power from the driven shaft of a gearbox or motor to the work machine. Common examples of driving elements are shafts, detachable shaft connections (couplings and clutches, clamping elements), bearings, traction drives (belts, bands, chains) and elements for converting rotary motion into linear motion (toothed belts, rack and pinions, leadscrews) or non-linear motion (thrust cranks, four-bar linkages).

Drive solution. A drive solution is the combination of drive components suitable to execute a drive task in a machine and usually comprises an electric motor for energy conversion, an inverter which drives the motor and contains software functions for drive control and motion control, a gearbox to adapt the operating point of the motor to the process and other driving elements for mechanical power transmission to the work machine. Some applications do not require an inverter or gearbox.

Electromagnetic compatibility. Electromagnetic compatibility (EMC) describes the interference between electrical components generated by e-lectromagnetic fields. As well as technical aspects, the term also covers the legal framework for limiting emitted inference from electromechanical and electronic devices and the immunity to such interference (conducted and radiated interference in the frequency range up to 1 GHz).

Engineering. Engineering describes all activities required to select a drive solution for an application and put it into operation so that it fulfils its intended purpose. The overall engineering process can be divided into several phases covering the selection of the drive system, its dimensioning, configuration (software function setup) and commissioning. Engineering activities are also carried out during the operation phase for diagnostics and service. Engineering activities are supported by engineering tools (computer programs).

Fans. Fans convey gases (e.g. air). They are usually driven by an electric motor.

Flying saw. Flying saws are machines which make saw cuts while the material is moving. In drive engineering and automation, motion control to synchronise the required positioning movements for this machining process is also referred to as flying saw.

Forming drives Forming drives drive intermittent or continuous forming processes (reforming, compression, extrusion, etc.). Application areas include for example extruders, injection moulding machines, presses and sheet metal folders.

Frequency inverter. Frequency inverters control power conversion for driving three-phase AC motors by modulating the frequency and output voltage for the generation of the rotating field. Frequency inverters usually work in open-loop operation without speed or angular position feedback from the motor. They are used in applications with low to average requirements in respect of the dynamic response or motion control accuracy (conveyor drives, travelling drives, hoist drives, main drives and drives for pumps and fans).

Gearbox. The term gearbox describes all mechanical transmissions used for speed and torque conversion. Gearboxes most commonly take the form of toothed gear units for mounting on motors. Other types include driving elements with a mechanical transmission (traction drives, rack and pinions, leadscrews, thrust cranks, four-bar linkages) and cams.

Geared motor. The term geared motor describes the combination of motors and toothed gearboxes to create a unit. This can be achieved using coupling elements or by means of direct mounting using common construction elements. This direct combination offers advantages in respect of size and weight.

Hoist drive. Hoist drives are used for the vertical transport of goods or persons. Operation against gravitational force results in specific requirements for the energy flow and the safety of hoist drives. Typical areas of application are cranes and lifts.

Industrial PC. An industrial PC is a computer system for use in automation which uses PC-based architectures and components. Its mechanical design and also its software are adapted to the requirements of automation. The software on an industrial PC can run control functions (soft PLC, soft CNC), visualisation functions, communication tasks and a wide variety of other tasks.

Intermittent production. In intermittent production, stations for discrete production steps are set up and arranged in line. The work piece is conveyed or transported at fixed intervals from one station to the next, where each production step in the manufacturing process is completed. Intermittent production is used in mass production.

Inverters. Inverters control the electrical power conversion from the power mains to the motor. They can set the output frequency and voltage for the appropriate generation of the rotating field of a three-phase AC motor. This book covers exclusively voltage source DC link inverters, comprising a rectifier and an inverter along with a DC link. The inverter is controlled by a signal processing unit that performs motor control and (if applicable) motion control algorithms. Frequency inverters are used for applications with reduced requirements in respect of dynamic response, setting ranges and motion control accuracy and servo inverters for applications with high to very high requirements in respect of these characteristics.

Life cycle. In respect of a machine or system, the term life cycle describes the total time from creation via operation (usage phase) to disposal. Costs are incurred throughout the life cycle. For a definitive assessment, drive systems have to be evaluated on the basis of the total costs throughout the whole life cycle.

Linear motor. Linear motors convert electrical energy directly into linear motion. Using them can offer advantages in respect of speed, acceleration and positioning accuracy, for example. They are used whenever conventional solutions with toothed belts or leadscrews cannot be implemented due to extremely high requirements.

Logistics. In this book, the term logistics is used to describe the in-house transport of goods and materials (intralogistics). It covers procurement, production, distribution and disposal logistics.

Main drive. The main drive is the central drive on a machine which sets the machining and process speeds. All movements are transferred via a shaft (line shaft). Main drives are used in all types of machine and in all sectors of industry.

Mass production. Mass production produces goods in high volumes (> 100,000 units/year). Mass production usually involves a high level of mechanisation and automation as well as a continuous flow of the produced goods.

Material flow systems. Material flow systems include in-house materials handling technology, storage technology and picking (whereby goods are grouped for consignment for specific orders). Conveying equipment (continuous and non-continuous conveyors, industrial trucks, overhead conveyors, storage and retrieval units), hoists and material handling systems for palletising and picking are used.

Mechatronics. The term "mechatronics" is used to describe the combination of mechanics, electronics and information technology. A mechatronic system uses software-controlled signal processing (open-loop control, closed-loop control) to control a mechanical process. Actuators are used to convert signals to forces and sensors to measure process states. A controlled drive which uses software to execute a drive function in a machine or system is referred to as a "mechatronic drive solution".

Motion control. Motion control converts commands from a sequence control (e.g. a PLC) into setpoints for drive control and thus generates the moves of the motor according to the requirements of the application. Setpoints are usually generated for the angular position or speed of the motor. Typical motion control functions are ramp function generators or the generation of a positioning profile. Motion control can be implemented in the inverter or in a separate control.

Motor. Motors generate mechanical energy from another energy source. This book deals exclusively with electric three-phase AC motors. They convert the energy from a three-phase electricity supply system or an inverter into mechanical rotation or linear motion. This book describes three-phase AC standard motors, synchronous and asynchronous servo motors as well as linear motors and torque motors.

Motor brake. Motor brakes are permanent magnetic brakes or spring-applied brakes built into or onto a motor. They are used as holding brakes or emergency stop brakes. Working or service brakes are not described in this book.

PLC. Programmable logic controllers (PLCs) execute Boolean operations to run programmable sequence controls on a machine or system. The PLC inputs the signals from the connected sensors, calculates the required output values in cyclic programs and controls the actuators – drives, for example – accordingly. Due to its mode of operation, a PLC is not perfectly prepared for motion control tasks.

Positioning drive. Positioning drives move material or a work piece to a target position with a defined positioning accuracy. Switch-off positioning techniques with frequency inverters without motor speed feedback and servo positioning techniques with feedback are used. The positioning drives described in this book are point-to-point positioning sequences with the positioning profile generated by motion control software. Just a few of the possible applications include the precise stopping of conveyors, automatic assembly machines and manipulators.

Process engineering. Process engineering describes the execution of processes designed to change the composition, properties or nature of materials without creating a defined geometry. A distinction is drawn between chemical, biological, thermal and mechanical processes. Processes in process engineering usually produce raw materials (liquids, gases, pellets, powders, etc.) for further processing in production facilities.

Production engineering. Production engineering describes the techniques for manufacturing and modifying work pieces to create a defined geometry. Distinctions are drawn between primary forming, forming, separating, joining, coating and the modification of material properties.

Project planning. Project planning describes the engineering activities for assigning the actual configuration to a drive system for a specific application. These include in particular software function setup by parameter setting, function block configuration or programming.

Pumps. Pumps convey liquids (e.g. water). There are various pumping techniques. Pumps are usually driven by an electric motor.

Reliability. "Reliability" describes a measure of the capacity of a machine or component to fulfil its function for a defined period of time. Within this period of time, which corresponds to the operating phase, random component failures may occur. The wear limit determines the end of the operating phase.

Robots. The industrial robots described in this book are material handling systems designated for universal use. They perform multidimensional coordinated movements in space with usually between three and six axes which can be freely programmed. In accordance with the area of application and prevailing requirements, robots are designed as articulated arm, gantry, SCARA or parallel robots, for example. The areas of application are quite numerous and can be found for example in the field of material handling and assembly technology; welding robots are also used in automotive production.

Safety technology. Safety technology ensures that automated machines and systems are not able to put the individuals in their vicinity at risk. For this purpose, specific concepts (for example redundancy techniques) are used to ensure that the machine is set to a defined status in the event of an error or malfunction. Today's mechatronic drive solutions feature on-board safety functions which support safety concepts and make them easier to implement. There are extensive standards governing safety engineering design.

Sensor. Sensors detect and measure the physical variables associated with a processes. These can be for example positions, speeds, forces, temperatures or pressures. There are also sensors for measuring chemical properties as well as for identifying goods. The output values of sensors are transferred to the control system by electrical signals (digital, analog, serial communication).

Servo drive. Servo drives are electronically-controlled drives for applications with very high requirements in respect of dynamic response, setting ranges and/or motion control accuracy. For such applications, servo inverters together with servo motors are used. The servo inverters evaluate an angle or position sensor and perform a very precise and dynamic motor control. Primary areas of application are positioning drives, co-ordinated drives, synchronised drives, cross cutters, flying saws and electronic cam profilers.

Servo inverter. Servo inverters perform a very precise and highly dynamic closed-loop motor control. The design of the inverter takes into account the special requirements of applications with high overload factors, high resolutions of the motor feedback system, fast control loop sampling times and the different concepts for the energy flow. Servo inverters are used as single-axis drives with high functionality for motion control or as multi-axis drives usually with a limited functional scope for motion control.

Servo motor. Servo motors are electric motors optimised for servo drives with high requirements in respect of dynamic response and accuracy. They have low mass inertias and high overload capabilities for fast acceleration. They include angle sensors (encoders or resolvers) for measuring the rotor position for the servo control in the inverter. Servo motors are provided as synchronous and asynchronous motors.

Standard three-phase AC motor. The term standard three-phase AC motor is used to describe robust asynchronous three-phase AC motors being operated either from a three-phase mains power supply or by an inverter. This type of motor is characterised by standardised main dimensions (standard motor). Standard three-phase AC motors are the most frequently used type of motor in industrial applications in the power range above 100 W.

Synchronised drive. On a synchronised drive, several drives work at a synchronous speed or with a synchronous angular position. As electrical shafts and electrical gearboxes, they replace mechanical interconnections via shafts. Areas of application for synchronised drives include the transport of webs in continuous production processes with continuous materials and printing units with single drives in printing machines.

Synchronous motor. Synchronous motors are three-phase AC motors with the magnetic field being generated in the rotor either by an excitation winding or by permanent magnets. In the power range covered by this book, only permanent magnet synchronous motors are used. Compared with asynchronous motors, synchronous motors are able to achieve better efficiency rates and have lower moments of inertia at the same rated torque. Slip is replaced by a load-dependent angular phase shift of the rotor position in relation to the rotating field of the supply voltage. Synchronous motors are used in industrial applications as servo motors in the power range up to approximately 10 kW.

Three-phase AC motors. Three-phase AC motors are electric motors which are driven electrically, usually by a three-phase AC system. This can be the three-phase mains supply or an inverter. Three-phase AC motors take the form of asynchronous motors and synchronous motors.

Tool drive. A tool drive provides the power required for the tool of a machining process. Most tool drives have high requirements in respect of constant speed and/or torque and some also run at very high speeds. Just some of the areas of application for tool drives are drills, milling cutters and grinding machines.

Torque motor. Torque motors are slow-running rotary direct drives. The majority of torque motors are designed as synchronous servo motors with a high number of pole pairs. Short designs with large diameters dominate.

Travelling drive. Travelling drives are used to transport goods or persons. Unlike conveyor drives, travelling drives move with the transport vehicle, thereby placing specific requirements on the power supply. Travelling drives are used for example on storage and retrieval units, monorail overhead conveyors and in cranes.

Visualisation. Visualisations are used for operator control and monitoring of automated machines and systems. They take the form of machine operator terminals with textual or graphical displays or PC-based systems.

Winding drive. Winding drives are used to wind or unwind continuous materials. They are characterised by the constantly changing diameter of the reel. Winding drives are characterised by high setting ranges for speed and torque and good smooth running performance. Typical areas of application are winding processes for foils, paper, threads, textile webs, wires, sheet metal, etc.

List of symbols

A .... Area, Probability of occurrence, Proportionality constant, Availability
c............ Rigidity
C1.......... Category 1 noise emission
C2.......... Category 2 noise emission
C3.......... Category 3 noise emission
C4.......... Category 4 noise emission
d .... Damping constant, Diameter
E............ Energy, Probability of discovery
f............. Frequency
F .... Force
h, H .... Gradient, Height
i .... Amperage, Ratio
I............. Amperage, Input signals
J............. Moment of inertia
L............ Inductance, Service life
m .... Mass
MTBF ... Mean time between failures

MTTF ....Mean time to failure
MTTR....Mean time to repair
n.............Speed
P.............Power
p.............Number of pole pairs
Q............Switching energy per switching cycle
R ....Reliability
r ....Radius
s ....Distance, Slip
T ....Time constant, Temperature
t.............Time
v, V ....Voltage, Number of revolutions
v............Velocity
W ....Work
α............Angular acceleration, Angle of tilt, Phase angle
λ............Failure rate
μ ....Coefficient of friction
η............Efficiency
ρ............Specific mass
τ............Moment, Pole pitch
φ ....Angle
Φ............Magnetisation
ω ....Speed, Angular velocity

Indices

a ....actual

AC......... Alternating Current

B............ breakdown

C............ Coil

d, D .... Drive, starting, Voltage rise

DC......... Direct Current

d, q .... Cartesian vector in field coordinates

dv .... DC link voltage, DC voltage

dyn .... Dynamic response

e, E .... Carrier, Exciter

el .... Electrical

fr............ Friction

G .... Weight

kin .... Kinetic

L, load... Load

M.......... Magnet

m, mean. Mean value

max .... Maxima

mech ...... Mechanical

meas....... Measured variables

min......... Minima

mot......... Motor

N ....Rated quantity, Normal value, Mains

O ....Operation

opt..........Optimum

P.............Pulse, Rotor

pot..........Potential

ref ....Reference

req..........Require

resn ....nth harmonic

rms.........R.m.s. value

S.............Spring

S.............Saddle, Stator, Terminal

st ....static, inclined

Sys.........System

Torsion ..Torsion

V ....Loss

Wn.........Inverter output variables

Bibliography

[Ap04]: Apfeld, R.: Neue Definition von Sicherheitsfunktionen für Antriebssysteme und die Berücksichtigung von SIL, Kat und Performance-Level. SPS/IPC/Drives Conference Proceedings, Nuremberg 2004, pp. 449-458.

[Ar+01]: Arlt, V.; Götz, O.; Nebuhr, K.; Ranze, C.: Der schnelle Weg zum richtigen Antrieb durch Produktkonfiguration Technologie, Anwendung, Nutzen. SPS/IPC/Drives Conference Proceedings, Nuremberg 2001, pp. 651-660.

[ArKo01]: Arlt, V.; Konieczny, F.: Funktionales Engineering modularer, verteilter Automatisierungsapplikationen auf Basis der IDA-Technologieplattform. SPS/IPC/Drives Conference Proceedings, Nuremberg 2001, pp. 83-91.

[BaDi06]: Baumann, F.; Dillig, R.: Einspeisungen für Mehrmotorenantriebe. SPS/IPC/Drives Conference Proceedings, Nuremberg 2006, pp. 141-150.

[Be03]: Beckmann, G.: Sicherheit vom Sensor zum Antrieb. SPS/IPC/Drives Conference Proceedings, Nuremberg 2003, pp. 699-706.

[Be05]: Bender, K.: Embedded Systems – qualitätsorientierte Entwicklung. Springer-Verlag Berlin Heidelberg, 2005.

[BePr06]: Berge, G.; Prangenberg, M.: Anwendungsgetriebene modulare Plattform für Kreiselpumpen. Elektrisch-mechanische Antriebssysteme Conference Proceedings, Böblingen 2006, pp. 217-228.

[Bl72]: Blaschke, F.: Das Prinzip der Feldorientierung zur Regelung der Asynchronmaschine. Siemens Forschungs- und Entwicklungsberichte, vol. 1 1972, pp. 184-193.

[Bl+03]: Bleisch, G.; Goldhahn, H.; Schricker, G.; Vogt, H.: Lexikon Verpackungstechnik. Hüthig GmbH & Co. KG, Heidelberg, 2003.

[Bo04]: Bohl, W.; Elmendorf, W.: Strömungsmaschinen 1. Vogel Buchverlag 2004.

[BrGe04]: Brandenburg, G.; Geißenberger, S.: Schnelle Schnittregister- und Bahnzugkraftregelung für Rollendruckmaschinen. SPS/IPC/Drives Conference Proceedings, Nuremberg 2004, pp. 435-448.

[BrGeKl02]:   Brandenburg, G.; Geißenberger, S.; Klemm, A.: Einfluss von Transport- und Leitwalzen mit Gleitschlupf auf die Bahndynamik von kontinuierlichen Fertigungsanlagen der Metall-, Kunststoff-, Papier- und Druckindustrie. SPS/IPC/Drives Conference Proceedings, Nuremberg 2002, pp. 679-690.

[Br99]:   Brosch, P. F.: Drehzahlvariable Antriebe für die Automatisierung. Vogel Buchverlag 1999.

[Bu06b]:   Bültmann,   U.:   Berührungslose   Energieübertragung. SPS/IPC/Drives Conference Proceedings, Nuremberg 2006, pp. 197-202.

[Bu06a]:   Statistisches Bundesamt: Statistisches Jahrbuch 2006 für die Bundesrepublik Deutschland. Statistisches Bundesamt, Wiesbaden, 2006.

[CzHe04]:   Czichos, H.; Hennecke, M.: HÜTTE Das Ingenieurwissen. Springer-Verlag, Berlin Heidelberg, 2004.

[De06]:   Deckers, J.: Drehmomentanalyse als unverzichtbare Methode zur Produktoptimierung und Sicherung der Anlagenverfügbarkeit – Ein Erfahrungsbericht. Elektrisch-mechanische Antriebssysteme Conference Proceedings, Böblingen 2006, pp. 455-470.

[DiSiRi99]:   Diedrich, C.; Simon, R.; Riedl, M.: Elektronische Gerätebeschreibungen. SPS/IPC/Drives Conference Proceedings, Nuremberg 1999, pp. 184-192.

[Do03]:   Doppelbauer, M.: Vergleich der Anforderungen an Wirkungsgrad und Normmotoren in Nordamerika und Europa. SPS/IPC/Drives Conference Proceedings, Nuremberg 2003, pp. 651-662.

[Dr01]:   Dresig, H.: Schwingungen mechanischer Antriebssysteme. Springer-Verlag, Berlin Heidelberg, 2001.

[Dub95]:   Dubbel, H.; Beitz, W.; Küttner, K.-H.: Taschenbuch für den Maschinenbau. Springer-Verlag, Berlin Heidelberg New York, 1995.

[EmKo02]:   Emde,   C.;   Köhler,   B.:   Energieeffiziente   Pumpenantriebe. SPS/IPC/Drives Conference Proceedings, Nuremberg 2002, pp. 480-572.

[Fr01]:   Fräger, C.: Servomotoren: Leistungsverbesserungen bei Servomotoren und Auswahl des richtigen Antriebs. SPS/IPC/Drives Conference Proceedings, Nuremberg 2001, pp. 518-526.

[Fr03]:   Fräger, C.: Kompaktservomotoren - höchste Dynamik und Leistungsdichte für moderne Antriebsaufgaben in der Handhabungstechnik. SPS/IPC/Drives Conference Proceedings, Nuremberg 2003, pp. 663-672.

[Fr06]:   Fräger, C.: Regelung mechatronischer Linearantriebe mit Zahnriemen. SPS/IPC/Drives Conference Proceedings, Nuremberg 2006, pp. 515-522.

[GaSc96]: Garbrecht, F. W.; Schäfer, J.: Das 1x1 der Antriebsauslegung. VDE-Verlag GmbH, Berlin and Offenbach, 1996.

[Ge88]: Gekeler, M. W.: Raumzeiger-Modulation bei Frequenzumrichtern. antriebstechnik 27, vol. 4 1988, pp. 131-133.

[GiHaVo03]: Giersch, H.-U.; Harthus, H.; Vogelsang, N.: Elektrische Maschinen. B.G. Teubner / GWV Fachverlage GmbH, Wiesbaden, 2003.

[Gi92]: Ginsbach, K.-H.: Leistungshalbleiter: Rückblick, Stand der Technik und Ausblick. ETG-Fachbericht 39, 1992, pp. 7-24.

[Go06]: Göpfrich, K.: F3E-Converter with Fundamental Freqeuency Front End. SPS/IPC/Drives Conference Proceedings, Nuremberg 2006, pp. 151-158.

[Gr03]: Grauer, J.: Automatisierungs- und Sicherheitstechnik wachsen zusammen - Standard und Sicherheitstechnik aus einem GusS. SPS/IPC/Drives Conference Proceedings, Nuremberg, 2003 pp. 343-348.

[GrGe06]: Greubel, K.; Gebert, K.: Asynchron- und Synchronmotoren für kompakte Motorspindeln. Elektrisch-mechanische Antriebssysteme Conference Proceedings, Böblingen 2006, pp. 201-216.

[GrHaWi06]: Groß, H.; Hamann, J.; Wiegärtner, G.: Elektrische Vorschubantriebe in der Automatisierungstechnik. Publicis Corporate Publishing, Erlangen 2006.

[Gr06]: Grosser, M.: Sicherheitsfunktionen im Antrieb integriert vereinfacht die Systemstruktur und bietet Potenziale zur Kosteneinsparung. SPS/IPC/Drives Conference Proceedings, Nuremberg 2006, pp. 135-140.

[GuLi99]: Gundelach, V.; Litz, L.: Moderne Prozessmesstechnik. Springer-Verlag Berlin Heidelberg, 1999.

[Ha04]: Habermann, S.: Standardisierte Antriebs- und Steuerungskomponenten für Elektrohängebahnsysteme. SPS/IPC/Drives Conference Proceedings, Nuremberg 2004, pp. 419-424.

[HaSt03]: Haj-Fraj, A.; Storath, A.: Torquemotoren, Aufbau und Anwendungen. SPS/IPC/Drives Conference Proceedings, Nuremberg 2003, pp. 833-840.

[Ha06]: Harms, M.: Modularisierte Anwendungslösungen in der elektrischen Antriebstechnik. SPS/IPC/Drives Conference Proceedings, Nuremberg 2006, pp. 593-602.

[He+04]: Hebbing, L.; Lepper, E.; Peitz, M.; Stach, T.: Systemoptimierung von Hochgeschwindigkeitsanwendungen durch Einsatz von Dreipunktwechselrichtern. SPS/IPC/Drives Conference Proceedings, Nuremberg 2004, pp. 391-398.

[HeGePo98]: Heimann, B.; Gerth, W.; Popp, K.: Mechatronik. Fachbuchverlag Leipzig at Carl Hanser Verlag, 1998.

[HeKa04]:   Heinemann, G.; Käsdorf, O.: Moderne Architektur für flexible und universell einsetzbare Antriebsregler. Elektrisch-mechanische Antriebssysteme Conference Proceedings, Fulda 2004, pp. 157-174.

[He03]:   Heyer-Reinfeld, A.: Ethernet - Just in Time. SPS/IPC/Drives Conference Proceedings, Nuremberg 2004, pp. 595-602.

[Hi02]:   Hilfert, S.; Thurn, M.: Synchron-Servomotor mit integrierter Spindel für den Einsatz in Punktschweißzangen. SPS/IPC/Drives Conference Proceedings, Nuremberg 2002, pp. 699-707.

[HoKi01]:   Hoene, E.; Kiel, E.: Ausbreitungswege elektromagnetischer Störungen in Umrichtern hoher Leistungsdichte. SPS/IPC/Drives Conference Proceedings, Nuremberg 2001, pp. 579-588.

[Ho92]:   Holtz, J.: Pulsewidth Modulation - A Survey. PESC Toledo, 1992, pp. 11-18.

[Is99]:   Isermann, R.: Mechatronische Systeme. Springer-Verlag, Berlin Heidelberg, 1999.

[JeWu95]:   Jenni, F.; Wüest, D.: Steuerverfahren für selbstgeführte Stromrichter. vdf Hochschulverlag AG at ETH Zürich and B.G. Teubner, Stuttgart, 1995.

[Jo00]:   Jochim, F.: Automatisieren von Verpackungsmaschinen mit Elektronischen Kurvenscheiben. SPS/IPC/Drives Conference Proceedings, Nuremberg 2000, pp. 591-599.

[Jo89]:   Jönnsen, R.: A new control scheme for AC induction motor, Power Conversion Conf., Tokyo 1989.

[Ki94a]:   Kiel, E.: Anwendungsspezifische Schaltkreise in der Drehstrom-Antriebstechnik. Dissertation at TU Braunschweig, 1994.

[Ki94b]:   Kiel, E.: Der analoge VeCon-Chip. elektronik industrie, vol. 6 1994, pp. 29-33.

[Ki00]:   Kiel, E.: Requirements and System Concepts for Industrial Converters in the Lower Power Range. CIPS Conference Proceedings, Bremen 2000, pp. 140-144.

[Ki04]:   Kiel, E.: Servoantriebe als mechatronische Maschinenautomatisierungselemente: Struktur, Software, Systemintegration. Elektrisch-mechanische Antriebssysteme Conference Proceedings, Fulda 2004, pp. 175-190.

[Ki05]:   Kiel, E.: Der Muskel der Automatisierung – Zukunft der Antriebstechnik. etz Elektrotechnik + Automation, vol. S2 2005, pp. 70-73.

[Ki06]:   Kiel, E.: Mechatronische Antriebslösungen für den Maschinenbau: Produktbaukästen und Entwurfsmethodiken. Elektrisch-mechanische Antriebssysteme Conference Proceedings, Böblingen 2006, pp. 229-240.

[KiSch92]:   Kiel, E.; Schierenberg, O.: Einchip-Controller für das SERCOS-Interface. Elektronik, vol. 6 1992, pp. 50-59.

[KiSch94a]:   Kiel, E.; Schumacher, W.: Der Servocontroller in einem Chip. 40
               Firmen entwickeln die Vecon-Chips für intelligente Antriebe. E-
               lektronik, vol. 8 1994, pp. 48-60.

[KiSch94b]:   Kiel, E.; Schumacher, W: Der VeCon-Prozessor: Signalverarbei-
               tung für Antriebe auf einem Chip. SPS/IPC/Drives Conference
               Proceedings, Nuremberg 1994, pp. 533-540.

[KiSch95a]:   Kiel, E.; Schumacher, W.: VeCon: A High-Performance Single-
               Chip-Servocontroller for AC Drives. 1995 pp. 1284-1289.

[KiSch95b]:   Kiel, E.; Schumacher, W.: VeCon: High-Performance Digital
               Control of AC Drives by One-Chip Servo Controller. 1995
               pp. 3001-3005.

[KiLo98]:   Kiel, E.; Loy, T.: Der digitale Antriebsregler auf dem Weg vom
               einfachen Aktor zur dezentralen intelligenten Bewegungssteue-
               rung. SPS/IPC/Drives Conference Proceedings, Nuremberg 1998,
               pp. 476-487.

[KiKrSch01]:   Kiel, E.; Krüger, M.; Schlichtermann, L.: Integration der digitalen
               und analogen Funktionen einer Servoregelung in den APMC-
               Motorcontroller. SPS/IPC/Drives Conference Proceedings, Nurem-
               berg 2001, pp. 748-756.

[KiBoDo03]:   Kimmich, R.; Bomke, U.; Doppelbauer, M.: Wirkungsgrade und
               Betriebseigenschaften von Asynchronmotoren mit Läuferkäfigen in
               Kupferdruckguss-Technologie. SPS/IPC/Drives Conference Pro-
               ceedings, Nuremberg, 2003 pp. 631-640.

[KiHeKo04]:   Kiel, J.; Hebinig, L.; Koch, U.: Feldschwächebetrieb von perma-
               nenterregten Synchronmaschinen - Ziele, Grenzen und Strategien.
               SPS/IPC/Drives Conference Proceedings, Nuremberg 2004,
               pp. 655-664.

[KiFr06]:   Kiel, E.; Fräger, C.: Positionierantriebe als mechatronische Syste-
               me. ASB Conference Proceedings, Stuttgart 2006.

[KiKu06]:   Kiel, E.; Kuppinger, S.: Zentral oder Dezentral – das ist nicht die
               Frage. IEE, vol. 1, 2006.

[KnSch06]:   Knopf, E.; Schreiber, S.: Aktive Drehschwingungsbeeinflussung
               bei Bogenoffsetdruckmaschinen. Elektrisch-mechanische Antriebs-
               systeme Conference Proceedings, Böblingen 2006, pp. 187-200.

[Ku05]:   Kummetz, J.: Positionserfassung in hochdynamischen Servoantrie-
               ben. SPS/IPC/Drives Conference Proceedings, Nuremberg 2005,
               pp. 451-462.

[Ku06]:   Kummetz, J.: Serielle Positionswertübertragung in geregelten An-
               trieben. SPS/IPC/Drives Conference Proceedings, Nuremberg
               2006, pp. 533-542.

[Le91]:   Leonhard, W.: 30 Years Space Vectors, 20 Years Field Orientation, 10 Years Digital Signal Processing with AC-Drives, a Review. EPE-Journal, vol. 1 / 2 1991 pp. 13-20 / 89-102.

[Le97]:   Leonhard, W.: Control of Electrical Drives. Springer-Verlag, Berlin Heidelberg, 1997.

[Le06]:   Lenze AG: Antriebslösungen. Lenze AG, Hameln, 2006.

[LiRo05]:   Liggesmeyer, P.; Rombach, D.: Software Engineering eingebetteter Systeme. Elsevier GmbH, Munich, 2005.

[Li07]:   Lindemann, U.: Methodische Entwicklung technischer Produkte. Springer-Verlag, Berlin Heidelberg, 2007.

[LoOsSp06]:   Lohrengel, A.; Ostertag, W.; Sprick, R.: Klassische Getriebetechnik oder servoelektrischer Direktantrieb? - Ein Vergleich aus der Medizin- und Hygieneindustrie. Elektrisch-mechanische Antriebssysteme Conference Proceedings, Böblingen 2006, pp. 23-36.

[LoKi05]:   Lorch, F.; Kiel, E.: Servoantriebssystem nach dem Baukastenprinzip. SPS/IPC/Drives Conference Proceedings, Nuremberg 2005, pp. 103-111.

[MaVo06]:   Manowarda, M.; Vollmer, R.: Direkt treibender Motor für Rollenfördersysteme - Neuartige Polspulenwicklung ermöglicht sehr hohe Drehmomentdichte. Elektrisch-mechanische Antriebssysteme Conference Proceedings, Böblingen 2006, pp. 59-70.

[Ma06a]:   Martin, H.: Transport- und Lagerlogistik. Friedr. Vieweg & Sohn Verlag/GWV Fachverlage GmbH, Wiesbaden, 2006.

[Mi06]:   Mirbach, S.: ETHERNET Powerlink im Antrieb - Weiterentwicklung des Systembus (CAN). SPS/IPC/Drives Conference Proceedings, Nuremberg 2006, pp. 233-242.

[No98]:   Nolte, R.: Schwingungsarme Bewegungssteuerungen bei elektronischen Kurven. SPS/IPC/Drives Conference Proceedings, Nuremberg, 1998 pp. 344-352.

[No99]:   Nolte, R.: Sollwertvorgaben für Servoantriebe mit Koppelmechanismen. SPS/IPC/Drives Conference Proceedings, Nuremberg 1999, pp. 464-473.

[No02]:   Nolte, R.: Genaue Servoantriebsauslegung mit Ungleichförmigkeiten und Antriebskennlinien. SPS/IPC/Drives Conference Proceedings, Nuremberg 2002, pp. 561-568.

[ObSaPo01]:   Obermeier, C.; Sattler, K.; Pommer, H.: Direkter oder indirekter Antrieb? Vergleichende Untersuchungen des Positionierverhaltens einer mit Linearmotor oder Zahnriemenantrieb realisierten Antriebsachse. SPS/IPC/Drives Conference Proceedings, Nuremberg 2001, pp. 528-536.

[PaJo02]:   Pauli, M.; Jochim, F.: Komplexe Bewegungsaufgaben einfach umsetzen. SPS/IPC/Drives Conference Proceedings, Nuremberg 2002, pp. 552-560.

[Po03]:   Pollmeier, S.: Antriebe im Automatisierungsverbund. SPS/IPC/Drives Conference Proceedings, Nuremberg 2003, pp. 519-538.

[PrSt03]:   Pritschow, G.; Staudt, S.: Kommunikation im Umbruch - Feldbusse versus Ethernet. SPS/IPC/Drives Conference Proceedings, Nuremberg 2003, pp. 549-560.

[Qu93]:   Quang, N. P.: Praxis der feldorientierten Drehstromantriebsregelungen. Expert Verlag, Ehningen bei Böblingen, 1993.

[Ri06]:   Rivoli, P.: Reisebericht eines T-Shirts. Econ, Berlin, 2006.

[RoStWi00]:   Rohde, M.; Stoll, L.; Willerich, K.: Innovative Komponenten für die Dezentralisierung von Antriebssystemen. SPS/IPC/Drives Conference Proceedings, Nuremberg 2000, pp. 576-575.

[Ro03]:   Rostan, M.: Ethernet bis in die Reihenklemme. SPS/IPC/Drives Conference Proceedings, Nuremberg 2003, pp. 301-310.

[Ro99]:   Roth-Stielow, J.: Antriebselektronik auf dem Weg vom Schaltschrank ins Feld. SPS/IPC/Drives Conference Proceedings, Nuremberg 1999, pp. 579-588.

[Sch06]:   Schäfers, E.; Denk, J.; Hamann, J.: Ein neues Verfahren zur aktiven Schwingungsdämpfung für Be- und Verarbeitungsmaschinen. Elektrisch-mechanische Antriebssysteme Conference Proceedings, Böblingen 2006, pp. 343-360.

[Sch03]:   Scheitlin, H.: ETHERNET Powerlink: kurz und knapp. SPS/IPC/Drives Conference Proceedings, Nuremberg 2003, pp. 283-290.

[Sch98]:   Schönfeld, R.: Bewegungssteuerungen. Springer-Verlag, Berlin Heidelberg, 1998.

[SchHo05b]:   Schönfeld, R.; Hofmann, W.: Elektrische Antriebe und Bewegungssteuerungen. VDE Verlag GmbH, Berlin and Offenbach, 2005.

[SchSt64]:   Schönung, A.; Stemmler, H.: Geregelter Drehstrom-Umrichter mit gesteuertem Umrichter nach dem Unterschwingungsverfahren. BBC-Mitteilungen, vol. 8 / 9 1964 pp. 555-577.

[Sch01a]:   Schröder, D.: Elektrische Antriebe - Regelung von Antriebssystemen. Springer-Verlag, Berlin Heidelberg, 2001.

[SchHe85]:   Schumacher, W.; Heinemann, G.: Fully digital control of induction motor. EPE Brussels 1985, vol. 2, pp. 191-196.

[SchRoLe85]:   Schumacher, W.; Rojek, R.; Letas, H.-H.: Hochauflösende Lage- und Drehzahlerfassung für schnelle Stellantriebe. Elektronik. vol. 10 1985 pp. 65-68.

[Sch+01b]:   Schürmann, U.; Peter, K.; Orlik, B.; Rudolph, C.; Lücken, R.: Parameteridentifikation an Asynchronmaschinen mit Hilfe von Anregungssignalen. SPS/IPC/Drives Conference Proceedings, Nuremberg 2001, pp. 589-597.

[Sch99]:   Schütte, F.: Robuste Lageregelung einer elektrisch angetriebenen Linearachse. SPS/IPC/Drives Conference Proceedings, Nuremberg 1999, pp. 793-803.

[St00]:   Strackenbrock, B.; Weiß, J.: Mensch, Maschinen, Mechanismen. F. A. Brockhaus GmbH, Leipzig, 2000.

[Ti99]:   Tinebor, M.: Sensorloser Hubwerksantrieb. SPS/IPC/Drives Conference Proceedings, Nuremberg, 1999 pp. 825-832.

[Wa04a]:   Wagner, W.: Kreiselpumpen und Kreiselpumpenanlagen. Vogel Buchverlag, 2004.

[Wa02]:   Wahler, M.: Vorbeugende Wartungsfunktionen durch intelligente Antriebe. SPS/IPC/Drives Conference Proceedings, Nuremberg 2002, pp. 542-551.

[Wa04b]:   Wahler, M.: Vorbeugende Wartung mit intelligenter Antriebstechnik. Elektrisch-mechanische Antriebssysteme Conference Proceedings, Fulda, 2004 pp. 561-570.

[WaBo00]:   Walter, A.; Born, H.: Optimierte Antriebe für die Mehrachsanwendung in Industrierobotern. SPS/IPC/Drives Conference Proceedings, Nuremberg 2000, pp. 734-741.

[We01a]:   Weck, M.: Werkzeugmaschinen Fertigungssysteme 3: Mechatronische Systeme: Vorschubantriebe und Prozessdiagnose. Springer-Verlag, Berlin Heidelberg, 2001.

[We01b]:   Weck, M.: Werkzeugmaschinen Fertigungssysteme 4: Automatisierung von Maschinen und Anlagen. Springer-Verlag, Berlin Heidelberg, 2001.

[WeWö06]:   Wenz, M.; Wörn, H.: Entwicklung einer selbstkonfigurierenden und selbstorganisierenden Bewegungssteuerung für Industrieroboter. SPS/IPC/Drives Conference Proceedings, Nuremberg 2006, pp. 619-628.

[Wi04]:   Wildemann, H.: Stabil und doch flexibel: Produktordnungssysteme. Harvard Business Manager, vol. 02 2004, pp. 37-42.

[Wi05a]:   Wiendahl, H.-P.: Betriebsorganisation für Ingenieure. Carl Hanser Verlag, Munich Wien, 2005.

[Wi05b]:   Witte, S.: Integration der Sicherheitstechnik in zukünftigen Servoumrichter-Generationen. SPS/IPC/Drives Conference Proceedings, Nuremberg 2005, pp. 141-149.

[Wr07]:   Wratil, P.; Kieviet, M.: Sicherheitstechnik für Komponenten und Systeme. Hüthig GmbH & Co. KG, Heidelberg, 2007.

[Za00]:   Zacher, S.: Automatisierungstechnik kompakt. Friedr. Vieweg &
              Sohn Verlagsgesellschaft mbH, Braunschweig/Wiesbaden, 2000.

[Zi99]:   Ziegler, M.; Hofmann, W.: Der Matrixumrichter - Kommutierung
              in nur zwei Schritten. SPS/IPC/Drives Conference Proceedings,
              Nuremberg 1999, pp. 512-530.

[Zw92]:   Zwanziger, P.: IGBT-Einsatztechnik in Umrichtern. ETG-
              Fachbericht 39, Bauelemente und ihre Anwendung 1992, pp. 265-
              280.

[ZV06a]:   ZVEI: Energiesparen mit elektrischen Antrieben. ZVEI-
              Zentralverband Elektrotechnik- und Elektronikindustrie e.V.,
              Frankfurt am Main 2006.

[ZV06b]:   ZVEI: Integrierte Technologie-Roadmap Automation 2015+.
              ZVEI-Zentralverband Elektrotechnik- und Elektronikindustrie e.V.,
              Frankfurt am Main 2006.

Acknowledgements for illustrations appearing in this book

We would like to thank the companies who allowed us to use the following illustrations:

Fig. 3-95:	Ruland Manufacturing Inc., Great Britain
Fig. 3-96:	KTR Kupplungstechnik GmbH, Rheine, Germany
Fig. 3-97:	Maschinenfabrik Mönninghoff GmbH & Co.KG, Bochum, Germany
Fig. 3-98:	SCHMIDT-KUPPLUNG GmbH, Wolfenbüttel, Germany
Fig. 3-99:	ETP Transmission AB, Sweden
Fig. 3-101:	Schaeffler KG, Herzogenaurach, Germany
Fig. 3-103:	SNR Wälzlager Deutschland GmbH, Bielefeld, Germany
Fig. 3-113:	Schaeffler KG, Herzogenaurach, Germany
Fig. 4-3:	TGW Transportgeräte GmbH, Austria
Fig. 4-6:	TGW Transportgeräte GmbH, Austria
Fig. 4-11:	Scheuch GmbH, Austria
Fig. 4-13:	TGW Transportgeräte GmbH, Austria
Fig. 4-14:	SIGMA Maschinenbau GmbH, Magdeburg, Germany
Fig. 4-15:	TGW Transportgeräte GmbH, Austria
Fig. 4-17:	EISENMANN AG, Böblingen, Germany
Fig. 4-18:	BLEICHERT Förderanlagen GmbH, Osterburken, Germany
Fig. 4-20:	BLEICHERT Förderanlagen GmbH, Osterburken, Germany
Fig. 4-23:	Brevetti Stendalto GmbH, Reutlingen, Germany
Fig. 4-24:	Paul Vahle GmbH & Co. KG, Kamen, Germany
Fig. 4-25:	Paul Vahle GmbH & Co. KG, Kamen, Germany
Fig. 4-26:	Paul Vahle GmbH & Co. KG, Kamen, Germany
Fig. 4-27:	Paul Vahle GmbH & Co. KG, Kamen, Germany
Fig. 4-32:	Demag Cranes & Components GmbH, Wetter/Ruhr, Germany
Fig. 4-33:	Demag Cranes & Components GmbH, Wetter/Ruhr, Germany
Fig. 4-36:	EISENMANN AG, Böblingen, Germany

Fig. 4-39:   Ziehl-Abegg AG, Künzelsau, Germany

Fig. 4-41:   Demag Cranes & Components GmbH, Wetter/Ruhr, Germany

Fig. 4-42:   BLEICHERT Förderanlagen GmbH, Osterburken, Germany

Fig. 4-63:   KUKA Roboter GmbH, Augsburg, Germany

Fig. 4-64:   RO-BER Industrieroboter GmbH, Kamen, Germany

Fig. 4-67:   Gebr. Kemper GmbH & Co. KG, Olpe, Germany

Fig. 4-68:   SKET - Verseilmaschinenbau GmbH, Magdeburg, Germany

Fig. 4-70:   A. GMonforts Textilmaschinen GmbH & Co. KG, Mönchengladbach, Germany

Fig. 4-78:   HERBERT OLBRICH GmbH & Co KG, Bocholt, Germany

Fig. 4-80:   MEGTEC Systems Inc., France

Fig. 4-82:   KIEFEL Extrusion GmbH, Worms, Germany

Fig. 4-88:   REISCH Maschinenbau GmbH, Austria

Fig. 4-113:   CLEXTRAL Group, France

Fig. 4-116:   Reifenhäuser EXTRUSION GmbH & Co. KG, Troisdorf, Germany

Fig. 4-117:   MEAF Machines b.v., Netherlands

Fig. 4-118:   Battenfeld Extrusionstechnik GmbH, Bad Oeynhausen, Germany

Fig. 4-120:   TOX® PRESSOTECHNIK GmbH Co. KG, Weingarten, Germany

Fig. 4-121:   TOX® PRESSOTECHNIK GmbH Co. KG, Weingarten, Germany

Fig. 4-122:   DIOSNA Dierks & Söhne GmbH, Osnabrück, Germany

Fig. 4-123:   Willy Degen Werkzeugmaschinen GmbH & Co KG, Schömberg-Schörzingen, Germany

Fig. 4-133:   Schmalenberger GmbH + Co. KG, Tübingen, Germany

Fig. 4-136:   Weser Kurier, Bremer Tageszeitungen AG, Bremen, Germany

Fig. 4-138:   Wemhöner Surface Technologies GmbH & Co. KG, Herford, Germany

Fig. 4-139:   KRONES AG, Neutraubling, Germany

Fig. 4-140:   WITRON Logistik + Informatik GmbH, Parkstein, Germany

Fig. 4-141:   KUKA Roboter GmbH, Augsburg, Germany

The following illustrations were provided by HartungDesign, Mühlheim an der Ruhr, Germany:

Fig. 2-13	Fig. 3-24	Fig. 3-28	Fig. 3-37
Fig. 3-38	Fig. 3-39	Fig. 3-40	Fig. 3-58
Fig. 3-83	Fig. 3-86	Fig. 3-87	Fig. 3-88
Fig. 3-90	Fig. 3-93	Fig. 3-100	Fig. 3-102
Fig. 4-10	Fig. 4-21	Fig. 4-29	Fig. 4-31
Fig. 4-35	Fig. 4-37	Fig. 4-43	Fig. 4-44
Fig. 4-57	Fig. 4-58	Fig. 4-59	Fig. 4-60
Fig. 4-61	Fig. 4-62	Fig. 4-65	Fig. 4-76
Fig. 4-89	Fig. 4-94	Fig. 4-96	Fig. 4-99
Fig. 4-100	Fig. 4-112	Fig. 4-119	Fig. 4-125

All other illustrations were provided by Lenze AG, Aerzen, Germany.

Index